T0122590

New Thinking in GIScience

Bin Li · Xun Shi · A-Xing Zhu · Cuizhen Wang ·
Hui Lin
Editors

New Thinking in GIScience

Editors
Bin Li
Department of Geography
and Environmental Studies
Central Michigan University
Mount Pleasant, SC, USA

A-Xing Zhu
Department of Geography
University of Wisconsin-Madison
Madison, WI, USA

Hui Lin
School of Geography and Environment
Jiangxi Normal University
Nanchang, Jiangxi, China

Xun Shi
Department of Geography
Dartmouth College
Hanover, NH, USA

Cuizhen Wang
Department of Geography
University of South Carolina
Columbia, SC, USA

ISBN 978-981-19-3818-4 ISBN 978-981-19-3816-0 (eBook)
https://doi.org/10.1007/978-981-19-3816-0

Jointly published with Higher Education Press
The print edition is not for sale in China (Mainland). Customers from China (Mainland) please order the print book from: Higher Education Press.

This Springer imprint is published by the registered company Springer Nature Singapore Pte Ltd.
The registered company address is: 152 Beach Road, #21-01/04 Gateway East, Singapore 189721, Singapore

Preface

Initially emerging from map-based spatial data inventories in the 1960s, Geographic Information Science (GIScience) has undergone decades of evolution characterized by a breadth of technical breakthroughs and new applied fields. At the 28th International Conference on Geoinformatics, Nanchang, China (in hybrid form), a panel of twenty-two scholars presented their theoretical perspectives on GIScience, which attracted a large audience worldwide. The enthusiastic response and constructive discussion prompted the idea of publishing a collection of short articles to reflect the new forms and frontiers of contemporary GIScience based on the panel discussions. Encouraged by the firmed support from the publishers, we decided to broaden the scope and invited an additional group of distinguished scholars to contribute thought-based essays in this book *New Thinking in GIScience.*

Over one hundred authors contributed forty manuscripts. With the majority from academia, the authors were leaders in GIScience research, development, and applications. Many lead authors are prolific writers well known in GIScience, among whom are editors and editorial board members of prominent journals in the discipline. Also included in the author team are junior scholars who are doing cutting-edge work that advances and expands the field.

Thus, an important and unique purpose of this book is to provide an uncommon outlet for a group of active thinkers in GIScience to present their most vanguard and sometimes early thoughts, ideas, and speculations about the discipline. In our invitation to the authors, we made it clear that what we need are seminal "position papers" rather than ordinary research papers.

The essays cover a wide range of topics and represent diverse perspectives on the future trend of GIScience. They are organized into four categories. The first category deals with conceptual topics, including the changing connotations of GIScience and a proposal to extend GIS to Holo Spatial Information System (HSIS), new representations in both GIS and remote sensing, elaboration on Virtual Geographic Environment (VGE) and the Neighborhood Effect Average Problem (NEAP), guidelines for presenting the relevance of research, and proposals of new law and novel approaches to GIS-based modeling. Topics in the second category concern the functional components of GIS, such as the management of unstructured geospatial data and Volunteered Geographic Information (VGI), implementation of new functions for spatial causal analysis, Bayesian modeling, domain knowledge-based modeling, GeoAI, deep learning, and energy-conscious cartographic design. Prospects on GIS software development in the new era of cloud computing are also included in this group. The third category highlights new research directions in remote sensing applications, namely retrieval of canopy structural and leaf biochemical parameters, LiDAR remote sensing of forest ecosystems, time series analysis of dense satellite image, and integration of earth remote sensing and social sensing. The fourth category covers domain-specific topics. In addition to new perspectives on the integration of GIS with humanities and social sciences as well as GIS-enabled urban science, the essays share authors' insights on specific application domains such as social governance, planning, transportation, crime, health, and land use.

We are indebted to the authors who demonstrated not only distinct scholarships but also a high level of professionalism. They submitted their first drafts in less than two months from early November 2021 and completed the revisions by the end of January 2022. The timeline was particularly demanding for many authors for whom English is not their working language. They gracefully responded to editors' suggestions for revisions in both content and language in a timely manner, which made it possible to produce this valuable volume on such a tight timeline.

Most of the authors, including ourselves as editors, are members of the International Association of Chinese Professionals in Geographic Information Sciences (CPGIS) which was founded in 1992. We proudly present the book as a tribute to the celebration of the 30th anniversary of CPGIS.

Mount Pleasant, USA — Bin Li
Hanover, USA — Xun Shi
Madison, USA — A-Xing Zhu
Columbia, USA — Cuizhen Wang
Nanchang, China — Hui Lin
March 2022

Contents

1 **From Representation to Geocomputation: Some Theoretical Accounts of Geographic Information Science** 1
May Yuan

2 **On Holo-spatial Information System** 9
Chenghu Zhou, Yixin Hua, Ting Ma, and Tao Pei

3 **The Virtual Geographic Environments: More than the Digital Twin of the Physical Geographical Environments** 17
Hui Lin, Bingli Xu, Yuting Chen, Qi Jing, and Lan You

4 **Big Remote Sensing Data as Curves** 29
Fang Qiu and Yunwei Tang

5 **GIScience from Viewpoint of Information Science** 41
Zhilin Li and Tian Lan

6 **Towards Place-Based GIS** 51
Song Gao

7 **The Bottom-Up Approach and De-mapping Direction of GIS** 59
Xun Shi, Meifang Li, and Xia Li

8 **The Geography of Geography** 67
Weihe Wendy Guan

9 **Classification and Description of Geographic Information: A Comprehensive Expression Framework** 75
Guonian Lv, Zhaoyuan Yu, Linwang Yuan, Mingguang Wu, Liangchen Zhou, Wen Luo, and Xueying Zhang

10 **On the Third Law of Geography** 85
A-Xing Zhu

11 Human Mobility and the Neighborhood Effect Averaging Problem (NEAP) 95
Mei-Po Kwan

12 How to Form and Answer the *So What* Question in GIScience 103
Lan Mu

13 Prospects on Causal Inferences in GIS 109
Bin Li

14 Bayesian Methods for Geospatial Data Analysis 119
Wei Tu and Lili Yu

15 GIS Software Product Development Challenges in the Era of Cloud Computing 129
Fuxiang Frank Xia

16 Spatial Thinking of Computational Intensity in the Era of CyberGIS 143
Shaowen Wang

17 GeoAI and the Future of Spatial Analytics 151
Wenwen Li and Samantha T. Arundel

18 Deep Learning of Big Geospatial Data: Challenges and Opportunities 159
Guofeng Cao

19 Towards Domain-Knowledge-Based Intelligent Geographical Modeling 171
Cheng-Zhi Qin and A-Xing Zhu

20 Mitigating Spatial Bias in Volunteered Geographic Information for Spatial Modeling and Prediction 179
Guiming Zhang

21 Dealing with Unstructured Geospatial Data 191
Huayi Wu and Zhaohui Liu

22 Green Cartography and Energy-Aware Maps: Possible Research Opportunities 197
Mingguang Wu, Guonian Lv, and Linwang Yuan

23 Next Step in Vegetation Remote Sensing: Synergetic Retrievals of Canopy Structural and Leaf Biochemical Parameters 207
Jing M. Chen, Mingzhu Xu, Rong Wang, Dong Li, Ronggao Liu, Weimin Ju, and Tao Cheng

24 LiDAR Remote Sensing of Forest Ecosystems: Applications and Prospects 221
Qinghua Guo, Xinlian Liang, Wenkai Li, Shichao Jin, Hongcan Guan, Kai Cheng, Yanjun Su, and Shengli Tao

25 Dense Satellite Image Time Series Analysis: Opportunities, Challenges, and Future Directions 233
Desheng Liu and Xiaolin Zhu

26 Digital Earth: From Earth Observations to Analytical Solutions 243
Cuizhen Wang

27 Spatial–Temporal Big Data Enables Social Governance 253
Jianya Gong and Gang Xu

28 Geo-computation for Humanities and Social Sciences 265
Kun Qin, Donghai Liu, Gang Xu, Yanqing Xu, Xuesong Yu, and Yang Zhou

29 Four Methodological Themes in Computational Spatial Social Science 275
Fahui Wang

30 Geosocial Analytics 283
Kai Cao, Yunting Qi, Mei-Po Kwan, and Xia Li

31 Defining Computational Urban Science 293
Xinyue Ye, Ling Wu, Michael Lemke, Pamela Valera, and Joachim Sackey

32 What Can We Learn from "Deviations" in Urban Science? 301
Fan Zhang and Xiang Ye

33 Variants of Location-Allocation Problems for Public Service Planning 309
Yunfeng Kong

34 Smart, Sustainable, and Resilient Transportation System 319
Zhong-Ren Peng, Wei Zhai, and Kaifa Lu

35 The "Here and Now" of HD Mapping for Connected Autonomous Driving 329
Liqiu Meng

36 Modelling Teleconnections in Land Use Change 341
Yimin Chen and Xia Li

37 Progresses and Challenges of Crime Geography and Crime Analysis 349
Lin Liu

38 GIS Empowered Urban Crime Research 355
Yijing Li and Robert Haining

39 GIS in Building Public Health Infrastructure 367
Ge Lin

40 Challenging Issues in Applying GIS to Environmental Geochemistry and Health Studies 375
Chaosheng Zhang, Xueqi Xia, Qingfeng Guan, and Yilan Liao

Chapter 1
From Representation to Geocomputation: Some Theoretical Accounts of Geographic Information Science

May Yuan

Abstract This essay discusses theoretical perspectives in GIScience in representing and computing geographic information. Grounding the discussion is the need for *new ways of thinking about new facts*. Information and geospatial technologies continue acquiring new facts of various kinds. New ways of thinking about these new facts are essential to theoretical advances. Geographic representation encodes new facts to evoke new ways of thinking about them. Geocomputation carries out analytical and modeling procedures to realize these new ways of thinking. Discussions follow the proposed object-field continuum and event-process continuum to capture the essence of geographic representation and computational thinking. While much progress has been made, theories in GIScience research mostly apply existing ones from other disciplines or surround conceptual, logical, or ontological arguments. The lack of a well-defined theory for geographic information presents an excellent research opportunity. Theories for statistics and machine learning are exemplars.

Keywords Thinking · Object-field · Event-process · Theory · Law

1.1 Introduction

"*The important thing in science is not so much to obtain new facts as to discover new ways of thinking about them*," remarked Sir William Lawrence Bragg, the youngest-ever Nobel laureate who received the 1915 Nobel prize in physics at age 25. Indeed, new facts without new thinking help grow an archive but contribute little to understanding, innovations, or predictions. In the era of data deluge, new facts are not only plentiful but overwhelming. The need to discover new ways of thinking about the ever-growing new facts intensifies. Scientific advances come from acquainting new facts with new thinking for novel perspectives and insights. This essay grounds Geographic Information Science (GIScience) in Bragg's premise to examine the synergistic convergence between new facts and new ways of thinking for theoretical

---

M. Yuan (✉)
The University of Texas at Dallas, Richardson, TX 75080, USA
e-mail: myuan@utdallas.edu

B. Li et al. (eds.), *New Thinking in GIScience*,
https://doi.org/10.1007/978-981-19-3816-0_1

developments in geographic information. The goal is not to have a comprehensive review of theories in GIScience but to highlight how new facts and new ways of thinking about them can scaffold new theories in GIScience.

New geospatial technologies acquire new facts, which stimulate new ways of thinking. Theories often follow innovations. Hans Lippershey invented the first telescope in 1608 physicists later grasped the theories of optics around 1650–1700. Charles Babbage created the first calculator and then the analytical engine (the precursor of computers) in the 1830s; theories of computing, algorithms, automation, formal language, and database rose during 1950–1970, over a century after. Likewise, GIS technologies emerged in the 1960s; thirty years later, Goodchild (1992) coined Geographic Information Science (GIScience) and pushed scientific questions about geographic information. Innovations acquire new facts or transform facts into new representations. Theories synthesize facts, drive intuitions, provide guidance to make exploration and experimentation more systematic and effective by reducing brutal-force search or serendipity happenstance, and enable us to make predictions.

This essay adopts Britannica's (2018) definition of *scientific theory*. Empirical laws express observed or posited regularities existing in objects and events. A scientific theory renders a structure suggested by empirical laws to explain why these laws obtain. Theories are imaginative constructions of the human mind resulting from philosophical and aesthetic judgments of observations. Hence, theories are only suggested by, not inductively generalized from, observational information. An empirical law may convey a unifying relationship among limited observations, but a scientific theory needs to explain a variety of known laws and predict laws yet known. In this context, laws are new facts generalized from empirical facts of observations; and theories are new ways of thinking about them.

1.2 Geographic Representation

Representation drives new ways of thinking about new facts. Geographic representation brings selected essentials before the mind to help simplify complexity for reasoning. Existing knowledge influences the selection, and the selection exploits current theories that structure the knowledge. Science is always a work in progress: theories change when the outcome of new thinking deems logically inconsistent or limiting with the existing knowledge or current theories. Geographic representation shapes computational methods and substrates theoretical developments in GIScience (Galton, 2001; Goodchild et al., 2007) and conceptually constitutes a continuum of objects and fields (Fig. 1.1).

Since the mid-1960s, geospatial technologies have proliferated two approaches of spatial thinking: GIS and Remote Sensing. GIS technologies provoke thinking that transforms paper maps into digital objects with coordinate strings and topological encoding. With objects (e.g., settlements, rivers, or mountains) in mind, geographic representation attends to what new facts that we can measure from identifiable objects and assess their relationships (e.g., the number of settlements along a

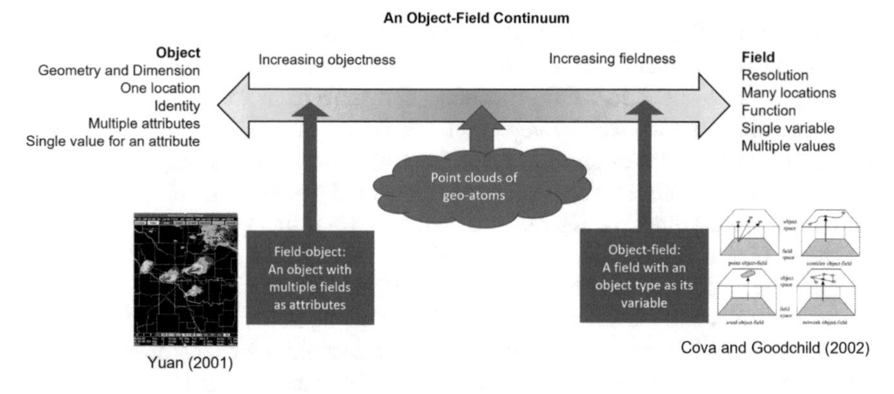

Fig. 1.1 A continuum of objects and fields in geographic representation

river) as shown in the left end of the object-field continuum in Fig. 1.1. Object-based thinking promotes inquiring about what objects are, where they are, how they relate spatially, and how to develop computational methods to derive and cogitate new facts. Examples of new thinking motivated by object-based facts include relational database theory (Maier, 1983), topological spatial relations (Egenhofer & Franzosa, 1991; Egenhofer & Mark, 1995), and the maptree (Worboys, 2012).

Complimentarily, remote sensing technologies digitize the landscape into discrete picture elements (i.e., pixels). On the right-hand side of the object-field continuum (Fig. 1.1), picture elements constitute a featureless space characterized by fields of properties with varying values across locations and subsequently invoke questions about and computational methods for what combinatory fields of properties are at locations, where properties of interest emerge, and how their spatial patterns and gradients evince underlying processes. Pixel-based thinking privileges locations over features; measurements are associated with locations, and spatial variations rise to feature emergence. New facts from pixels may be finer, more frequent, and richer. Consequently, correspondent emerging features may be in greater detail over space and time. Examples of new thinking based on rasters of pixel-based facts include map algebra (Tomlin, 1990), digital terrain modeling (Li et al., 2004), and scale ramifications (Cohen et al., 2016; Goodchild, 2011).

GIScience literature commonly adopts the following convention: real-world *entities* are represented by database *objects* and symbolized by cartographic *features*. People manipulate objects but cultivate fields (Couclelis, 1992). Fields provide geographic contexts for objects (Gold, 2010). Most GIS technologies implement vector and raster models to differentiate between such object- and field-based representations. Yet, such differentiation is flawed. Voronoi diagram (or Voronoi tessellation) partitions geography into spatially exhaustive object-centric locations and spatially arranges locations and neighborhoods in a scale-dependent hierarchy. Changes to features (e.g., movement or densification) alter the spatial structure of locations, which hints at underlying geographic processes responsible for the changes (Gold, 2016).

The rise of GPS, SmallSats, and machine learning also pushes for vector-raster integration rather than differentiation. As GPS technologies become ubiquitous, geotagged facts about real-time locations and movement trajectories open new ways of thinking for spatial interactions, activity space, contextual complexity, and geographic dynamics among vector features over raster landscapes. A marked increase in GIScience research fashions Hägerstrand's time geography in modeling movements, exposures, or interactions (Yuan, 2021), timely for COVID-19 risk mapping (Li et al., 2021), contact tracing (Dodge et al., 2021), and mobility analytics (Toger et al., 2021). The continuing growth of the research interest drives efforts towards establishing movement science (Demšar et al., 2021; Miller et al., 2019).

SmallSats, CubeSats, and NanoSats expand opportunities for earth observations at higher spatial, temporal, and spectral resolutions. Unmanned Aerial Vehicles (UAV) customize geospatial data acquisitions on-demand at fine spatial resolutions down to millimeters. Besides fine resolutions, these new remote sensing technologies acquire geospatial data in multiple spectra across numerous elevations and generate dense point clouds of geo-atoms that can be assembled into objects, fields, or anything in-between (Fig. 1.1). These new facts cannot be fully ingested without innovative algorithms with scalability, crosslinks to accelerate data routing, self-replan capabilities for broader collaboration among satellites and hierarchical normalization frameworks to produce and organize analysis-ready data. New data processing and analytical approaches prevail, such as structure from motion, data fusion and machine learning methods for applications. Image processing migrates from pixel- and object-based classification to convolutional, self-supervised semantic segmentation (Dong et al., 2020). Learning algorithms alternate the perspectives of objects and fields throughout computational pipelines to extract micro-objects from fields and assemble them to form higher-order objects.

An event-process continuum expands the object-field continuum from space to space–time (Fig. 1.2). Space–time considerations are now common in the de facto approach to many GIS applications. Analogous to objects and fields, events and processes anchor the two ends of the discrete–continuous continuum and serve as the basis for space–time analysis and reasoning (Yuan, 2020a, 2020b). Despite plentiful real-world examples, research in geographic representation has yet much attended

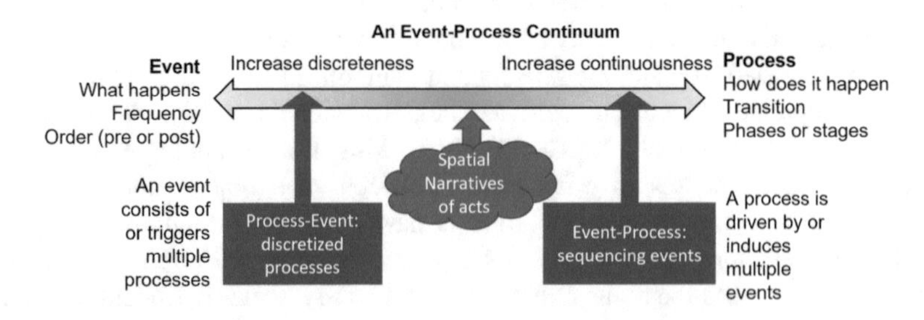

Fig. 1.2 A continuum of events and processes in geographic representation

to events encompassing multiple processes (e.g., a coastal flood resulted from overland runoff and storm surge), and processes driven by multiple events (e.g., inflation caused by a pandemic and supply chain failure). Spatial narratives of acts, analogous to point clouds of geo-atoms, serve as the space–time fundamentals to form any space–time constructs along the event-process continuum. A narrative connects happenings and situations meaningfully (i.e., acts by biotic or abiotic beings) to tell a story. For spatial narratives, acts must be spatially relevant and geographically contextualized. A spatial act associates a geo-atom with an action in a situation and enables GIS to automatically generate spatial narratives (Yuan et al., 2015).

These new ways of thinking, while providing systematic frameworks to reason or analyze geographic information, fall short of being scientific theories. GIScience research has made great strides in spatial analysis and reasoning on spatial autocorrelation, spatial dependency, spatial heterogeneity, spatial relationships and how they vary spatially. Much attention directs to search for geographic laws but not much on developing theories that give explanations of why the laws obtain and enable predictions of new laws. As such, GIScience studies often suffer limited validity and generalizability.

1.3 Geocomputation

Innovations multiply. New facts lead to new ways of thinking about them. New thinking in geographic representation promotes new geocomputation. Spatial autocorrelation or neighborhood effects push for neighborhood and proximity measures. For example, different clustering methods reflect various perspectives to consider distance, density, reachability, centrality, and similarity. With field-based data, GIS analyses elicit patterns, states, and state transitions. Integration of objects and fields enables studies of movements, lifelines, activity spaces, and many individualized geographic contexts.

There are many ways to compute events. Figure 1.3 summarizes four examples of common thinking in building computational methods. Sequence analysis arranges events over time and identifies periodicity or survival intervals, commonly used in political science. Lifeline analysis and trajectory analysis follow events on individuals to uncover accumulative effects. Spatial studies typically consider humans or animals as individuals (Demšar et al., 2021). With locations as individuals, events endow experiences to locations and, as such, transform a space into a place (Cho & Yuan, 2019).

Location analysis and spatial analysis bring geographic perspectives to many disciplines. As geospatial technologies proliferate, many social sciences and humanities take a spatial turn for new perspectives in the geographic dimension. Social scientists and humanists georeference objects and events and examine patterns and relationships in space and time. Environmental phenomena, physical infrastructure or barriers, or social and political conditions give the geographic context for interpretations.

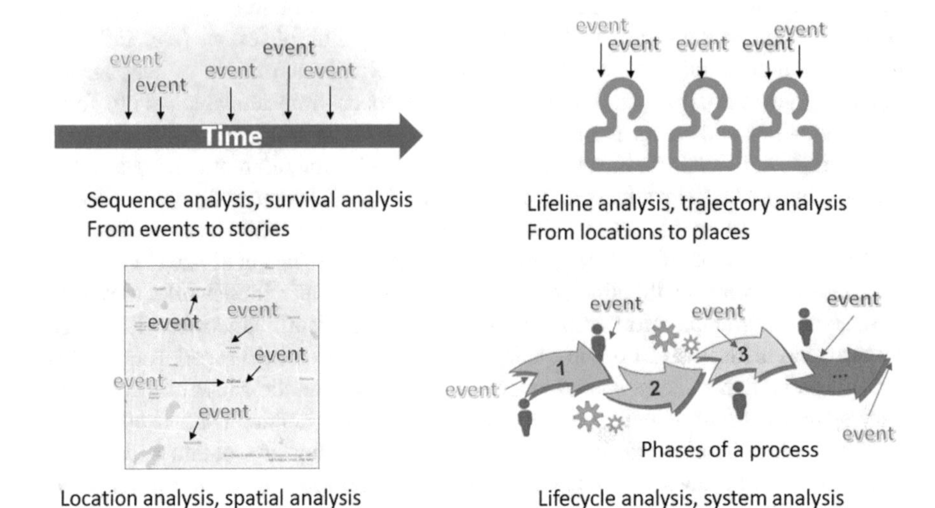

Fig. 1.3 Four example ways to analyze events

The fourth example shows events associated with multiple phases of a process. These events may be external or internal to the process. External events may influence the course of the process, such that a vehicle crash alters the traffic flow. Internal events signal mechanistic or compositional changes in the process. A mesoscale thunderstorm that produces hailstones signals the formation of strong updrafts and growing severity. Environmental impact assessment adopts lifecycle analysis to evaluate effects on resources and human health throughout the manufacture, usage, and disposal of a product. Likewise, system analysis applies to the entire business process to assure efficiency and meet the set goals. Regardless of nature or social processes, the key is to identify phases of a process and analyze external or internal events taking place in each phase.

The four ways to organize facts about events build upon temporal ordering, accumulating to individuals, georeferencing, and phasing along a process (Fig. 1.3). There are many other possibilities, such as functional or semantical hierarchies, spatial or temporal scales, and dependency connections. Agent-based modeling, genetic algorithms, and various simulation methods are popular methods to decipher the space–time dynamics of objects, fields, events, and processes. The emphasis is that what we can analyze and conclude depends on what new fact is represented and how we think about them.

1.4 Concluding Remarks

Theoretical advances in GIScience have been made mainly on two fronts. One is applying existing theories to GIS modeling. Examples are location theory, decision theory, and Dempster-Shafer theory of evidence. The other is bringing forward theoretical considerations in GIS research. Examples are essays on scale, space–time, conceptual models, formal logic, and ontologies. This essay falls in this group.

Geographic representation and geocomputation are at the heart of theoretical development for geographic information. Geographic representation encodes and organizes new facts and gives rise to new ways of thinking about them. Geocomputation carries out analytical or modeling procedures to realize new ways of thinking about the new facts. This essay presents an object-field continuum and an event-process continuum to discuss the spectrum of geographic representation and ensued GIS methods. Moreover, the essay presents four examples showing how different ways of thinking about events correspond to other analyses and findings. Innovations build upon new ways of thinking about new facts. Theories arise when we can meaningfully structure new facts to invoke new ways of thinking about them and making predictions.

Innovations in GIScience have led to new algorithms, new modeling techniques, and new applications. To date, theories of geographic information are yet available, although GIScience research has utilized existing theories from geography and other disciplines. Like GIScience, Statistics and Machine learning are method-centric disciplines. Statistical theories ground study design, data analysis, and statistical inference. Computational learning theory scopes learning tasks and quantifies the learning difficulty of a problem and the learning capacity of an algorithm. What will a GIScience theory say?

References

Cho, S., & Yuan, M. (2019). Placial analysis of events: A case study on criminological places. *Cartography and Geographic Information Science, 46*(6), 547–566.

Cohen, J. M., Civitello, D. J., Brace, A. J., Feichtinger, E. M., Ortega, C. N., Richardson, J. C., Sauer, E. L., Liu, X., & Rohr, J. R. (2016). Spatial scale modulates the strength of ecological processes driving disease distributions. *Proceedings of the National Academy of Sciences, 113*(24), E3359–E3364.

Couclelis, H. (1992). People manipulate objects (but cultivate fields): Beyond the raster-vector debate in GIS. In A. U. Frank, I. Campari, & U. Formentini (Eds.), *Theories and methods of spatio-temporal reasoning in geographic space* (Vol. 639, pp. 65–77). Springer.

Cova, T. J., & Goodchild, M. F. (2002). Extending geographical representation to include fields of spatial objects. *International Journal of geographical information science, 16*(6), 509–532.

Demšar, U., Long, J. A., Benitez-Paez, F., Brum Bastos, V., Marion, S., Martin, G., Sekulić, S., Smolak, K., Zein, B., & Siła-Nowicka, K. (2021). Establishing the integrated science of movement: Bringing together concepts and methods from animal and human movement analysis. *International Journal of Geographical Information Science, 35*(7), 1273–1308.

Dodge, S., Su, R., Johnson, J., Simcharoen, A., Goulias, K., Smith, J. L. D., & Ahearn, S. C. (2021). ORTEGA: An object-oriented time-geographic analytical approach to trace space-time contact patterns in movement data. *Computers, Environment and Urban Systems, 88*, 101630.
Dong, H., Ma, W., Wu, Y., Zhang, J., & Jiao, L. (2020). Self-supervised representation learning for remote sensing image change detection based on temporal prediction. *Remote Sensing, 12*(11), 1868.
Egenhofer, M. J., & Franzosa, R. D. (1991). Point-set topological spatial relations. *International Journal of Geographical Information Systems, 5*(2), 161–174.
Egenhofer, M. J., & Mark, D. M. (1995). Modelling conceptual neighbourhoods of topological line-region relations. *International Journal of Geographical Information Systems, 9*(5), 555–565.
Galton, A. (2001). A formal theory of objects and fields. In *International conference on spatial information theory* (pp. 458–473).
Gold, C. (2016). Tessellations in GIS: Part II–making changes. *Geo-spatial Information Science, 19*(2), 157–167.
Gold, C. (2010). The dual is the context: Spatial structures for GIS. In *2010 International symposium on Voronoi diagrams in science and engineering* (pp. 3–10).
Goodchild, M. F. (1992). Geographical information science. *International Journal of Geographical Information Science, 6*(1), 31–45.
Goodchild, M. F. (2011). Scale in GIS: An overview. *Geomorphology, 130*(1–2), 5–9.
Goodchild, M. F., Yuan, M., & Cova, T. J. (2007). Towards a general theory of geographic representation in GIS. *International Journal of Geographical Information Science, 21*(3), 239–260.
Li, J., Wang, X., He, Z., & Zhang, T. (2021). A personalized activity-based spatiotemporal risk mapping approach to the COVID-19 pandemic. *Cartography and Geographic Information Science, 48*(4), 275–291.
Li, Z., Zhu, C., & Gold, C. (2004). *Digital terrain modeling* (1st ed.). CRC Press.
Maier, D. (1983). *The theory of relational databases*. Computer Science Press. https://web.cecs.pdx.edu/~maier/TheoryBook/TRD.html
Miller, H. J., Dodge, S., Miller, J., & Bohrer, G. (2019). Towards an integrated science of movement: Converging research on animal movement ecology and human mobility science. *International Journal of Geographical Information Science, 33*(5), 855–876.
Toger, M., Kourtit, K., Nijkamp, P., & Östh, J. (2021). Mobility during the COVID-19 pandemic: A data-driven time-geographic analysis of health-induced mobility changes. *Sustainability, 13*(7), 4027.
Tomlin, D. (1990). *Geographic information systems and cartographic modeling*. Prentice-Hall.
Worboys, M. (2012). The Maptree: A fine-grained formal representation of space. In N. Xiao, M.-P. Kwan, M. F. Goodchild, & S. Shekhar (Eds.), *Geographic information science* (Vol. 7478, pp. 298–310). Springer.
Yuan, M. (2020a). *FC-09—Relationships between space and time | GIS&T body of knowledge*. Relationships between Space and Time. https://gistbok.ucgis.org/bok-topics/relationships-between-space-and-time
Yuan, M. (2020b). Why are events important and how to compute them in geospatial research? *Journal of Spatial Information Science, 21*, 47–61.
Yuan, M. (2021). GIS research to address tensions in geography. *Singapore Journal of Tropical Geography, 42*(1), 13–30.
Yuan, M., McIntosh, J., & DeLozier, G. (2015). GIS as a narrative generation platform. In D. J. Bodenhamer, J. Corrigan, & T. M. Harris (Eds.), *Deep maps and spatial narratives* (pp. 176–202). Indiana University Press.

Chapter 2
On Holo-spatial Information System

Chenghu Zhou, Yixin Hua, Ting Ma, and Tao Pei

Abstract Geographic entities exist in a spatio-temporal continuum, yet traditional geographic information system represents this dynamic geographic world using a map model which is often static in nature. This discrepancy between the dynamitic nature of the real world and the static representation scheme calls for a new system which can capture and represent the dynamic nature of the spatio-temporal continuum. This chapter presents such a system, referred to as Holo-Spatial Information System (HSIS). HSIS consists of two major components: an object-oriented representation scheme, which captures and represents the spatial entities in the dynamic world as multi-granular spatio-temporal objects (MGSTO), and the information management framework, which comprehensively manages the multi-dimensional information in adaptive transformation under different objectifications and granularity abstractions. The development of HSIS will drive innovations in theoretical, methodological, technological and system framework perspectives, which should lead to a drastic change in the landscape of geographic information science.

Keywords Holo-spatial information systems · Pan-spatial information systems · Geographic information systems · Spatial information system · Geographic information science

C. Zhou (✉) · T. Ma · T. Pei
State Key Laboratory of Resources and Environmental Information System, Institute of Geographical Sciences and Natural Resources Research, Chinese Academy of Sciences, Beijing 100101, China
e-mail: zhouch@lreis.ac.cn

T. Ma
e-mail: mting@lreis.ac.cn

T. Pei
e-mail: peit@lreis.ac.cn

Y. Hua
Institute of Geographic Space Information, The PLA Information Engineering University, Zhengzhou 450052, China
e-mail: chxyhyx@163.com

B. Li et al. (eds.), *New Thinking in GIScience*,
https://doi.org/10.1007/978-981-19-3816-0_2

2.1 Introduction

Sixty years have flown by since Dr. Roger Tomlinson put forward Geographical Information System (GIS) in 1962. As a modern extension of traditional cartography, GIS has continuously evolved with the development of related scientific fields and application requirements. The late twentieth century saw the rapid development of GIS and the glories. GIS has been extensively applied in many fields, ranging from natural resources, environment management, ecosystem studies, to public health, urban planning and design, demographics, education, socioeconomics research, and policy making. The scale of these fields spans over a wide range from a small, closed space to a planetary one. The growing diversity of applications and fields involved boosts the development of GIS both in depth and in breath. Present-day GIS is a broad discipline which consists of information science, information technology, statistics, civil engineering, mathematics, and other inter-discipline fields. To some extent, GIS not only made Geography and surveying/mapping flourishing and invigorating, but also significantly promoted the entire field of Geosciences.

Classical GIS, starting with the raster and vector models, represents objects and processes on Earth surface under geographic reference systems. However, emergences of innovative information science and new technologies, such as observation network, mobile internet, internet of things, big data, and cloud computing, increasingly present challenges for the development of GIS. These challenges will lead to great changes in the theories, methods, technologies, platforms, and applications with spatial information. The traditional geo-spatial information modeling approach using the map as a primary representational model is difficult to meet the demands of acquiring, managing, analyzing, expressing, and applying information in the multi-dimensional and dynamic space. These difficulties are mainly reflected in the following aspects:

1. The dilemma of geometric abstraction. Simple plane/solid geometry and abstraction using grid can hardly describe the complex shapes and status of objective entities (entities that exist in the world as they are), as well as objective existence and processes like flows and networks.
2. The dilemma of static modeling. Objects in the real world have dynamic characteristics as reflected in location, shape, relationship, and properties. The static data model and traditional map are difficult to describe the dynamic real world, which exists in a spatio-temporal continuum.
3. The dilemma of discretization in modeling. It is hard to describe the complex composition structures and functions of the entities in the objective world with highly discrete abstraction and discrete organization.
4. The dilemma of behaviors and cognitions. The traditional modeling approach are limited in handling the behaviors of objective entities in learnings and cognition.
5. The dilemma of symbolization and expression. The map is suitable for symbolic and layer-based expression methods, but it is limited for use with new technologies such as AR/VR.

6. The dilemma of data organization and management. The map-based spatial data model can hardly adapt to the demands for managing and analyzing spatio-temporal big data.

2.2 The Concept of Holo-spatial Information System

Traditional GIS is designed to handle static georeferenced objects and processes; what would happen when the locations and the attributes of these objects and processes become dynamic? Mobile objects and flows, for example, obviously become dominant concepts in GIS nowadays. Thus, the domain and scope of GIS need to expand into diverse ranges fully covering all observational spaces including geographical and non-geographical, or more generally 'Holo-Spatial Information System' (HSIS), which uses object-based model to describe this world (Zhou, 2015). The idea of HSIS is driven by the developments of both theoretical fundamentals and applications, and contains four salient aspects:

1. The extension of observations and research scope. From earth surface measurement space to the universe, from indoor space to outdoor space, from macro-space to micro-space, from geographical entity space to virtual network space, from natural element space to social-economic space.
2. The extension of spatial and temporal dimensions of geographic data. From measurement data to various types of perceptual big data, from discrete sampling data to continuous scene data, from regional representation data to individual object data, from independent variable data to high-dimensional data, and from simple relational data to complex network data.
3. The extension of data analysis methods. From single point computing to cluster and cloud computing, from independent variable mining to massive text and complex network mining, from spatial analysis to big data feature analysis, and from single geographic process computing to multi-geographical process analysis.
4. The extension of geographical law discovery and cognitive perspectives. From the multi-variable correlation to pan-variable correlation, from the understanding of multi-factor influence to pan-factor effect, from multi-process to pan-process geographical events analysis, from the recognition of independent geographical phenomenon to the recognition of collective geographical phenomenon, and from idiographic (qualitative and quantitative) to nomothetic.

Several extensive and innovative manners are highlighted in HSIS in the context of theory, method, techniques, and system framework (Jiang et al., 2017). In theory, mapping mechanism among physical space, social space and cyber space should be deeply investigated for constructing universal expression and logical semantics of pan-spatial big data. In the methodology domain, pan-spatial sampling and assessing, pan-spatial field-based modelling, curved pan-spatial surface construction, multi-scales stationary and non-stationary processes discrimination, and machine learning and mining, should receive much more attentions. In the development of software systems, the combination of GIS system and distributed file system, as well as parallel computing system and spatial database system are required for pan-spatial big data computations. In the computing technology front, high performance geo-computation and analytical techniques of big data are crucial for promoting extensive applications in broad fields: urban dynamics, Earth's surface processes, environment and ecosystem evolutions, social network analysis. Moreover, new modelling and analysis approaches should be developed to meet the needs of specific HSIS like indoor, underground, underwater, and social space domains that cannot be covered by traditional GIS.

2.3 Object-Oriented Modeling for HSIS

The main goal of HSIS is aimed at breaking away from the indirect modeling method that based on map model in traditional GIS and focuses on describing the real world from micro-level to macro-level in a straightway manner with multi-granular spatio-temporal object (MGSTO) and establishing an innovative information modeling and management framework to meet the needs of HSIS (Hua & Zhou, 2017). Both the description and the expression models for The MGSTO model is established through cognitive abstraction and contains formal definition, characterization, and functional expression of object entities over time and space, which facilitates the realization of multi-granular object-oriented modeling of the world in a pan-spatial framework.

Based upon multi-scale, multi-dimensional and multi-perspective spatial knowledge and requirements of specific applications, the description and expression model for multi-granular spatio-temporal objects are used for the multi-granular feature description and expression of objective entities in the pan-space, in terms of semantics, scale, dimensionality, temporality, and cognition. Considering the object expression, life cycle and attribute structure, the design of this pan-spatial spatio-temporal variation description method for multi-granularity objects is based on multi-granularity, multi-dimensional and life-cycle characteristics. According to the correlation, association, causality, logic and other relationships between different object elements, the relationship expression model of multi granularity spatio-temporal object elements forms the grammatical, semantic, and pragmatic rules of object relational expression. A multi-granularity object association network model is constructed based on the research of the evolution law of the characteristics, behavior, semantics, and knowledge information of objects with different spatio-temporal granularity.

The MGSTO model M can be conceptually expressed as:

$$M = (E, O)$$

where E is the representative set, which is used to describe and model the relationship between objects; O is the operation set, which used to model the process of dynamic evolution and interaction of objects.

E consists of eight expression elements:

$$E = \{R, S, D, A, \Sigma, G, \aleph, f\}$$

where R is spatio-temporal reference system, including temporal reference and spatial reference, which is used to describe the spatial coordinate and temporal coordinate reference information of the multi-granular spatio-temporal object; S is the spatio-temporal position, which is used to describe the location information of the object under different spatio-temporal granularity levels; D is the spatial morphology, which is used to express the geometric and morphological characteristics of multi-granular spatio-temporal object; A is the attribute feature, which is used to describe the dynamic attributes of the object under the multi granularity cognitive system, Σ is the expression of the composition structure, which is used to describe the relationship between the part and the overall spatio-temporal objects, including logical composition and spatial composition; G is the association relationship, which is used to express and model the association and action relationships among the multi-granular spatio-temporal objects; $\aleph$ is the behavior expression, which is used to model the action occurrence and consequences of spatio-temporal objects with behavior ability; f is the cognitive expression, which is used to model the cognitive process of spatio-temporal objects with cognitive ability.

O is the operational set defined on E:

$$O = \{\alpha, \beta, \gamma, \delta, \varepsilon\}$$

where α contains the construct and deconstruct operations, the operation and calculation of the rebirth and extinction of the multi-granular spatio-temporal objects in the life cycle; β is the decomposition and combination operation, realizing the segmentation and combination of different expression modes of the multi-granular spatio-temporal object sets; γ is the transform and evolve operation, the dynamic operation of the state and composition of the multi-granular spatio-temporal object set is realized; δ is the relationship dynamic operation, the operation model is developed to realize the change of association relationship between the multi-granular spatio-temporal objects; ε is the learning and decision operation, which is for the process of the multi-granular spatio-temporal objects generating decision output through self-learning for external input.

MGSTO is the result of direct object modeling and mapping of geographical elements or entities in the objective world on the knowledge system. Compared with the general object-oriented modeling method, the main characteristics are as follows: (1) Modeling method is still based on objectification, but produces the multi-granular objects as the observation scale, knowledge system, and description dimension changes; (2) Multi-granularity means that in the same modeling scenario, the mapped objects may have different scale and morphology, and different levels of attributes, and different description dimensions; (3) MGSTO are dynamic, with life cycles, morphological transformation, attribute evolution, composition, and relationship changes corresponding to objective processes that correspond to real-world processes; but are independent of observational scale and modeling expression categories. (4) Partial MGSTO have behavioral and cognitive features and can produce results on external inputs. These modeling operations are implemented inside the multi-granular object model rather than driven by external functions.

Compared with traditional GIS modeling methods, the main advantages of the pan-spatial information modeling based on abstract expression of MGSTO are as follows: (1) In geographic information theory, combining spatial modeling and geographic cognition, defining, and describing the perceptive objective entities through MGSTO are effective means to break through the limitations of traditional modeling methods. (2) Through the integrated and comprehensive modeling of MGSTO on the object space and cognition system, the multi-dimensional attributes, complex behaviors, and cognition of a larger range of complex observation objects, can be comprehensively described and expressed, thus providing a new basis for multimodality spatio-temporal analysis and visualization of geographic information (Li et al., 2020). (3) With MGSTO as the core, research activities in this area can include, but not limited to the interrelated processes of pan-spatial abstract method, description and expression model establishment, spatio-temporal object modeling implementation, as well as calculation and operation method. Outcomes from these research activities will form a comprehensive theoretical basis for HSIS modeling.

2.4 Information Management Framework of HSIS

HSIS is an information system that obtains, processes, manages, analyzes, and visualizes MGSTO. Therefore, *holo-spatial* is not only the coverage of all scales and contents, but also a multi-dimensional information comprehensive expression and management framework with an adaptive spatial scale transformation under objectification and granularity abstraction. It is a new form of geographical information science. The framework of MGSTO organization and management for HSIS are based upon the spatio-temporal domain in the holo-spatial digital world, a comprehensive data body composed of spatio-temporal objects describing various entities and elements in the real/virtual worlds in a computing environment (Hua et al., 2021).

The spatio-temporal domain is a data set composed of some spatio-temporal objects in a spatio-temporal continuum in the pan-spatial digital world, that is, the spatio-temporal domain formed by constraining spatio-temporal range and subject content. The pan-spatial digital world can be regarded as a combination of multiple spatio-temporal domains (Xiao et al., 2017). The spatio-temporal domain is a collection of objects divided according to the spatio-temporal scope and is the organization unit of the pan-spatial digital world. The spatio-temporal domain can meet the requirements of the pan-spatial digital world for spatio-temporal object organization and management. In order to organize and manage the spatio-temporal objects in the spatio-temporal domain, we also need to record the organization data of the spatio-temporal objects in the spatio-temporal continuum. These organizational data include: the sub-spatio-temporal domain, spatio-temporal object class, spatio-temporal object relations and spatio-temporal object life cycle sequence (Li et al., 2017). The management of the pan-spatial digital world is to manage the spatio-temporal continuum and its spatio-temporal objects, which is divided into two parts. The first part is the management of the spatio-temporal domain; the second part is the management of the spatio-temporal objects in the temporal space. The management includes the creation, deletion and maintenance of the spatio-temporal domain and the collective operation of several of the spatio-temporal domains. There are three management modes, including top-down hierarchical management mode, cross-reference network management mode, and linear management mode based on space–time object life cycle sequence.

2.5 Conclusion and Discussion

Boosted by the rapid developments in related scientific fields, a new generation of GIS should meet the needs of all-around sensing, broad connections and comprehensive understandings of spatial objects, phenomena, processes, interactions, and evolutions with several innovative information techniques in terms of data collecting, computing, mining, and visualization. Extensions in spatio-temporal domain, information content, analysis and visualization methods are inevitable, and crucial to the evolution from "geo-spatial information system" to "holo-spatial information system" (HSIS). HSIS is an innovative spatial information system designed to capture, store, manipulate, analyze, manage, and represent multi-granular spatio-temporal object, which plays the central role in context of the direct modeling of real world into computer systems. In HSIS, everything is an object represented by MGSTO, which depends on the observation scale, application strategy and dynamics under a new spatial modeling framework. HSIS is designed to fully support dynamics modeling of MGSTOs over the observation period in the context of changes in locations, shapes, patterns, components, relationship, and behaviors. We thus argued HSIS as a direct mapping of real world to a cyber world. HSIS also implies a remarkable extension in the context of the spatio-temporal domain into spatio-temporal continuum, information content and analysis methodology aiming at a complete coverage of dynamics in a changing world, which are crucially related to a new research paradigm of geographical information science.

Acknowledgements We are appreciated very much for Professors A-Xing Zhu, Bin Li, Xun Shi, Cuizhen Wang, and Hui Lin for their invitation to the participation in this book project. During last 35 years, we have benefitted greatly from the support and helps from CPGIS. Over the same time our friendship also grew significantly. The work reported here was supported by National Key Research

Program (2021YFB3900901), and the National Natural Science Foundation of China (Grant Nos: 41421001, 42050101).

During the preparation of this chapter, memories on how CPGIS significantly contribute to GIS development in China came to alive. Many CPGIS members participated in a variety of projects in GIS research, technology development, and education. Two events were worth mentioning. One is CPGIS Go-West lecture series initiated in June 2002. The other is the domestic software evaluation from 1996 to 2009. To some extent the glories of China GIS today would not have taken place without the supports from CPGIS.

References

Hua, Y. X., & Zhou, C. H. (2017). Description frame of data model of multi-granularity spatio-temporal object for Pan-spatial information system. *Journal of Geo-Information Science, 19*(9), 1142–1149.

Hua, Y. X., Zhang, J. S., & Cao, Y. B. (2021). Research on organization and management of spatio-temporal objects in pan-spatial digital world based on spatio-temporal domain. *Journal of Geo-information Science, 23*(1), 75–82.

Jiang, N., Fang, C., & Chen, M. J. (2017). Initial exploration of pan-spatial cognition and representation. *Journal of Geo-information Science, 19*(9), 1150–1157.

Li, D. S., Liu, Y., Shi, G. G., et al. (2017). The expression and modeling of relationship evolution of spatio-temporal objects of multi-granularity based on time-dependent network. *Journal of Geo-information Science, 19*(9), 1171–1177.

Li, Y., Zhu, Q., Fu, X., et al. (2020). Semantic visual variables for augmented geovisualization. *The Cartographic Journal, 57*(1), 43–56.

Xiao, S. J., Zong, Z., Xiang, L. Y., et al. (2017). The unified expression and calculation of spatial relationships of spatio-temporal object of multi-granularity. *Journal of Geo-information Science, 19*(9), 1178–1184.

Zhou, C. H. (2015). Prospects on pan-spatial information system. *Progress in Geography, 34*(2), 129–131.

Chapter 3
The Virtual Geographic Environments: More than the Digital Twin of the Physical Geographical Environments

Hui Lin, Bingli Xu, Yuting Chen, Qi Jing, and Lan You

Abstract With past more than 20 years of development, virtual geographic environments (VGE) had gradually matured and formed its own supporting theories and remarkable characteristics. During this period, the remarkable steps forwards of VGE were often inseparable from the promotion of new technologies. Recently, the term of digital twins has emerged and attracted researchers from the community of geographic information sciences to discuss what the digital twins of the physical geographic environments should be alike. This chapter focuses on discussing the conceptual connotations and typical characteristics of both virtual geographic environments and digital twins, analyzes the basic requirements for building digital twins of physical geographic environments, and summarizes whether VGE can match the framework of digital twins of physical geographic environments. The final conclusions of this chapter declare that: The concepts and framework of VGE are essentially consistent with those of digital twins; The characteristics of VGE can absolutely meet

H. Lin
School of Geography and Environment, Jiangxi Normal University, Nanchang 330022, Jiangxi, China
e-mail: huilin@cuhk.edu.hk

B. Xu (✉)
Department of Information and Communication, The Academy of Army Armored Forces, Beijing 100000, China
e-mail: xublmail@126.com

Y. Chen
School of Urban Planning and Design, Peking University Shenzhen Graduate School, Shenzhen 518052, Guangdong, China
e-mail: yutingchen@link.cuhk.edu.hk

Q. Jing
Institute of Space and Earth Information Science, The Chinese University of Hong Kong, Hong Kong 999077, China
e-mail: jingqi@link.cuhk.edu.hk

L. You
Faculty of Computer Science and Information Engineering, Hubei University, Wuhan 430062, Hubei, China
e-mail: yoyo@hubu.edu.cn

B. Li et al. (eds.), *New Thinking in GIScience*,
https://doi.org/10.1007/978-981-19-3816-0_3

the basic requirements of digital twins of physical geographic environments; What's more, VGE has been more than a digital twin of the physical geographic environments, for instance, it can extensively fit well with the conceptual framework of metaverse of geographic environments which have eight characteristics including identity, friends, immersive, low friction, variety, anywhere, economy, civility.

Keywords Virtual geographic environments · Digital twins · Physical geographic environments

3.1 Introduction

A digital twin, which was first applied by Michael Grieves (Jones et al., 2020), is defined as a virtual representation that serves as the digital counterpart of a physical object or process (Wikipedia, 2021). Digital twins are now widely used in diverse applications, such as smart factories/manufacturing (Qi & Tao, 2018), Industry 4.0 (Yang et al., 2017), and smart cities (Farsi et al., 2020). Geographic information science researchers are especially interested in defining and applying digital twins in geographical environments.

Virtual Geographic Environments (VGE) is a new and important direction of geographic research, particularly for Geo-Information Science. Salient features of VGE include three-dimensional reconstructions of the geographic environments, geographic process modeling and simulation, multi-dimensional geo-visualization, multi-disciplinary knowledge sharing, distributed collaborations, establishing connections between virtual environments and physical environments, and many others (Qi et al., 2004; Lin & Batty, 2009; Xu et al., 2011; Chen et al., 2016; Lü et al., 2018). Such features have a significant overlap with the characteristics of digital twins. Thus, there are many critical questions facing geo-information scientists about how to integrate both concepts: What are the characteristics of digital twins for geographic environments? Can VGE match the characteristics of digital twins in geographic environments? What is the relationship between VGE and digital twins? In this chapter, we will discuss the evolution and development, basic conception, and main characteristics of VGE. We also compare and contrast the similarities, differences, and relations between VGE and digital twins. Finally, we will discuss the questions posed above and draw conclusions.

3.2 Virtual Geographic Environments

3.2.1 The Definition and Concepts of Virtual Geographic Environments

VGE refers broadly and collectively to all geographic environments which are not "real". More specifically, virtual geographic environments are computer-generated digital representation of geographic environments in which complex geographic systems are realized via multi-channel human–computer interactions, distributed geographic modeling and simulation, and cyberspace geographic collaboration (Lin & Gong, 2001; Chen et al., 2016). Natural laws represented by spatial and temporal distribution patterns, evolutionary laws of geographic processes, and interaction mechanisms between geographic elements are the core concepts of virtual geographic environments (Wan et al., 2021) and the driving force behind the research and development of VGE (Lü, 2011). Based on above definitions, the structure of a complete VGE was designed to have four components, which were the data, the modeling and simulation, the interaction, and the collaboration. The four components were responsible for geographic data organization, the implementation of geographic modeling and simulations, interactive channel construction, and collaborative tool design, respectively (Chen et al., 2016).

3.2.2 The Evolution of Virtual Geographic Environments

The embryonic period of virtual geographic environments is during 1997 to 2002. In 1997, Michael Batty firstly used the term "virtual geography" in his paper (Batty, 1997) to study geography in a digital world. Then in 2001, Lin and Gong formally proposed the term of virtual geographic environments which was described as environments for exploring the relationship between humans and 3D virtual worlds (Lin & Gong, 2001).

During the period from 2003 to 2008, VGE underwent a continuous exploration stage. 2003 was a crucial year for the development of VGE. In that year, the first VGE International Conference was held at the Chinese University of Hong Kong. A collection of the conference papers had since been published and distributed in China and the United States. Later, the collection had also been translated into Russian and Czech for distribution in Europe. In terms of theoretical research, researchers at this stage are tried to seek what were the characteristics of VGE themselves. From the view of communication, Lin and Gong regards VGE as the third generation of geographic language comparing to the traditional maps, computer aided mapping, and GIS and remote sensing (Lin et al., 2003). Meanwhile, Lin and Gong pointed out that VGE could be a basic tool to represent geospatial information for both physical environments and social environments. Technically, many searches carried out to build VGE in high dimensions, which vastly benefited understanding and communicating geo-phenomena and geo-processes among researchers (Lin & Zhu, 2005; Xu et al., 2005; Zhu et al., 2007). Those technical achievements absolutely proved that the VGE could be competent as the new generation of geo-language.

From 2009 to 2020, VGE was in the stage of explosion. Firstly, two cores of geo-data and geo-models were emphasized to differentiate VGE from GIS (Lin & Batty, 2009; Xu, 2009). At the same time, geo-collaboration among multi-users form distributed locations, various domains of knowledge, and different departments was added as another key feature to support VGE to be outstanding (Lin et al., 2013; Lü et al., 2018; Xu et al., 2011). Secondly, VGE was theoretically promoted to be a geographic experiment tool for geographic analysis (Chen et al., 2015; Lin et al., 2013) based on a lot of outstanding researches, for instance air pollution simulation (Xu et al., 2011), dam break simulation (Zhu et al., 2016), crowd evacuation simulation (Gong et al., 2018), socio-environmental modeling (Voinov et al., 2018), and so on. Third, in the recent years, knowledge engineering was proposed for managing and sharing geographic knowledge in VGE (Lin & Chen, 2015).

Nowadays, driven by new technologies especially in 5G communication, virtual reality, the internet of things, big data, artificial intelligence, and blockchain, concepts of digital twins and metaverses have been rekindled. These new concepts will surely drive VGE to step into a new stage by improving its principles, methodologies, technologies, applications, and other aspects.

3.2.3 Features of Virtual Geographic Environments

1. **Three-dimensional (3D) reconstructions of geographic environments**

   Virtual reality is one of the important premises of VGE. VGE constructs virtual environments which matches the real environment and can reconstruct the geographic environment in three dimensions. There are many current technical methods for 3D reconstruction of VGE, including traditional 3D modeling methods (Zhao et al., 2011), 3D modeling with remote sensing images (Zomer et al., 2002), 3D modeling with laser point cloud scanning (Moon et al., 2019), and 3D modeling with UAV tilt photogrammetry (Guo et al., 2021). Whatever the method employed, these three-dimensional reconstructions must be based on accurate geographic spatial–temporal reference and must follow the principle of spatial–temporal scale.

2. **Geographic process modeling and simulation**

   Geographic modelling and simulation is used to better understand geographic environments and support decision making. VGE emphasizes the integration of data and models. Geographical process modeling and simulation are important features that distinguish VGE from GIS. The study of geographic process modeling and simulation is currently a critical direction of VGE researches. Geographic models have been applied to many applications such as air pollution diffusion simulations (Xu et al., 2011), dam breach simulations (Yu et al., 2021; Zhu et al., 2014), crowd evacuation simulations (Song et al., 2013), and many others. Additionally, geoscientific workflow management methods in VGE using version management can allow geographic simulation and computational results to be easily reproducible and extendable (Chen et al., 2020).

3. **Multi-dimensional geo-visualization**

   VGE visualization can be multi-dimensional, i.e., it can employ traditional two-dimensional methods, can be three-dimensional, or may even involve higher dimensional visualization. Visualization methods may involve a traditional desktop mode, an immersive mode using VR headsets, or an augmented reality mode (Xu et al., 2011; Chen & Lin, 2018; Lü et al., 2018). Visualization of virtual geographic environments is realized in a geo-referenced space, meaning that every entity and every process in the VGE needs to be geo-coded with relevant time and space constraints and on the correct geographic temporal and spatial scales. VGE visualization can be established on a macro scale, or it can involve a fine-scale micro-expression which is nearly consistent with the real environment.

4. **Multi-disciplinary knowledge sharing**

   VGE can also assist in the realization of Geographic Knowledge Engineering (GKE), the extension of the concept and connotation of geographic knowledge (Laurini, 2014). This refers to knowledge closely related to geographic things or processes, which have typical temporal and spatial geographic characteristics. GKE can be used in geographic problem analysis, process simulation, phenomenon prediction, and multiple other applications. Knowledge engineering attempts to express the hidden mathematical parameters of geographic knowledge. A combination of knowledge engineering and VGE can help to solve the integration, evolution, innovation of different forms of geographical knowledge in different fields (You et al., 2016), and can integrate emerging computer technologies such as artificial intelligence, cloud computing, mixed reality, and knowledge graphs.

5. **Distributed collaboration**

   VGE pursues the organic integration of data, knowledge, and resources distributed in different regions, different professions, and different departments, and forms a comprehensive platform for solving complex geographic problems through distributed collaboration. It is often possible to form a "synergy mode" of different temporal and spatial parameters such as for the same place at the same time, different place at the same time, different time at the same place, or different time and different place. The development of Internet of Things (IoT) technology can be combined with virtual geographic environments to synthesize virtual environments and real environments and yield effective data intercommunication, mutual information integration, and operation interaction between the virtual geographic environments and the physical geographic environments.

6. **Interactions between virtual environments and physical environments**

   The fusion and interaction of virtual and real refer to the interoperability and two-way interaction between digital spaces and physical spaces for specific objects or services, such as video-based virtual and real integration, dynamic loading of tilted photography, or cross-terminal human–computer interaction. It can reproduce and influence the real world in digital space and can also enter virtual spaces in the real world. This kind of visualization possesses real-time, dynamic, automatic, and interactive attributes, as well as the capability for automatic real-time dynamic evolution of a digital twin scene, automatic real-time dynamic restoration of a digital twin operating situation, reverse intervention of a digital twin system in the physical world, and a multi-entry physical world to reach a digital twin system.

3.3 Digital Twins

3.3.1 Concepts and Definitions of Digital Twin

There are many definitions of digital twins. Digital twins were first proposed in the manufacturing field, and subsequently adopted by other fields. Therefore, the definition of digital twins in manufacturing is relatively rich, while other fields are relatively less well defined. In this section, we will focus on collating the varied definitions of digital twins, starting with the manufacturing field, and then expanding the conceptual understanding of digital twin in a geographic information science context.

The term “digital twin” was first coined by Grieves in a 2003 presentation and later documented in a white paper (Fuller et al., 2020). A digital twin was described as a digital informational construct about a physical system created as an entity on its own and linked with physical system through the entire lifecycle of the system. Later, in 2017, Grieves refined the definition of digital twin as: “a set of virtual information constructs that fully describes a potential or actual physical manufactured product from the micro atomic level to the macro geometrical level”. In 2012, NASA released a paper detailing future designs of aerospace vehicles which defined a digital twin as: “an integrated multi-physics, multi-scale, probabilistic simulation of an as-built vehicle or system that uses the best available physical models, sensor updates, fleet history, etc., to mirror the life of its corresponding flying twin” (Glaessgen & Stragel, 2012). This definition is regarded as key milestone in the definition of digital twins (Fuller et al., 2020). Since then, there have been many permutations on the definition. For instance, Negri defined it as: “The DT consists of a virtual representation of a production system that is able to run on different simulation disciplines that is characterized by the synchronization between the virtual and real system, thanks to sensed data and connected smart devices, mathematical models and real time data elaboration”.

Within the field of geo-information science, the concept of a digital twin is still searching for a suitable definition. At the city level, Deren et al. defined a digital twin city as: "a digital twin city aims at constructing a complex giant system between the physical world and the virtual space that can map each other and interact with each other in both directions" (Li et al., 2021). Stefano defines the concept of digital twin of the Earth as a "digital replica of an earth system component, structure, process, or phenomenon obtained by merging digital modelling and real-world observational continuity—i.e., remote, in-situ, and synthetic data streams" (Nativi et al., 2021). In the book of *Manual of Digital Earth*, the digital twin for the Earth is associated with the concept of Digital Earth which is widely accepted by scientists in geo-information science (Guo et al., 2020).

3.3.2 Characteristics of Digital Twins

Grieves characterized a digital twin concept by three main components which were "Physical products in real space, virtual products in virtual space, and the connections of data and information that ties the virtual and real products together" (Grieves, 2003; Holler et al., 2016). In 2017, Grieves re-defined digital twins and emphasized four elements: real space, virtual space, the link for data flow from real space to virtual space, and the link for information flow from virtual space to real space (Grieves & Vickers, 2017). These four characteristics distinguish digital twins from other similar concepts, such as digital models, digital shadows, and others. In 2020, a systematic literature review titled with "Characterising the Digital Twin—A systematic literature review" conducted a thematic analysis of 92 Digital Twin publications from the previous ten years (Jones et al., 2020). This work broadened the characteristics of digital twins into thirteen aspects: Physical Entity/Twin, Virtual Entity/Twin, Physical Environment, Virtual Environment, State, Realization, Metrology, Twinning, Twinning Rate, Physical-to-Virtual Connection/Twinning, Virtual-to-Physical Connection/Twinning, Physical Processes, and Virtual Processes.

Digital twins are characterized differently in technology applications. Liu et al. contended that digital twins should be: individualized, high-fidelity, real-time, and controllable (Liu et al., 2021). Tao and Qi (2019) outlined the main problems of digital twins which are technically closed to data and models. For data problems, the types of data collection, the optimal number and placing of sensors for data collection, the merging of disparate data types, and the scattered ownership of data are identified as key difficulties. For model problems, standards and guidelines for verifying the accuracy of the model results, model combination, and model integration are highlighted as primary technical challenges.

3.4 Discussion

1. **What characteristics should the digital twin of a geographical environment have?**

Following the previous definitions of digital twins, we here define them for geographical environments as follows: The digital twin of a physical geographical environment is constructed as a virtual counterpart using geo-data, geo-models, geo-visualization, and other related technologies in a virtual geographic environment and realizes the information interactions and mutual control between the physical geographic environment and the virtual geographic environment. The digital twin of the geographical environment needs to have at least the following three aspects:

First, the physical geographic environment (PGE), i.e., the true geographic space, the geographic entities, and the geographic processes must exist.

Second, the virtual geographic environment organizes geographic entities and geographic processes in a geo-referenced digital environment, acting as a counterpart of the physical geographic environment.

Third, interaction and collaboration between the PGE and the VGE requires that the data and information of the PGE can be accurately monitored, transmitted with a standard format, passed to the VGE, and can be used to update the VGE in real time. At the same time, the data and information generated in the VGE via operation and simulation can provide feedback to and control the PGE.

2. **Can the current framework of VGE match the requirement of digital twins in geographic environments?**

First, digital twins need to have a physical entity and a mirrored virtual entity, and the two entities should be well-matched. The physical entity and the virtual entity should be consistent in terms of composition structure and action process. In essence, the virtual geographic environment maps the geographic entities and geographic processes of the physical world into digital counterparts which are represented in geo-referenced virtual environments through three-dimensional reconstructions of geographic entities and modeling/simulation of geographic processes. The mapping of VGE to PGE follows the principle of geographic space–time scaling. Consequently, the fidelity of the mapping meets the requirements of space–time scaling which depends on the relevant research problems and has multiple dimensions. Overall, the virtual geographic environment acts as a mirror of the physical geographic environment.

Second, the digital twin emphasizes the information interactions between real entities and their counterparts as virtual entities. Key components of virtual geographic environments are the distributed collaboration among multiple users, multiple domains of knowledge, and multiple departments. The collaboration of VGE is formulated according to pace and time and across physical geographic environments and virtual geographic environments. The technical methodology of VGE and PGE collaboration is implemented via data intercommunication, information sharing, operational interaction, and consistent expression between the two types of environments. Therefore, the collaboration between VGE and PGE is consistent with the information interaction between virtual and real matching entities emphasized by previous digital twin definitions.

Third, data and models are the foundation for realizing digital twins. The data is mainly used for the construction of the digital twin, and the models are used for the representation of the processes conducted by the twin. At its inception, a virtual geographic environment synthesizes geographic data and geographic models to present and simulate geographical phenomena and processes. In this way, a VGE is a suitable twin for a PGE in the context of data and models.

3. **Is VGE more or less than a digital twin of the physical geographic environment**

The characteristics and basic functions of VGE include not only virtual mapping for the physical geographic environment and the interaction between VGE and PGE, but also knowledge collaboration, multi-person collaboration, multiple visualization, spatial–temporal expression, etc. VGE can couple people, geographic entities, and geographic processes together to produce a comprehensive integrated seminar environment for complex geographic problems solving in the scopes of physical geography, human geography, and other sciences overlapped with geography. Therefore, it can be inferred that VGE is fully capable of playing the role as the digital twin of physical geographic environments. What is more, the concept and framework of VGE have gone beyond the basic scope of digital twins of physical geographic environments and can fit well with the conceptual framework of metaverse in geographic environments which have eight characteristics including identity, friends, immersive, low friction, variety, anywhere, economy, civility.

4. **Who should lead the research on digital twins of geographic environments**

The geographic environment digital twin involves and integrates diverse disciplines such as geography, information science, environmental science, humanities, economics, medicine, etc. Effective integration of these many perspectives to form ideal digital twins of geographical environments requires geography, especially geographic information science, to take the lead in future development of this methodology. Geographic information science itself uses information technology to study the spatial patterns and spatial processes under appropriate time and space constraints. GIS also studies the interactions between humanity and environments, explores the interaction and coupling of natural and social processes, and interacts closely with many disciplines. Therefore, continued research on digital twins of geographic environments should be a primary direction of geographic information scientists in the near future, and will necessitate effective communication with information scientists to make full use of the latest information technology.

3.5 Conclusions

The basic concept of virtual geographic environments has existed for more than 20 years. The concepts and framework of VGE are essentially consistent with those of digital twins and also meet the basic requirements of digital twins of physical geographic environments. This is still a conceptual construct, as there is no digital twin of a geographic environment in the full sense at present. Digital twins, as currently applied in information or industrial fields, are insufficient as" true" digital twins of geographic environments. However, the digital twin framework outlines a promising new research direction for building digital twins of physical geographic environments. Virtual geographic environments, which we have demonstrated here in theory, qualify as digital twins of physical geographic environments. What is more, VGE has been more than a digital twin of the physical geographic environments, it can extensively fit well with the conceptual framework of metaverse of geographic environments which have eight characteristics including identity, friends, immersive, low friction, variety, anywhere, economy, civility.

Acknowledgements Support for this chapter was partially funded by the National Natural Science Foundation of China (No. 41771442 and No. 41271402).

References

Batty, M. (1997). Virtual geography. *Futures, 29*(4–5), 337–352.

Chen, M., & Lin, H. (2018). Virtual geographic environments (VGE): Originating from or beyond virtual reality (VR)? *International Journal of Digital Earth, 11*(4), 329–333.

Chen, M., Lin, H., Kolditz, O., & Chen, C. (2015). Developing dynamic virtual geographic environments (VGE) for geographic research. *Environmental Earth Sciences, 74*(10), 6975–6980.

Chen, M., Lin, H., & Lu, G. (2016). Virtual geographic environments. *International Encyclopedia of Geography: People, the Earth, Environment and Technology: People, the Earth, Environment and Technology* (pp. 1–11).

Chen, Y., Lin, H., Xiao, L., Jing, Q., You, L., Ding, Y., Hu, M., & Devlin, A. T. (2020). Versioned geoscientific workflow for the collaborative geo-simulation of human-nature interactions—A case study of global change and human activities. *International Journal of Digital Earth*, 1–30.

Farsi, M., Daneshkhah, A., Hosseinian-Far, A., & Jahankhani, H. (2020). *Digital twin technologies and smart cities*. Springer.

Fuller, A., Fan, Z., Day, C., & Barlow, C. (2020). Digital twin: Enabling technologies, challenges and open research. *IEEE Access, 8*, 108952–108971.

Glaessgen, E., & Stargel, D. (2012). The digital twin paradigm for future NASA and US Air Force vehicles. In *53rd AIAA/ASME/ASCE/AHS/ASC structures, structural dynamics and materials conference 20th AIAA/ASME/AHS adaptive structures conference 14th AIAA*.

Gong, J., & Lin, H. (2006). A collaborative virtual geographic environment: Design and development. In S. Balram, & S. Dragicevic (Eds.), *Collaborative geographic information systems* (pp. 186–206). IGI Global.

Gong, J., Li, W., Zhang, G., Shen, S., Huang, L., & Sun, J. (2018). An augmented geographic environment for geo-process visualization—A case of crowd evacuation simulation. *Acta Geodaetica Et Cartographica Sinica, 47*(8), 1089.

Grieves, M. (2003). *Digital twin: Manufacturing excellence through virtual factory replication.*
Grieves, M., & Vickers, J. (2017). *Digital twin: Mitigating unpredictable, undesirable emergent behavior in complex systems* (pp. 85–113). Springer.
Guo, H., Goodchild, M. F., & Annoni, A. (2020). *Manual of digital earth.* Springer Nature.
Guo, Q., Liu, H., Hassan, F. M., Bhatt, M. W., & Buttar, A. M. (2021). Application of UAV tilt photogrammetry in 3D modeling of ancient buildings. *International Journal of System Assurance Engineering and Management, 13*(1), 424–436.
Holler, M., Uebernickel, F., & Brenner, W. (2016). Digital twin concepts in manufacturing industries—A literature review and avenues for further research. In *18th International Conference on Industrial Engineering and Engineering Management (IJIE).*
Jones, D., Snider, C., Nassehi, A., Yon, J., & Hicks, B. (2020). Characterising the digital twin: A systematic literature review. *CIRP Journal of Manufacturing Science and Technology, 29*, 36–52.
Laurini, R. (2014). A conceptual framework for geographic knowledge engineering. *Journal of Visual Languages & Computing, 25*(1), 2–19.
Li, D., Yu, W., & Shao, Z. (2021). Smart city based on digital twins. *Computational Urban Science, 1*(4), 11.
Lin, H., & Batty, M. (2009). *Virtual geographic environments.* Science Press.
Lin, H., & Chen, M. (2015). Managing and sharing geographic knowledge in virtual geographic environments (VGE). *Annals of GIS, 21*(4), 261–263.
Lin, H., Chen, M., & Lu, G. (2013). Virtual geographic environment: A workspace for computer-aided geographic experiments. *Annals of the Association of American Geographers, 103*(3), 465–482.
Lin, H., & Gong, J. (2001). Exploring virtual geographic environments. *Geographic Information Sciences, 7*(1), 1–7.
Lin, H., Gong, J., & Shi, J. (2003). From maps to GIS and VGE—A discussion on the evolution of the geographic language (in Chinese). *Geography and Geo-Information Science, 19*(4), 6.
Lin, H., & Zhu, Q. (2005). Virtual geographic environments. In S. Zlatanova, & D. Prosperi (Eds.), *Large-scale 3D data integration: Challenges and opportunities* (pp. 211–231). CRC Press.
Liu, M., Fang, S., Dong, H., & Xu, C. (2021). Review of digital twin about concepts, technologies, and industrial applications. *Journal of Manufacturing Systems, 58*, 346–361.
Lü, G. (2011). Geographic analysis-oriented virtual geographic environment: Framework, structure and functions. *Science China Earth Sciences, 54*(5), 733–743.
Lü, G., Chen, M., Yuan, L., Zhou, L., Wen, Y., Wu, M., Hu, B., Yu, Z., Yue, S., & Sheng, Y. (2018). Geographic scenario: A possible foundation for further development of virtual geographic environments. *International Journal of Digital Earth, 11*(4), 356–368.
Moon, D., Chung, S., Kwon, S., Seo, J., & Shin, J. (2019). Comparison and utilization of point cloud generated from photogrammetry and laser scanning: 3D world model for smart heavy equipment planning. *Automation in Construction, 98*, 322–331.
Nativi, S., Mazzetti, P., & Craglia, M. (2021). Digital ecosystems for developing digital twins of the earth: The destination earth case. *Remote Sensing, 13*(11), 2119.
Qi, M., Chi, T., & Zhang, X., & Huang, J. (2004). Collaborative virtual geographic environment: concepts, features and construction. In *IEEE International Conference on Geoscience and Remote Sensing Symposium* (Vol. 7, pp. 4866–4869).
Qi, Q., & Tao, F. (2018). Digital twin and big data towards smart manufacturing and industry 4.0: 360 degree comparison. *IEEE Access, 6*, 3585–3593.
Shao, G., & Helu, M. (2020). Framework for a digital twin in manufacturing: Scope and requirements. *Manufacturing Letters, 24*, 105–107.
Song, Y., Gong, J., Li, Y., Cui, T., Fang, L., & Cao, W. (2013). Crowd evacuation simulation for bioterrorism in micro-spatial environments based on virtual geographic environments. *Safety Science, 53*, 105–113.
Tao, F., & Qi, Q. (2019). *Make more digital twins.* Nature Publishing Group.

Voinov, A., Çöltekin, A., Chen, M., & Beydoun, G. (2018). Virtual geographic environments in socio-environmental modeling: A fancy distraction or a key to communication? *International Journal of Digital Earth, 11*(4), 408–419.

Wan, G., Lin, H., Zhu, Q., & Liu, Y. (2021). Virtual geographical environment. *Advances in cartography and geographic information engineering* (pp. 443–477). Springer.

Wikipedia. (2021). Digital twin.

Xu, B. (2009). *A prototype of collaborative virtual geographic environments to facilitate air pollution simulation.* The Chinese University of Hong Kong.

Xu, B., Gong, J., Lin, H., Li, W., Zhang, J., Zhu, J., & Wu, X. (2005). Virtual geographic environment database design and collaboration. In *IEEE International Conference on Geoscience and Remote Sensing Symposium* (Vol. 2, pp. 25–29).

Xu, B., Lin, H., Chiu, L., Hu, Y., Zhu, J., Hu, M., & Cui, W. (2011). Collaborative virtual geographic environments: A case study of air pollution simulation. *Information Sciences, 181*(11), 2231–2246.

Yang, W., Tan, Y., Yoshida, K., & Takakuwa, S. (2017). Digital twin-driven simulation for a cyber-physical system in Industry 4.0. DAAAM International Scientific Book (pp. 227–234).

You, L., & Lin, H. (2016a). A conceptual framework for virtual geographic environments knowledge engineering. *International Archives of the Photogrammetry, Remote Sensing & Spatial Information Sciences, 41.*

You, L., & Lin, H. (2016b). Towards a research agenda for knowledge engineering of virtual geographical environments. *Annals of GIS, 22*(3), 163–171.

Yu, D., Tang, L., Ye, F., & Chen, C. (2021). A virtual geographic environment for dynamic simulation and analysis of tailings dam failure. *International Journal of Digital Earth, 14*(9), 1194–1212.

Zhao, H., Bai, R., & Liu, G. (2011). 3D modeling of open pit based on AutoCAD and application. *Procedia Earth and Planetary Science, 3*, 258–265.

Zhu, J., Gong, J. H., Liu, W. G., Song, T., & Zhang, J. Q. (2007). A collaborative virtual geographic environment based on P2P and Grid technologies. *Information Sciences, 177*(21), 4621–4633.

Zhu, J., Yin, L., Wang, J., Zhang, H., Hu, Y., & Liu, Z. (2014). Dam-break flood routing simulation and scale effect analysis based on virtual geographic environment. *IEEE Journal of Selected Topics in Applied Earth Observations and Remote Sensing, 8*(1), 105–113.

Zhu, J., Zhang, H., Yang, X., Yin, L., Li, Y., Hu, Y., & Zhang, X. (2016). A collaborative virtual geographic environment for emergency dam-break simulation and risk analysis. *Journal of Spatial Science, 61*(1), 133–155.

Zomer, R., Ustin, S., & Ives, J. (2002). Using satellite remote sensing for DEM extraction in complex mountainous terrain: Landscape analysis of the Makalu Barun National Park of eastern Nepal. *International Journal of Remote Sensing, 23*(1), 125–143.

Chapter 4
Big Remote Sensing Data as Curves

Fang Qiu and Yunwei Tang

Abstract The latest improvement of sensor resolutions has led to the emergence of various big remote sensing data, which provides great potentials for extracting valuable geospatial information. However, traditional perceptions of these data could not fully capitalize their benefits. In pursuing better solutions to the challenges posed by the volume and variety of new big remote sensing data, we processed the data from different perspectives by transforming the data into curves in frequency domain. The derived curves enabled us to develop a new thinking about big remote sensing data, based on which a theoretical framework was established with new directions on forming commensurate variable types, compatible processing units, as well as matching strategies and algorithms to process and fuse big remote sensing data. The new thinking bears significant theoretical implication and will have widespread adoption with its easiness to be applied to different features of the emerging data.

Keywords Curves · Matching · Big remote sensing data · Data fusion · Frequency distribution

4.1 Introduction

How you see a target determines how you treat it because thinking always precedes doing. In the same sense, how you perceive your data often defines how you will process the data. For a long period, remotely sensed images have been perceived as a 2-dimensional (2D) or 3D array of pixels. As a result, pixel-based approaches have dominated digital image processing. When the focus is changed from an individual pixel to also its neighboring pixels, various kernel- or convolution-based approaches were developed, introducing the new spatial concept into processing.

F. Qiu (✉)
The University of Texas at Dallas, Richardson, TX 75080, USA
e-mail: ffqiu@utdallas.edu

Y. Tang
Aerospace Information Research Institute, Chinese Academy of Sciences, Beijing 100094, China
e-mail: tangyw@aircas.ac.cn

B. Li et al. (eds.), *New Thinking in GIScience*,
https://doi.org/10.1007/978-981-19-3816-0_4

Since the turn of last century, the advancement of sensor technologies has triggered the emergence of many new remote sensors, constantly acquiring data from satellites, aircrafts, unmanned aerial vehicles (UAV), mobile and terrestrial platforms. These emerging sensors have significantly improved the spectral, spatial, vertical, and temporal resolutions of traditional remote sensing data, resulting in hyperspectral, high spatial resolution (HSR), LiDAR and multiple temporal remotely sensed data. The high resolutions of big remote sensing data have provided unprecedent potentials for extracting more precise, more accurate and much richer geospatial information not possible before. The volume of these big data, however, have also posed tremendous challenges for image analysts to process them, especially if they still perceive the data from the perspectives they had with traditional data.

Each sensor type possesses advantages that are not available in other types. Ideally, the complementary information from data acquired by different types of sensors should be integrated through data fusion, allowing the shortcomings of one type to be compensated by others. Data fusion is expected to achieve better results than using data from a single sensor alone. However, the fusion of sensor data with a variety of higher resolutions is even more challenging due to the disparate data structures involved.

To meet these challenges, it is desirable to have a new kind of thinking about the emerging big remote sensing data, so that commensurate variables can be extracted from disparate data sources and fused in compatible processing units using a unified methodology as a total solution to the problems faced in current algorithms. To this end, this chapter proposes a new theoretic framework based on a novel concept of "curve" to enable data processing and fusion of disparate big remote sensing data much more intuitive and efficient. A set of "matching" algorithms have also been proposed based on the new concept, which can be uniformly applied to different types of remote sensing data and allow the combination of the advantages of disparate data sources.

4.2 Traditional Perceptions of Big Remote Sensing Data

To realize the potentials offered by the big remote sensing data, various algorithms have been developed to process a specific type of data. Based on traditional perception of remote sensing data, these algorithms achieve limited success and face great challenges in taking full advantage of the emerging data.

The spatial resolution of remotely sensed data has been significantly improved with the advent of many commercial sensor systems that can acquired data with 1–4-m multispectral bands and a sub-meter panchromatic band since 1999. The reduction of pixel size from Landsat TM's 30 m to WorldView-3's 30 centimeters increases the data volume by 10,000 folds, while their temporal resolution is also greatly improved with their revisit frequency shortening from 16 days to 1 day through satellite constellation, a tremendous leap in data acquiring velocity. A single pixel in a HSR image now occupies only a small portion of a geographic object. Consequently, the spectral information of a single pixel becomes less representative of the target class (Arroyo et al., 2006). As a result, pixel-based image analysis methods are no longer appropriate, and object-based image analysis (OBIA) was proposed to perform image classification using image objects or segments as processing units. For classification, object-level statistical summaries (such as mean and standard deviation) are first derived from image segments and are used as inputs to a conventional image classifier. In this case, each object is treated as if it was a pixel, and the object-level summaries are treated as if they were pixel values.

LiDAR is an active sensor using visible or near-infrared laser beams, offering improved vertical resolution by providing multiple readings for the same location (Fowler, 2000; Zhou & Troy, 2008). There are two types of LiDAR data based on how the signal is recorded: discrete-return LiDAR and full-waveform LiDAR (Ussyshkin & Theriault, 2011). Currently, discrete-return LiDAR data with up to 6 laser returns are the mainstream product of commercial LiDAR (Ussyshkin & Theriault, 2011). Full waveform, a relatively new product of LiDAR, records the quasi-continuous time-varying measurements of signal strengths from the illuminated area (i.e., footprint) using small time intervals (e.g., 1 ns), resulting in thousands of returns for each transmitted laser pulse (Ussyshkin & Theriault, 2011). With the resulting fine vertical resolution (15 cm), a waveform is big data that offers an enhanced capability to reflect the 3D structures of geographical objects than the discrete-return LiDAR that only provides a few measurements, often separated by meters in vertical distance (Ussyshkin & Theriault, 2011; Zaletnyik et al., 2010)

LiDAR data can detect elevated objects and the ground. However, LiDAR data in the form of point cloud may need to be transformed into image format for traditional remote sensing algorithms to be applied. This pixel perception of the LiDAR data causes the degradation of the original precision during the transformation process. As a result, point cloud-based classification (PCBC) for LiDAR data has been developed, which are usually much more computationally intensive (Zhou, 2013). For the full-waveform LiDAR, a waveform may exhibit many echoes, corresponding to multiple objects or one object with multiple components that are vertically separable within the footprint. Consequently, the shapes of a waveform can be complex. To utilize the waveforms, existing studies have focused on simplifying the complex full-waveform data. One approach is to discretize full-waveform into discrete returns. Another is to derive shape-related metrics from the full-waveforms. Shape-related metrics, such as the number of echoes, echo amplitude, echo width, skewness, and kurtosis, are extracted by Gaussian decomposition to represent the important characteristics of the waveform shapes (Zaletnyik et al., 2010), so that dealing with the complex shape of a waveform itself is avoided. This is similar to the perception of traditional object-based image analysis which treats each object as a pixel and uses the statistical summaries as pixel values. In this case, each shape (or Gaussian decomposition) is equivalent to a discrete return and its shape-related metrics as its return values.

With hundreds of contiguous bands with narrow bandwidth, *hyperspectral* image is a big data cube made up of a massive 3D array of pixels. Compared to traditional multispectral imagery such as Landsat TM with 7 bands, the dimensionality of a hyperspectral image such as AISA sensor with 492 bands are 70 times more in dimension. By measuring subtle absorption features related to physical, chemical, and biological properties of ground materials, hyperspectral data provide improved discrimination capability (Cochrane, 2000; Ustin et al., 2004). The presence of a large number of spectral bands in hyperspectral data, however, results in unsatisfactory classification when classical methods such as maximum likelihood classification (MLC) are applied. This is known as Hughes phenomenon, where with a fixed number of training samples, the predictive power reduces as the dimensionality increases (Duds & Hart, 1973). A popular solution for this problem is to utilize spectra signatures of pure materials, referred to as endmembers (Bajorski, 2004; Hestir et al., 2008). Several endmember-based classification (EMBC) algorithms have been developed. For example, spectral angle mapper (SAM) assigns a class value to each pixel based on how it resembles an endmember in the shapes of their spectral signature (Yuhas et al., 1992). However, obtaining representative endmember is difficult. Limiting a single endmember to one class simplifies the classification process but weakens the generalization ability for the EMBC algorithms.

It is observed from the above that current methodologies for analyzing high resolution sensors have a common limitation. They all try to convert big remote sensing data into small data first to meet the computational challenges posed by the improved resolution. For HSR data it is the use of object-level summary instead of all pixel values in the object; for LiDAR waveform, it is the discretized returns or derived metrics instead of the original full waveforms; and for hyperspectral, it is the use of one endmember instead of multiple training samples per class. Converting to small data causes loss of tremendous valuable information that big data offer and thus is not an ideal solution.

Data fusion in remote sensing is generally defined as integrating the information acquired with sensors of different spatial, spectral, vertical, and temporal resolutions to produce fused data. The fused data is expected to contain more detailed and comprehensive information than each individual data source (Zhang & Qiu, 2012; Zhang, 2010). The challenges posed by the variety of big remote sensing data lie in the fact that HSR and hyperspectral imagery are raster grid data with different pixel size, while LiDAR data are vector point cloud of varied footprints with either discrete returns or continuous waveform returns. Therefore, these fusion methodologies suffer from the following limitations: (1) Conversion of data from one format to another, causing loss of precision during the conversion process; (2) Merge of inputs of different types (e.g. reflectance and elevation), which are not commensurate variables; (3) Inclusion of processing units that are incompatible in size and shape due to variety in pixel and footprint dimensions, and resulting in loss of accuracy in the resampling process; and (4) Application of ad hoc solutions that are not designed for high resolution sensor data and failing to take full advantage of information that big data provide.

4.3 Novel Perceptions of Big Remote Sensing Data

To exploit the benefits of emerging big remote sensing data and to overcome the aforementioned challenges, we have developed many specialized algorithms. Initially the algorithms have been developed independently without connection, and later we found they all operated on similar novel perceptions of remote sensing data.

HSR as histogram: Object-level statistical summary is a valid representation of the pixel values of an image object only when they follow a normal distribution. We observed that the spectral histograms of image objects of high spatial resolution imagery were often not normal and different object classes demonstrated unique patterns in their histograms. In this case, why not to classify image objects based on their histogram instead of object-level statistical summaries? Spectral histograms encompass all pixel values of image objects to depict their full frequency distribution instead of only the central tendency or the dispersion of the distribution and contains much richer information about the objects. Graphically, we can use Q-Q plot to compare the histogram of an unknown image objects with that of a reference image object to perform classification. Quantitatively, we developed an algorithm based on Kolmogorov-Smirnov (KS) distance by matching object-level cumulative histogram and achieved significant improvement over the statistical summary based results (Sridharan & Qiu, 2013). Adopting a similar spectral histogram view of the images, Stow et al. (2012) and Toure et al. (2013) also proposed to use spectral angle mapper to perform OBIA, treating the object-level histogram as if it is the spectral signature of an image object.

LiDAR as waveforms: We explored the full LiDAR waveforms to differentiate objects having different vertical structures without simplifying the intensive waveform data through either discretization or derivation of shape-based metrics. Conceptually a waveform can be perceived as a time-varying frequency distribution of returning impulses (Zaletnyik et al., 2010). Therefore, we similarly used the KS distance to classify the full waveform of ICESAT data and achieved results substantially better than the widely adopted rule-based classification using Gaussian decomposition derived metrics (Zhou et al., 2015). Under this perception, the full waveform has been utilized for classification instead of the shape metrics.

Fusion of HSR and LiDAR by pseudo waveform: To reap the maximum benefit from the complementary strengths of HSR and high-density discrete return LiDAR, we performed object-level data fusion. We first generated object-level spectral histogram of pixels from HSR imagery. However, the discrete-return LiDAR does not have waveform. To achieve the data fusion, we innovatively synthesized an object-level pseudo-waveforms of discrete-returns LiDAR data by constructing height-based histogram of the LiDAR points within the boundary of the object. It was observed that the height-based pseudo-waveform was nearly identical to the corresponding full waveform of the ICESAT data. Therefore, the height based pseudo-waveforms were then fused with the object-level spectral histograms from HSR imagery to perform classification using a Kullback–Leibler divergence-based histogram matching approach and achieved an improvement of 7.61% in overall classification accuracy over the use of HSR imagery alone (Zhou et al., 2015).

Unlike most of existing approaches that process high resolution sensor data by simplifying the complexity of the big data, the novel algorithms that we have developed above attempted to fully utilize the complete information provided by the emerging sensors.

4.4 New Thinking of Big Remote Sensing Data and New Theoretic Frame for Data Processing and Fusion

Big remote sensing data as Curves: These novel algorithms were originally developed to specifically process and fuse HSR and LiDAR, but they all have something important in common by coincidence. The new perceptions on the emerging remote sensing data are all in the form of a certain type of "*curve*", which laid the foundation of their processing algorithms. For HSR data, pixel values of an object are transformed into spectral histograms (Stow et al., 2012; Toure et al., 2013) and cumulative histogram (Sridharan & Qiu, 2013) to provide a more comprehensive description of various components of an image object. Histogram and cumulative histogram are "*curves*" of spectral frequency distribution of the pixels. For LiDAR data, waveform is a better 3D profile of an object's structure compared to the discrete LiDAR returns. Waveform is again a "*curve*" of quasi-continuous time-varying measurements of signal strengthen from the illuminated area. High density discrete LiDAR returns were transformed to pseudo waveforms by synthesizing height-based histograms, which are also essentially "curves" of aggregated signal strength of the returned laser pulses across height (i.e., time also). They are also similar to the spectral signature of hyperspectral data, which is a "curve" of reflectance aggregation across the spectrum. Since "*curves*" in a certain form can be obtained from the high-resolution data acquired by different types of sensors, "*curve*" is proposed in this chapter as a common concept that established the theoretic framework for defining commensurate variables, compatible processing units, and matching strategies and algorithm for processing and fusing HSR, LiDAR and hyperspectral data.

Commensurate variable types: To process and fuse data from disparate sources with heterogeneous structures, it is imperative to find commensurate variable types that can be obtained or derived from the individual sources to be fused, whether the fusion is to happen at the pixel or the object level. For fusion involving only imagery, such variables can be the reflectance or radiance of the pixels. When data fusion involves imagery and LiDAR whose value is point elevation, the issue becomes more complex because pixel value and point elevation are not commensurate with each other, which is a hurdle for traditional perception of remote sensing data to overcome.

Under the new thinking about big remote sensing data as curves, this hurdle is much easier to overcome. For example, a waveform not only is a curve but also can be considered as a time-varying frequency distribution function of return signal strength (i.e., histogram) (Muss et al., 2011; Zaletnyik et al., 2010). The fact that they are of the same mathematical nature provides the theoretic basis for conceptualizing LiDAR waveform, synthesized pseudo-wave forms, and object level spectral histogram as commensurate variables. When these histogram "curves" are normalized, they all become probability distribution "curves", which make their subsequent integration with each other legitimate.

Compatible processing units: With commensurate variables defined, compatible processing units are the next important issue to solve. The processing units for imagery with moderate spatial resolution (such as 30 m resolution Hyperion hyperspectral data) are regularly pixels, while those for imagery with high spatial resolution (such as WorldView-2, 3 data) are segments or objects. The footprints of LiDAR data, however, are of varied sizes and shapes, be it large, medium, or small, and with irregular spacing between them, which match neither the regularly gridded pixels nor the boundary of image objects that vary dramatically in shape and size. This obstacle makes it difficult to fuse LiDAR and imagery data. We propose to still make pixels as compatible processing units for fusing coarse and moderate imagery with high density LiDAR, and objects as compatible processing units for fusing HSR imagery with LiDAR. To fuse them, however, we will need to first synthesize the pixel level or object-level pseudo-waveforms with the new footprint corresponding to the boundary of a pixel cell or an image object respectively using high density discrete-return or full waveform LiDAR by the method mentioned above. For object level fusion, segmentation of high-resolution data is needed to generate object boundaries first. The fusion can be achieved then by simply integrating the image "curve" layer and the LiDAR pseudo waveform as an additional "curve" layer to form a composite dataset. Each processing unit, either a pixel or an object, now contain "curves" made of array of values forming spectral signatures, object-level spectral histograms, or object-level pseudo-waveforms, instead of a single pixel value or an object-level statistical summary. The fused curve layers will then be subject to the curve matching classifiers to be discussed below for classification.

Matching strategy and algorithms: The curve matching classifiers to be developed operate on the idea that assigns the curve of an unknown pixel or object to a known class that has the best resemblance based on the degree of matching of the curves involved. We propose two different matching strategies. (1) Curve-to-curve matching, (2) Curve-to-surface matching. For both matching strategies, innovative methods will be developed to improve their generalization ability. Generalization is the ability to predict correct outputs for unseen inputs. Statistical classifiers such as MLC have generalization ability with the probability density surface derived from the training data. With good generalization, a pixel can be assigned to its desired class based on the highest probability determined by the probability density surface of the trained model no matter it is seen or unseen in the training data. Curve-to-curve matching operates on the direct comparison of an unknown curve with reference curves of known classes and assigning it to the reference with highest matching. Like EMBC algorithms, curve-to-curve matching may have low generalization ability because no trained model is involved. To ensure the generalization ability for the curve-to-curve matching, it is necessary to have multiple reference curves for each class. To reduce redundancy and maximize within-class variability, it is necessary to select a representative subset of references from the training samples.

Many algorithms that were designed to match spectral hyperspectral signatures, such as spectral angel mapper and root sum squared difference, can be extended, and reprogrammed as curve-to-curve matching algorithms. A rich collection of statistical methods used to compare two histogram or cumulative histogram distributions such as Kullback-Leibler (KL) divergence and Kolmogorov-Smirnov (KS) distance can also be generalized as curve-to-curve matching algorithms. The curve-to-surface matching algorithms, on the other hand, may have better generalization ability using training samples to define probability density surface. The neuro-fuzzy classifier that we designed originally for hyperspectral data (Qiu, 2008) by utilizing multiple endmembers of each class to form fuzzy membership surface is an option that is free from the curse of Hughes phenomenon. It can be easily extended as a general-purpose curve-to-surface matching algorithm to achieve generalization ability. Other machine learning methods, such as the one-dimensional deep learning algorithms can also be adopted to perform curve-to-surface matching.

Theoretical implication and widespread adoptions: The "curves" obtained or derived from the big remote sensing data above were conceptualized as histograms or cumulative histograms, which depict the frequency distribution of pixel values or returned pulses over spectral or height intervals. The transform of big remote sensing data into histogram is similar to the thinking of Fourier transform. This similarity has rendered significant implication for our new thinking. Fourier transform converts a whole image from the 2D spatial domain to the 2D frequency domain, resulting in a new set of algorithms for tasks impossible or very difficult to solve in the spatial domain. Our new thinking, on the other hand, transforms big remote sensing data from the 2D spatial domain to the 1D frequency domain for individual portions of the data (e.g., image objects or LiDAR footprints), giving rise to a new set of matching algorithms that could better capitalize the potentials of the emerging data.

Transforming data into 1D frequency domain is a simple aggregation calculation of data values across multiple intervals of certain measurement (e.g., spectral reflection, height, and time). Using this simple process, virtually all features obtained or derived from remote sensing data can be transformed as a certain type of curve, which promises a widespread adoption of this new thinking. Guided by this new thinking, we have already seen many innovative applications in our recent research, a few of which are exemplified below.

The increasing availability of new sensors provides large volume of multi-temporal data from which time series trajectories can be extracted as temporal-based frequency distribution "curves". Time series trajectory can reveal the distinctive phenological cycles of vegetation species, which allowed us to better identify winter wheat distribution (Zhang et al., 2019) and perform tree species classification of forest stands (Wan et al., 2021) through data fusion. We envision that time series trajectory can be also employed to conduct change detection analysis, where different types of changes are represented as curves with unique patterns.

It is well-known that the same histogram can have a very different spatial distribution of the input values. Curves that can capture spatial autocorrections such as empirical semivariogram, model-fitted semivariogram, and binary spatial covariogram, can be also included as additional curve layers to represent spatial distribution of pixels. By combining object-level spectral histogram and binary spatial covariogram curve layers, we were able to integrate within-object spectral variability and spatial distribution of the pixels to improve OBIA using curve matching algorithms (Tang et al., 2020). Additionally, we further integrated the within-object spectral variability with the between-object spatial association, which is represented by frequency curves of pairwise classes in four main directions (Tang et al., 2021). A new recurrent curve matching algorithm is also developed to facilitate the integration through deep relearning.

Recently, convolution neural networks (CNN) have been increasingly used to classify high spatial resolution images, because the deep features they generate during the training process can served as additional spatial texture layers for classification. These deep features can also be used to synthesize curves for each image object and used by curve matching algorithms for classification, which is anticipated to be much more effective than the current pixel based deep learning classification.

4.5 Conclusions

New big remote sensing data demand new algorithms to analyze them effectively. New algorithms should be derived from new thinking about the data instead of being confined by traditional perception. Adopting new artificial intelligence algorithms from other discipline could improve performance, but human intelligence from image analysts can still play an even more important role, which is what a geospatial researcher should contribute through creative thinking. In this chapter, we explored different ways to fully capitalize the potentials provided by emerging big remote sensing data, which led to a new thinking about remote sensing data based on curves. This new thinking has given rise to a new theoretical framework for processing and fusing various remote sensing data in an effective and consistent manner, which may fundamentally change how remote sensing data are being analyzed.

References

Arroyo, L. A., Healey, S. P., Cohen, W. B., Cocero, D., & Manzanera, J. A. (2006). Using object-oriented classification and high-resolution imagery to map fuel types in a Mediterranean region. *Journal of Geophysical Research: Biogeosciences, 111*(G04S04), 1–10.

Bajorski, P. (2004). In simplex projection methods for selection of endmembers in hyperspectral imagery. In *IEEE International on Geoscience and Remote Sensing Symposium. IGARSS'04* (pp. 3207–3210).

Cochrane, M. (2000). Using vegetation reflectance variability for species level classification of hyperspectral data. *International Journal of Remote Sensing, 21*, 2075–2087.

Duds, R. O., & Hart, P. E. (1973). *Pattern classification and scene analysis.* Wiley Interscience Press.

Fowler, R. A. (2000). The lowdown on LIDAR. *Earth Observation Magazine, 9*(3), 5.

Hestir, E. L., Khanna, S., Andrew, M. E., Santos, M. J., Viers, J. H., Greenberg, J. A., Rajapakse, S. S., & Ustin, S. L. (2008). Identification of invasive vegetation using hyperspectral remote sensing in the California Delta ecosystem. *Remote Sensing of Environment, 112*, 4034–4047.

Muss, J. D., Mladenoff, D. J., & Townsend, P. A. (2011). A pseudo-waveform technique to assess forest structure using discrete lidar data. *Remote Sensing of Environment, 115*, 824–835.

Qiu, F. (2008). Neuro-fuzzy based analysis of hyperspectral imagery. *Photogrammetric Engineering and Remote Sensing, 74*(10), 1235–1247.

Sridharan, H., & Qiu, F. (2013). Developing an object-based HSR image classifier with a case study using WorldView-2 data. *Photogrammetric Engineering & Remote Sensing, 79*(11).

Stow, D. A., Toure, S. I., Lippitt, C. D., Lippitt, C. L., & Lee, C. R. (2012). Frequency distribution signatures and classification of within-object pixels. *International Journal of Applied Earth Observation and Geoinformation, 15*(1), 49–56.

Tang, Y., Qiu, F., Jing, L., Shi, F., & Li, X. (2020). Integrating spectral variability and spatial distribution for object-based image analysis using curvematching approaches. *ISPRS Journal of Photogrammetry and Remote Sensing, 169*, 320–336.

Tang, Y., Qiu, F., Jing, L., Shi, F., & Li, X. (2021). A recurrent curve matching classification method integrating within-object spectral variability and between-object spatial association. *International Journal of Applied Earth Observation and Geoinformation., 101*, 102368.

Toure, S. I., Stow, D. A., Weeks, J. R., & Kumar, S. (2013). Histogram curve matching approaches for object-based image classification of land cover and land use. *Photogrammetric Engineering and Remote Sensing, 79*(5), 433–440.

Ussyshkin, V., & Theriault, L. (2011). Airborne lidar: Advances in discrete return technology for 3D vegetation mapping. *Remote Sensing, 3*(3), 416–434.

Ustin, S. L., Roberts, D. A., Gamon, J. A., Asner, G. P., & Green, R. O. (2004). Using imaging spectroscopy to study ecosystem processes and properties. *BioScience, 54*, 523–534.

Wan, H., Tang, Y., Jing, L., Li, H., Qiu, F., & Wu, W. (2021). Tree species classification of forest stands using multisource remote sensing data. *Remote Sensing., 13*, 144.

Yuhas, R., Goetz, A., & Boardman, J. (1992). Discrimination among semi-arid landscape endmembers using the Spectral Angle Mapper (SAM) algorithm, JPL. *Summaries of the Third Annual JPL Airborne Geoscience Workshop.*

Zaletnyik, P., Laky, S., & Toth, C. (2010). LIDAR waveform classification using self-organizing map. In *American Society for Photogrammetry and Remote Sensing Annual Conference 2010: Opportunities for Emerging Geospatial Technologies* (pp. 1055–1066).

Zhang, C., & Qiu, F. (2012). Mapping individual tree species in an urban forest using airborne lidar data and hyperspectral imagery: AAG remote sensing specialty group 2011 award winner. *Photogrammetric Engineering and Remote Sensing, 78*(10), 1079–1087.

Zhang, J. (2010). Multi-source remote sensing data fusion: Status and trends. *International Journal of Image and Data Fusion, 1*(1), 5–24.

Zhang, X., Qiu, F., & Qin, F. (2019). Identification and mapping of winter wheat by integrating temporal change information and Kullback-Leibler divergence. *International Journal of Applied Earth Observation and Geoinformation, 76*, 26–39.
Zhou, W. (2013). An object-based approach for urban land cover classification: Integrating LiDAR height and intensity data. *IEEE Geoscience and Remote Sensing Letters, 10*(4), 928–931.
Zhou, W., & Troy, A. (2008). An object-oriented approach for analysing and characterizing urban landscape at the parcel level. *International Journal of Remote Sensing, 29*(11), 3119–3135.
Zhou, Y., Qiu, F., Al-Dosari, A., & Alfarhan, M. (2015). ICESat waveform-based land cover classification using a curve matching approach. *International Journal of Remote Sensing, 36*(1), 36–60.

Chapter 5
GIScience from Viewpoint of Information Science

Zhilin Li and Tian Lan

Abstract The term "Geographical Information Science" (GIScience) was formally introduced in 1992, after 30-year development of Geographical Information Systems (GIS). The authors believe that it is the appropriate time to reexamine what GIScience should actually be, as it has reached an age of 30 years. In this article, it is noted that GIScience at its current content is focused on the "G" aspect and deals with the theoretical aspect of spatial data handling. However, it is argued that GIScience should also be a type of specialized information science (or a branch of information science) as geographical information is a special type of information. Then, it is pointed out that the foundation of developing GIScience as a branch of information science has been laid down already and it is time to develop theories behind such a science. This article provides an insight into the future development of GIScience.

Keywords GIScience · Spatial data handling · Specialized information science

5.1 Introduction

It is well-known that the first computerized GIS (Geographical Information Systems) in the world, i.e., "Canada Geographic Information System", was developed in 1963 by Roger Tomlinson to store, analyze, and manipulate data collected for the Canada Land Inventory. After nearly 30 years of development, (that is, when it reached an age of 30—the age of establishment in Chinese culture), the term GIScience (Geographical Information Science) was proposed by Michael Goodchild in 1992, in a paper (Goodchild, 1992) published at the "International Journal of Geographical Information Systems", which was based on his keynotes at the Fourth International Symposium on Spatial Data Handling and EGIS 91.

Z. Li (✉) · T. Lan
State-Province Joint Engineering Laboratory in Spatial Information Technology for High-Speed Railway Safety and Faculty of Geosciences and Environmental Engineering, Chengdu 611756, China
e-mail: dean.ge@home.swjtu.edu.cn

T. Lan
e-mail: lantian@my.swjtu.edu.cn

B. Li et al. (eds.), *New Thinking in GIScience*,
https://doi.org/10.1007/978-981-19-3816-0_5

In the past 30 years, lots of discussions on the contents of GISceince have been conducted (see next section for details) and a number of research agenda has been set (e.g., Elmes, 2004; McMaster & Usery, 2004; UCGIS, 2006). As a consequence, a range of accomplishments has been achieved and a lot of changes have happened. For example, the "International Journal of Geographical Information Systems" was renamed as "International Journal of Geographical Information Science" and "Cartography and Geographic Information Systems" as "Cartography and Geographic Information Science", in 1997. It is worthy of noting that the University Consortium for Geographic Information Science (UCGIS) was formed in 1995, to serve as "an effective, unified voice for the geographical information science research community; to foster multidisciplinary research and education; and to promote the informed and responsible use of geographical information science and geographic analysis for the benefit of society" (Elmes, 2005).

As GIScience is reaching the age of establishment, it is the time to reexamine what GIScience is meant currently, what GIScience should be from different perspectives and what is the future direction of development.

The remainder of this article is as follows. Section 5.2 examines the contents of GIScience with its current definitions; Sect. 5.3 argues that GIScience should be a specialized type of information science (or a branch of information science); Sect. 5.4 argues that the foundation of developing GIScience as a branch of information science has been laid down already and it is the time to develop theories behind such a science; and Sect. 5.5 presents an outlook.

5.2 GIScience in Its Current Definitions

Goodchild in his paper on GIScience tried to distinguish GIScience and GIS. He stated that "spatial data handling may describe what we do, but give no sense of why we do it" (Goodchild 1992). That is, while GIS answers "what" and "where", GIScience is concerned with the "how" (GISGeography, 2021). This implies that GIScience is concerned with the "G" aspect and deals with the theoretical aspect of spatial data handling.

Since 1992, a number of definitions have been produced by various organizations and individual researchers. But the contents are very similar although emphasis might be placed on different aspects.

In the NCGIA Core Curriculum in Geographic Information Science (Goodchild, 1997), GIScience is defined as "the science behind the technology, which considers the fundamental questions raised by the use of systems and technologies, and is the science needed to keep technology at the cutting edge".

In the article entitled "Geographic information science: Critical issues in an emerging cross-disciplinary research domain", which was a report on a workshop held in January 1999 at the National Science Foundation, Mark (2000) defined GIScience as "the basic research field that seeks to redefine geographic concepts and their use in the context of geographic information systems".

The University Consortium for Geographic Information Science provided an indirect definition of GIScience. It states "The University Consortium for Geographic Information Science is dedicated to the development and use of theories, methods, technology, and data for understanding geographic processes, relationships, and patterns. The transformation of geographic data into useful information is central to geographic information science" (UCGIS, 2002).

The Wikipedia (2021) defines GIScience as the "scientific discipline that studies the techniques to capture, represent, process, and analyze geographical information".

Examples of other definitions by individual researchers are as follows:

- "Any aspect of the capture, storage, integration, management, retrieval, display, analysis, and modeling of spatial data" (Wilson & Fotheringham, 2007).
- "The science associated with developing and advancing GIS and technologies" (Masucci, 2008).
- "The general knowledge and important discoveries that have made GI systems possible" (Longley et al., 2015).
- "A multi-disciplinary and a multi-paradigmatic field, where 'spatial thinking' is fundamental" (Cabrera-Barona, 2017).
- "The scientific discipline that studies data structures and computational techniques to capture, represent, process, and analyze geographic information" (Granell-Canut & Aguilar-Moreno, 2018).

From these definitions, it can be observed that GIScience has been defined as a scientific discipline dealing with geographical information, with emphasis on "G". However, we would like to argue that these definitions correctly reflect the contents of current GIScience, which places much emphasis on spatial data handing but pay too little attention to the properties and flow of information, as pointed out by Li (2002, 2021).

5.3 GIScience from the Viewpoint of Information Science

Indeed, in the "Introduction" to the book "Foundations of Geographic Information Science" edited by Duckham et al. (2003), it was stated that "Information science can be defined as the systematic study according to scientific principles of the nature and properties of information. Geographic information science is the subset of information science that is about geographic information". But there is no alternative definition of GIScience provided, from this perspective of information science.

We argue that geographical information science is a branch of information science, or a specialized type of information science (Li, 2002, 2021), because geographical information is a special type of information, just like physical geography is a branch of geography; and that a new definition of GIScience from the perspective of information science should be provided. The definition may be given by an analogy to that of information science.

"Information science is a discipline that investigates the properties and behavior of information, the forces governing the flow of information, and the means of processing information for optimum accessibility and usability. More precisely, such a field is concerned with that body of knowledge relating to the origination, collection, organization, storage, retrieval, interpretation, transmission, transformation, and utilization of information" (Borko, 1968). With an analogy to this definition of information science, GIScience can be described as a discipline that investigates the properties and behavior of geographical information, the forces governing the flow of geographical information, and the means of processing geographical information for optimum accessibility and usability. More precisely, such a field is concerned with that body of knowledge relating to the origination, collection, organization, storage, retrieval, interpretation, transmission, transformation, and utilization of geographical information.

By comparing this definition with the definitions described in Sect. 5.2, it is found that current GIScience particularly lacks theories for the properties and behavior of geographical information and the forces governing the flow of geographical information. These theories may be built upon the theories in information science.

5.4 GIScience as a Branch of Information Science

As information science studies the properties and behavior of information and the forces governing the flow of information, there must be a quantitative measure of information, first of all. The entropy introduced by Shannon (1948), also called Shannon entropy, serves for such a purpose (Bristow & Kennedy, 2015). In Shannon's work for digital communication, the information content contained in a message is defined as follows:

$$H(X) = -\sum_{i=1}^{n} P(x_i) \log_2 P(x_i) \tag{5.1}$$

where X is a discrete random variable with possible values of $\{x_1, x_2, \ldots, x_i, \ldots, x_n\}$, $P(x_i)$ is the probability of X taking the value of x_i and $H(X)$ is the Shannon entropy.

With entropy as the foundation, Shannon established the mathematical theory of communication, or simply as information theory (Shannon & Weaver, 1949). Since then, information theory has found wide applications in various fields, such as biology, cognitive science, econometrics, linguistics, medical science, neural computation, psychology, social sciences, remote sensing, and telecommunication (e.g., Deco & Obradovic, 1996; Judge & Mittelhammer, 2011; Uda, 2020). On the other hand, researchers do encounter difficulties in applying information theory to spatial sciences such as cartography and urban studies. Although the concept of spatial entropy has been proposed by some researchers (e.g., Batty, 1974), they are however

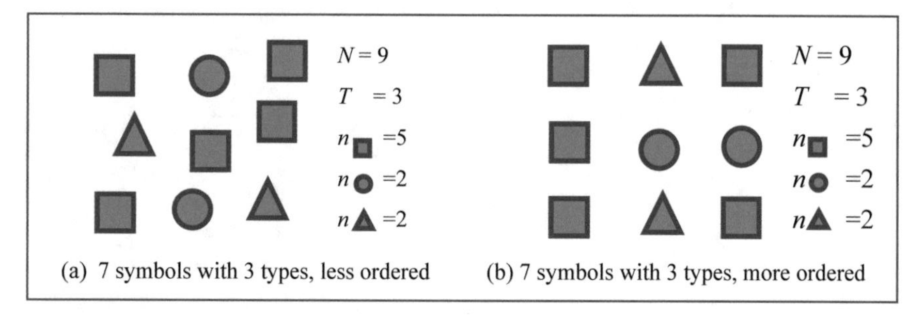

Fig. 5.1 Two maps with same number and types of symbols

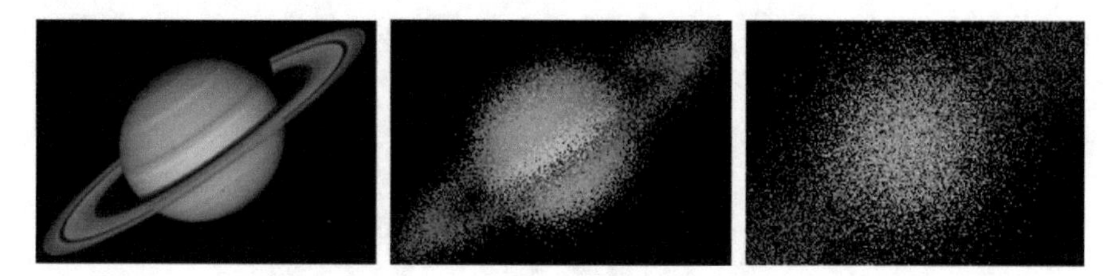

Fig. 5.2 Images with different spatial configurations but the same Shannon entropy

essentially statistics-based and have not taken spatial configurations into consideration. This is perhaps the reason why the theories of the GIScience as a branch of information science have not been developed yet.

The reason why information theory has difficulties in its applications to spatial science is that Shannon entropy is a statistical measure, which does not take the spatial configurations into consideration (Li & Huang, 2002). This can be illustrated by the two examples shown in Figs. 5.1 and 5.2. In Fig. 5.1, the patterns of the two maps appear to be very different. However, they indeed possess the same amount of Shannon entropy because both of them have the same number of symbol types and same number of symbols in each type. In Fig. 5.2, the left image is randomized into different degrees, leading to different spatial configurations, however they still possess the same amount of Shannon entropy because the number of pixels for each grey level is identical.

To provide an insight into this problem, Li and Huang (2002) pointed out that (a) there should be four types of information for a set of spatial data, i.e., statistical information, metric information, thematic information, and topological information; and (b) entropies for the other three types of information should be clearly defined, apart from the current Shannon entropy for statistical information. As a consequence, they developed mathematical models of entropies for metric information, thematic information, and topological information. Figure 5.3 and Eq. (5.2) show the definition of metric information. The map space is tessellated by Voronoi diagram of map symbols and the Voronoi region of each symbol is regarded as the zone of influence of the symbol. Then the entropy for metric information, donated as $H(M)$, is defined

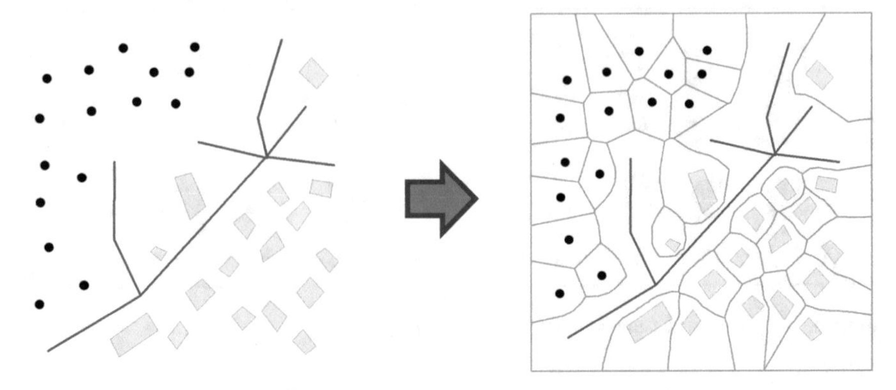

Fig. 5.3 A map and its Voronoi diagram (modified from Li & Huang, 2002)

as follows:

$$H(M) = -\sum_{i=1}^{N}\left(\frac{S_i}{S}\right) \times \log\left(\frac{S_i}{S}\right) \tag{5.2}$$

where $S_i(i = 1, 2, \ldots, N)$ is the Voronoi region of the ith map symbol; S is the whole map space (i.e. $S = \sum S_i$); and N is the total number of map symbols.

This set of information works well for spatial data in vector mode (e.g., maps) but not convenient for raster data (e.g. images and land-cover maps). For the latter, Boltzmann entropy (Perrot, 1998) is the solution. The basic equation was proposed by physicist Ludwig Boltzmann in the 1870s as follows:

$$S = k_B \log(W) \tag{5.3}$$

where W is the number of microstates that belongs to a defined macrostate (macroscopic state) for a thermodynamic system, k_B is a constant, equal to $1.38 * 10^{-23}$ J/K.

Although Boltzmann entropy is fundamentally important, and researchers have made efforts to apply this measure to different disciplines, yet it remains largely at a conceptual level. This difficulty has been emphasized by Bailey (2009), that is, "quite problematic when the notion of entropy is extended beyond physics, and researchers may not be certain how to specify and measure the macrostate/microstate relations". Fortunately, Gao et al. (2017) have solved the problem by using the multi-scale images to define these two concepts. Figure 5.4 illustrates the relationship between a given macrostate and its microstates of a (2 × 2) image. Figure 5.5 illustrates the relationship between a given macrostate and its microstates of a (2 × 2) categorical maps.

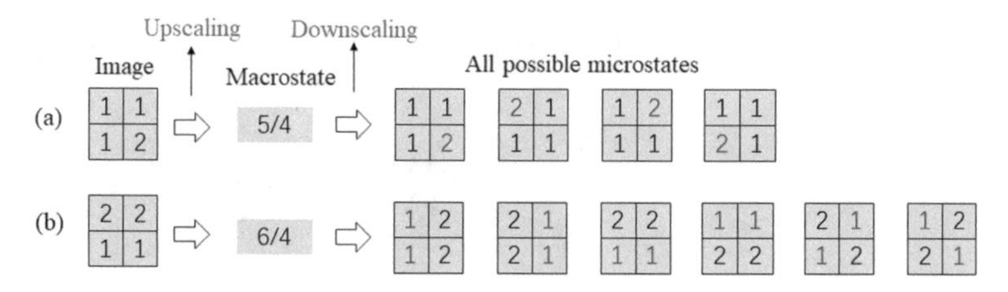

Fig. 5.4 Macrostates of two simple images and corresponding possible microstates (modified from Gao et al., 2017)

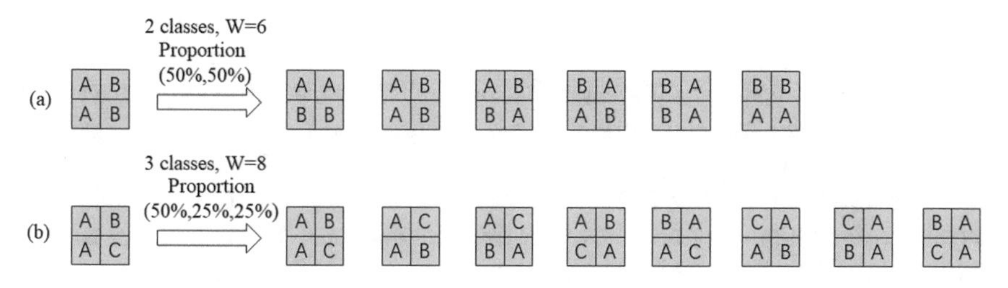

Fig. 5.5 Macrostates of two simple categorical maps and corresponding possible microstates

With the establishment of these mathematical models, it is now possible to build theories of GIScience upon such a foundation. Indeed, information theory of cartography has recently been built based on such a basis (Li et al., 2021).

5.5 Outlook

In this article, we argued that (a) existing GIScience is focused on the theories behind spatial data handling and pays too little attention to the information aspect; (b) GIScience is a branch (or specialized type) of information science as geographical information is a special type of information, thus theories for GIScience from the perspective of information science must be developed; (c) now it is possible to build such theories as the foundation for spatial information since the entropies for geographical information have been developed to overcome the shortcomings of Shannon entropy.

It is noted that the GIScience theories from the perspective of information science would be as important as those from the perspective of spatial data dandling because information flow is essential to our world. According to the triangular relations among matter, energy, and information (Somma, 2009), information is one of the three major elements (the other two are matter and energy) of the real world (see Fig. 5.6). It is believed by many that, without matter, nothing would exist; without energy, nothing would happen; without information, nothing would be meaningful.

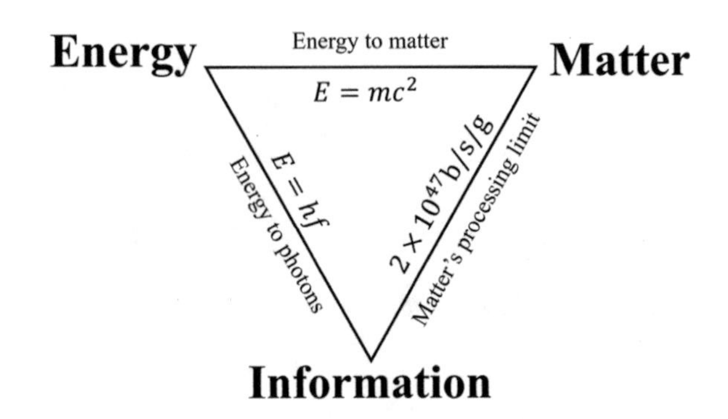

Fig. 5.6 Resource triangle of matter, energy, and information modified from Somma, 2009)

It is believed that the GIScience theories built upon information science will become more and more important as information itself will become more and more important. Indeed, a growing number of researchers (Wheeler, 1990, Seife, 2007, Vedral, 2012, Becker, 2020, Davies & Gregersen, 2014) have been wondering whether information may be primary, or more fundamental than matter and energy. For example, Wheeler (1990) advocated the "physical world as mode of information, with energy and matter as incidental". Therefore, it is expected that more efforts will be made in building GIScience theories upon information science.

Acknowledgements This work is supported by a NSFC key project (41930104, 41971330 and 42101442) and a Hong Kong RGC GRF project (#152219/18E).

References

Bailey, K. D. (2009). Entropy systems theory. *Systems science and cybernetics*. Eolss Publishers/UNESCO.

Batty, M. (1974). Spatial entropy. *Geographical Analysis*, *6*(1), 1–31

Becker, K., Is information fundamental? https://www.pbs.org/wgbh/nova/article/is-information-fundamental/. Accessed on February 8, 2020.

Borko, H. (1968). Information science: What is it? *American Documentation, 19*(1), 3–5.

Bristow, D., & Kennedy, C. (2015). Why do cities grow? Insights from nonequilibrium thermodynamics at the urban and global scales. *Journal of Industrial Ecology, 19*(2), 211–221.

Cabrera-Barona, P. (2017). From the 'Good Living' to the 'Common Good': What is the role of GIScience? Joint Urban Remote Sensing Event (JURSE) (pp. 1–4).

Davies, P., & Gregersen, N. H. (2014). *Information and the nature of reality: From physics to metaphysics*. Cambridge University Press.

Deco, G., & Obradovic, D. (1996). *An information-theoretic approach to neural computing*. Springer.

Duckham, M., Goodchild, M. F., & Worboys, M. F. (2003). *Foundations of geographic information science*. Taylor and Francis Press.

Elmes, G. (2004). *The UCGIS Research Agenda—Driving forward by locating challenges*. www.directionsmag.com/entry/the-ucgis-research-agenda-driving-forward-by-locating-challenges/123681. Accessed on January 1, 2022.

Elmes, G. (2005). The university consortium for geographic information science: Shaping the future at ten years. *Transactions in GIS, 9*(3), 273–276.

Gao, P. C., Zhang, H., & Li, Z. L. (2017). A hierarchy-based solution to calculate the configurational entropy of landscape gradients. *Landscape Ecology, 32*(6), 1133–1146.

GISGeography. (2021). What is GIScience (Geographic Information Science)? https://gisgeography.com/giscience-geographic-information-science/. Accessed on January 1, 2022.

Goodchild, M. F. (1992). Geographical information science. *International Journal of Geographical Information Systems, 6*(1), 31–45.

Goodchild, M. (1997). *NCGIA Core Curriculum in Geographic Information Science.* http://www.ncgia.ucsb.edu/giscc/

Goodchild, M. F. (2011). Challenges in geographical information science. *Proceedings of the Royal Society a: Mathematical, Physical and Engineering Sciences, 467*(2133), 2431–2443.

Granell-Canut, C., & Aguilar-Moreno, E. (2018). Geospatial influence in science mapping. *Encyclopedia of information science and technology* (4th ed., pp. 3473–3483).

Judge, G. G., & Mittelhammer, R. C. (2011). *An information theoretic approach to econometrics.* Cambridge University Press.

Li, Z. L., & Huang, P. Z. (2002). Quantitative measures for spatial information of maps. *International Journal of Geographical Information Science, 16*(7), 699–709.

Li, Z. L. (2002). *The "I" in G.I.S., Invited Talk, Central South University, China.*

Li, Z. L. (2021). From spatial data handling to GIScience. Theoretical Perspectives on GISci: Panel Discussion. In *The 28th International Conference on Geoinformatics (On-line conference),* November 1, 2021.

Li, Z. L., Gao, P., & Xu, Z. (2021). Information theory of cartography: An information-theoretic framework for cartographic communication. *Journal of Geodesy and Geoinformation Science, 4*(1), 1–16.

Longley, P., Goodchild, M. F., Maguire, D. J., & Rhind, D. W. (2015). *Geographic information science and systems.* Wiley.

Mark, D. M. (2000). Geographic information science: Critical issues in an emerging cross-disciplinary research domain. *URISA Journal, 12*(1), 45–54.

Masucci, M. (2008). Interrelationships between Web-GIS and E-collaboration research. *Encyclopedia of E-Collaboration* (pp. 405–410). IGI Global.

McMaster, R. B., & Usery, E. L. (2004). *A research agenda for geographic information science.* CRC Press.

Perrot, P. (1998). *A to Z of thermodynamics.* Oxford University Press.

Seife, C. (2007). *Decoding the universe: How the new science of information is explaining everything in the cosmos, from our brains to black holes.* Penguin Publishing Group.

Shannon, C. E. (1948). A mathematical theory of communication. *The Bell System Technical Journal, 27*(3), 379–423.

Shannon, C. E., & Weaver, W. (1949). *The mathematical theory of communication.* University of Illinois Press.

Somma, R. (2009). Matter-energy and information. *Ideonexus.* https://ideonexus.com/2009/05/05/matter-energy-and-information/

UCGIS. (2002). *UCGIS bylaws.* Available from http://www.ucgis.org/fByLaws.html. Last accessed on June 14, 2002.

UCGIS. (2006). *UCGIS Research Agenda.* https://gistbok.ucgis.org/publication/research-priorities. Last accessed on July 10, 2018.

Uda, S. (2020). Application of information theory in systems biology. *Biophysical Reviews, 12*(2), 377–384.

Vedral, V. (2012). *Decoding reality: The universe as quantum information* (p. 2012). Oxford University Press.
Wheeler, J. (1990). *Information, physics, quantum: The search for links*. Addison-Wesley.
Wikipedia. (2021). *The definition of geographic information science (GIScience)*. https://en.wikipedia.org/wiki/Geographic_information_science. Last accessed on January 1, 2022.
Wilson, J., & Fotheringham, A. S. (2007). *The handbook of geographic information science*. Wiley.

Chapter 6
Towards Place-Based GIS

Song Gao

Abstract The space-place dichotomy has long been discussed in human geography, digital humanity, and more recently in cartography and geographic information science. Place-based GIS are not yet well developed, although there is an increasing interest in semantic and ontological approaches. In this chapter, I present the technological building blocks towards the implementation of an operational place-based GIS that requires the input of platial data from crowdsourced data streams, the understanding of place characteristics and associated human activities and cognition, the support of representation and computational models of place, and the development of platial analysis and visualization. Based on the literature review, I found that the platial analysis functionalities with regard to their spatial counterparts were not sufficiently implemented yet. Therefore, more researches are needed into the development of platial operators for place-based GIS.

Keywords Place · Place-based GIS · Platial analysis

6.1 Introduction

The space-place dichotomy has long been discussed in human geography, digital humanity, and more recently in cartography and geographic information science (GIScience) (Couclelis, 1992; Goodchild, 2011; Janowicz, 2009; Jones et al., 2008; MacEachren, 2017; Merschdorf & Blaschke, 2018; Pezanowski et al., 2018; Purves et al., 2019; Tang & Painho, 2021; Tuan, 1977; Winter et al., 2009). Place names are usually mentioned in human conversations while locations with underlying coordinate information (latitude and longitude) are used in digital navigation systems to answer the "where" questions. In the past decades' development of geographic information systems (GIS) and spatial analysis methods, there exist rich studies about the role of space but only a few about the role of place due to the challenges on conceptualization, digital representations, computational modeling and analysis

S. Gao (✉)
Geospatial Data Science Lab, Department of Geography, University of Wisconsin-Madison, Madison, USA
e-mail: song.gao@wisc.edu

B. Li et al. (eds.), *New Thinking in GIScience*,
https://doi.org/10.1007/978-981-19-3816-0_6

of place in GIS. The typical spatial perspective in GIS is based on geometric reference systems that include coordinates, objects/fields, distances, and directions; while the alternative place-based perspective is characterized by place descriptions and semantic relationships extracted from human discourses, experiences, and activities (Gao et al., 2013; Goodchild, 2011; Westerholt et al., 2020; Winter & Freksa, 2012). The concepts of place (e.g., neighborhoods, vague cognitive regions, and sense of place) are complex and difficult to handle in GIS. One gap lies between the vagueness and richness of place in the human mind and the formalization need for place-based representations and analytical operations in place-based GIS (or platial information systems). Therefore, one of the main goals of place-based GIS research is to integrate the concepts and characteristics of place into platial (or placial) data, operational and analytical standards in GIS (Purves et al., 2019; Tang & Painho, 2021). There have been recent discussions and reviews on the advancements towards place-based GIS. For example, Merschdorf and Blaschke (2018) discussed the role of place in various research branches of GIScience including critical GIS, participatory GIS, crowdsourced geographic information, semantics, and ontologies, etc. Giordano and Cole (2018) argued for a place-based GIS that can integrate quantitative spatial analysis and qualitative methods and data such as social networks and textual corpus. Purves et al. (2019) reviewed the key challenges in representation and modeling of place for information science. Westerholt et al. (2020) organized a special issue on place-based GIS in the journal of *Transactions in GIS* and argued the need for representational models, analytical approaches, and visualization methods for place in GIScience. Tang and Painho (2021) conducted a comprehensive literature review and bibliographic analysis on the topic of place-based research in GIScience. There should also be humans in the loop for place-based GIS. For example, Scheider and Janowicz (2014) showed how place references can be identified and localized by involving participants. Blaschke et al. (2018) addressed the importance of human language and culture differences reflected in place-based GIS. Shaw and Sui (2020) proposed a smart space-place (splatial) framework to synthesize multidimensional information of space and place to study human dynamics. Although the existence of intensive conceptual reviews, few studies have addressed the key technological issues for the development of an operational system for place-based GIS.

To this end, in this chapter, I focus on the discussion of technological building blocks towards the development and implementation of place-based GIS from a systematic perspective.

6.2 Building Blocks Towards Place-Based GIS

As shown in Fig. 6.1, the key technological building blocks for an operational place-based GIS mainly include the input data about various characteristics of place, the representation, and computational models of place, and platial analysis and visualization functionalities.

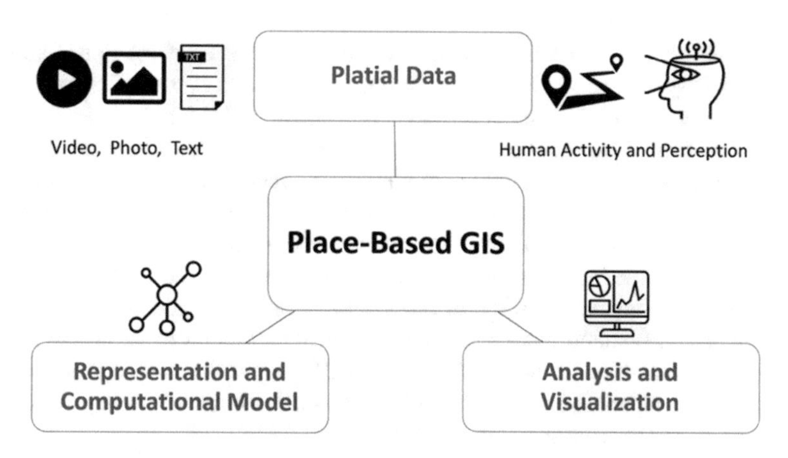

Fig. 6.1 Key building blocks for operational place-based GIS

6.2.1 Platial Data and Characteristics

Place can serve as a function between location and people (Mennis & Mason, 2016), a function of location, activity, and time (McKenzie & Adams, 2017), and a function of social relations (Giordano & Cole, 2018). Traditionally, data about places were collected through mapping agency (e.g., *gazetteer* usually includes place names and related entities) and survey-based narratives. The emergence of geospatial big data brings new opportunities to extract fine spatiotemporal resolution of human-place interaction data and understand the place semantics from large-scale volunteered geographic information and crowdsourced data streams, such as social media posts (including texts, photos, and videos), GPS trajectories, location-based social networks, comments and reviews on points of interest (POIs) or neighborhoods, and other Web documents (Gao et al., 2014, 2017; Hu et al., 2019; Kruse et al., 2021; McKenzie et al., 2015; Zhang et al., 2020). Those multiple data sources and advanced (geospatial) data science and machine learning methods provide a great opportunity to understand and extract the characteristics of place as well as associated human activities, experiences, emotions, and movements in different contexts.

6.2.2 Representation and Computational Models of Place

In order to effectively process, manage, analyze, and visualize the platial data, different approaches have been proposed for formalization, representation, and modeling of place in GIScience. Some key questions to ask include: What is a place? What are the core attributes and methods for a class of "place" in object-oriented programming and system design? Gao et al. (2017) represented a place as a field-object in which the degree of a location belongs to uses a membership function (e.g.,

Southern/Northern California) and demonstrated that place representations (thematic and culture aspects) might be relaxed in human cognition compared to the metric representation in GIS. Purves et al. (2019) demonstrated the linkage between the ontology of spatial information and the social and cognitive aspects of place and argued that places should include both names and geometries as well as relations. Hierarchical relationship is common to both physical systems and to human cognition (Golledge, 2002), e.g., river networks and administrative divisions. The hierarchical and other semantic relationships between places stored in the information systems support human cognition of places and their affordance in the real world. Recently, user-generated content has been used in extracting place characteristics and representations. For example, Wu et al. (2019) proposed a fuzzy formal concept analysis-based approach to uncovering spatial hierarchies among vague places (local toponym) extracted from social media data.

Digital gazetteers (i.e., dictionary of places) usually contain three core elements of geographic features: place names, place types, and spatial footprints (Hill, 2009). Place names are often used in human conversations and link to entities in gazetteers. A place may have more than one feature type based on different levels of categorization or using different schemata. The footprint of a place may be simply represented as a point in the information systems. However, it is challenging to select such a point for different types of places. The geometric center may not always be the best representative point. For example, one would not use the geometric center for a national park but use the entry points along the road networks. In addition, places may also be represented as areal objects (fields or polygons). Some places such as "downtown" are valuable in nature. Fuzzy-set-based methods and kernel density-based representations are usually used to model the intermediate boundaries of vague places (Burrough & Frank, 1996; Jones et al., 2008). Based on the assumption that a vague object can be viewed as the conceptualization of a field, a categorization framework including five distinct categories to formalize the semantic differences between vague objects using the fuzzy set theory is proposed by Liu et al. (2019), which can be used to model vague places.

Places can be modeled using graphs where nodes represent place entries and edges represent semantic relationships among places (e.g., part-of, directions, nearby). Patterns and relations between places can be computed and extracted from place graphs. For example, Chen et al., (2018a, 2018b) proposed a computation procedure to georeference textual place descriptions to gazetteer entries based on string and semantic similarities and qualitative spatial relationships using place graphs and natural language processing techniques. Zhu et al. (2020) analyzed place characteristics in geographic contexts through graph-based convolutional neural networks. Mai et al. (2019) proposed to represent places as Linked Data and demonstrated how to utilize Semantic Web reasoning and ontologies to extract and represent additional properties of places. A Linked Data connector was also developed as a set of ArcGIS's toolboxes to enable the retrieval, integration, and analysis of Linked Data from Web resources within GIS (Mai et al., 2019). In addition to the structured Linked Data, future developments of place-based GIS need to further integrate un-structured data about subjective human narratives about their experiences on places and dynamic relations among places over time.

6.2.3 Platial Analysis and Visualization

The spatial analysis and statistical functions are key capabilities of current GIS and based on the concepts of space, location, distance, and direction. Regarding the characteristics of place, what are the equivalent platial operation functionalities for their spatial counterparts? Gao et al. (2013) designed two platial analysis functions using semantics, namely *platial join* and *platial buffer*. Analogous to spatial join, the purpose of platial join is to attach the properties from the join entities to the target place using semantics (e.g., part-whole relation, qualitative spatial relations) rather than geometric constructs (e.g., geographic distance). The platial buffer might be able to mitigate the uncertainty issue of using spatial joins when objects are closed to border regions (Gao et al., 2013). In addition, platial buffer is to infer places or derive place-based knowledge through the connectivity, hierarchical relations, or other semantic relations between places. For example, using the semantic predicts between subway lines and shared transit stations as well as the station-to-station connectivity information, one can automatically generate a platial configuration on the subway system without accurate geometric information of the subway lines. The fundamental principles of platial operations rely on semantic relations between places rather than geometry. However, the operators for places in GIS are not sufficiently implemented yet. More researches are needed into the development of functionalities for place-based GIS. Some of the research directions may include platial associations, platial focal/zonal/global analyses on place graphs.

Recent advancements in geospatial artificial intelligence (GeoAI) and geospatial data science provide new opportunities for place-based analysis (Janowicz et al., 2020). For example, A quantitative measurement framework for place locale was developed using urban scene elements obtained from street-level images using a deep learning model (Zhang et al., 2018). A data-driven approach was proposed to uncover the inconspicuous-nice places in cities using street view images and social media check-in records combined with deep convolutional neural networks (Zhang et al., 2020). Using AI-powered facial expression detection techniques, a computational framework for extracting human emotions from over 6 million georeferenced photos at different places was proposed to enrich the understanding of sense of place (Kang et al., 2019).

Geospatial semantic queries and visualizations are also important functions in place-based GIS. Yan et al. (2017) proposed a novel approach to reasoning about place type similarity and relatedness by learning embeddings of places from augmented spatial contexts. Papadakis et al. (2020) developed a rule-based framework to support functional queries of a place (e.g., shopping areas). Hu et al. (2015) constructed thematic and geographic matching features from the textual descriptions of places and implemented semantic search for linked-data-driven geoportals on ArcGIS Online using geospatial and semantic expansion operators. As for visualization, MacEachren (2017) addressed the importance of leveraging geospatial big data with visual analytics to understand places and their inter-connectedness. A geovisual analytics framework *SensePlace3* for place–time–attribute information is proposed and implemented by Pezanowski et al. (2018). Moreover, graph visualization and interactive visual analytics techniques would be useful for place-based knowledge discovery and supporting human decision making.

6.3 Conclusion

In this chapter, I discussed the notion of place in GIScience and presented the technological building blocks towards the development and implementation of an operational place-based GIS, which requires the input of platial data, the understanding of its characteristics, the support of representation and computational models of place, and platial analysis and visualization. Joint efforts from multiple disciplines such as human geography, computer science, cartography and GIScience can facilitate the design and development process towards future place-based GIS.

References

Blaschke, T., Merschdorf, H., Cabrera-Barona, P., Gao, S., Papadakis, E., & Kovacs-Györi, A. (2018). Place versus space: From points, lines and polygons in GIS to place-based representations reflecting language and culture. *ISPRS International Journal of Geo-Information, 7*(11), 452.

Burrough, P. A., & Frank, A. U. (1996). *Geographic objects with indeterminate boundaries*. Taylor & Francis.

Chen, H., Winter, S., & Vasardani, M. (2018a). Georeferencing places from collective human descriptions using place graphs. *Journal of Spatial Information Science, 2018*(17), 31–62.

Chen, H., Winter, S., & Vasardani, M. (2018b). Georeferencing places from collective human descriptions using place graphs. *Journal of Spatial Information Science, 17*, 31–62.

Couclelis, H. (1992). Location, place, region, and space. *Geography's Inner Worlds, 2*, 15–233.

Gao, S., Janowicz, K., McKenzie, G., & Li, L. (2013). Towards Platial joins and buffers in place-based GIS. In *COMP'13: Proceedings of The First ACM SIGSPATIAL International Workshop on Computational Models of Place*, November 2013 (pp. 42–49).

Gao, S., Janowicz, K., Montello, D. R., Hu, Y., Yang, J. A., McKenzie, G., Ju, Y., Gong, L., Adams, B., & Yan, B. (2017). A data-synthesis-driven method for detecting and extracting vague cognitive regions. *International Journal of Geographical Information Science, 31*(6), 1245–1271.

Gao, S., Li, L., Li, W., Janowicz, K., & Zhang, Y. (2014). Constructing gazetteers from volunteered big geo-data based on Hadoop. *Computers, Environment and Urban Systems, 61*, 172–186.

Giordano, A., & Cole, T. (2018). The limits of GIS: Towards a GIS of place. *Transactions in GIS, 22*(3), 664–676.

Golledge, R. G. (2002). The nature of geographic knowledge. *Annals of the Association of American Geographers, 92*(1), 1–14.

Goodchild, M. F. (2011). Formalizing place in geographic information systems. In L. Burton, S. Kemp, M.-C. Leung, S. Matthews, & D. Takeuchi (Eds.), *Communities, neighborhoods, and health* (pp. 21–33). Springer.

Hill, L. L. (2009). *Georeferencing: The geographic associations of information*. MIT Press.

Hu, Y., Deng, C., & Zhou, Z. (2019). A semantic and sentiment analysis on online neighborhood reviews for understanding the perceptions of people toward their living environments. *Annals of the American Association of Geographers, 109*(4), 1052–1073.

Hu, Y., Janowicz, K., Prasad, S., & Gao, S. (2015). Metadata topic harmonization and semantic search for linked-data-driven geoportals: A case study using ArcGIS online. *Transactions in GIS, 19*(3), 398–416.

Janowicz, K. (2009, September). The role of place for the spatial referencing of heritage data. In *The Cultural Heritage of Historic European Cities and Public Participatory GIS Workshop* (Vol. 9, p. 57).

Janowicz, K., Gao, S., McKenzie, G., Hu, Y., & Bhaduri, B. (2020). GeoAI: Spatially explicit artificial intelligence techniques for geographic knowledge discovery and beyond. *International Journal of Geographical Information Science, 34*(4), 625–636.

Jones, C. B., Purves, R. S., Clough, P. D., & Joho, H. (2008). Modelling vague places with knowledge from the Web. *International Journal of Geographical Information Science, 22*(10), 1045–1065.

Kang, Y., Jia, Q., Gao, S., Zeng, X., Wang, Y., Angsuesser, S., Liu, Y., Ye, X., & Fei, T. (2019). Extracting human emotions at different places based on facial expressions and spatial clustering analysis. *Transactions in GIS, 23*(3), 450–480.

Kruse, J., Kang, Y., Liu, Y. N., Zhang, F., & Gao, S. (2021). Places for play: Understanding human perception of playability in cities using street view images and deep learning. *Computers, Environment and Urban Systems, 90*(101693), 1–15.

Liu, Y., Yuan, Y., & Gao, S. (2019). Modeling the vagueness of areal geographic objects: A categorization system. *ISPRS International Journal of Geo-Information, 8*(7), 306.

MacEachren, A. M. (2017). Leveraging big (geo) data with (geo) visual analytics: Place as the next frontier. *Spatial data handling in big data era* (pp. 139–155). Springer.

Mai, G., Janowicz, K., Yan, B., & Scheider, S. (2019). Deeply integrating linked data with geographic information systems. *Transactions in GIS, 23*(3), 579–600.

McKenzie, G., & Adams, B. (2017). Juxtaposing thematic regions derived from spatial and platial user-generated content. In *13th international Conference on Spatial Information Theory (COSIT 2017)*. Schloss Dagstuhl-Leibniz-Zentrum fuer Informatik.

McKenzie, G., Janowicz, K., Gao, S., Yang, J. A., & Hu, Y. (2015). POI pulse: A multi-granular, semantic signature–based information observatory for the interactive visualization of big geosocial data. *Cartographica: The International Journal for Geographic Information and Geovisualization, 50*(2), 71–85.

Merschdorf, H., & Blaschke, T. (2018). Revisiting the role of place in geographic information science. *ISPRS International Journal of Geo-Information, 7*(9), 364.

Mennis, J., & Mason, M. J. (2016). Modeling place as a relationship between a person and a location. In *International Conference on GIScience Short Paper Proceedings* (Vol. 1, No. 1).

Westerholt, R., Mocnik, F. B., & Comber, A. (2020). A place for place: Modelling and analysing platial representations. *Transactions in GIS, 24*(4), 811–818.

Papadakis, E., Resch, B., & Blaschke, T. (2020). Composition of place: Towards a compositional view of functional space. *Cartography and Geographic Information Science, 47*(1), 28–45.

Pezanowski, S., MacEachren, A. M., Savelyev, A., & Robinson, A. C. (2018). SensePlace3: A geovisual framework to analyze place–time–attribute information in social media. *Cartography and Geographic Information Science, 45*(5), 420–437.

Purves, R. S., Winter, S., & Kuhn, W. (2019). Places in information science. *Journal of the Association for Information Science and Technology, 70*(11), 1173–1182.

Scheider, S., & Janowicz, K. (2014). Place reference systems: A constructive activity model of reference to places. *Applied Ontology, 9*, 97–127.

Shaw, S. L., & Sui, D. (2020). Understanding the new human dynamics in smart spaces and places: Toward a splatial framework. *Annals of the American Association of Geographers, 110*(2), 339–348.

Tang, V., & Painho, M. (2021). Operationalizing places in GIScience: A review. *Transactions in GIS, 25*(3), 1127–1152.

Tuan, Y. F. (1977). *Space and place: The perspective of experience*. University of Minnesota Press.

Yan, B., Janowicz, K., Mai, G., & Gao, S. (2017). From ITDL to Place2Vec: Reasoning about place type similarity and relatedness by learning embeddings from augmented spatial contexts. In *Proceedings of the 25th ACM SIGSPATIAL International Conference on Advances in Geographic Information Systems* (pp. 1–10).

Winter, S., & Freksa, C. (2012). Approaching the notion of place by contrast. *Journal of Spatial Information Science, 5*, 31–50.

Winter, S., Kuhn, W., & Krüger, A. (2009). Guest editorial: Does place have a place in geographic information science? *Spatial Cognition & Computation, 9*(3), 171–173.

Wu, X., Wang, J., Shi, L., Gao, Y., & Liu, Y. (2019). A fuzzy formal concept analysis-based approach to uncovering spatial hierarchies among vague places extracted from user-generated data. *International Journal of Geographical Information Science, 33*(5), 991–1016.

Zhang, F., Zhang, D., Liu, Y., & Lin, H. (2018). Representing place locales using scene elements. *Computers, Environment and Urban Systems, 71*, 153–164.

Zhang, F., Zu, J., Hu, M., Zhu, D., Kang, Y., Gao, S., Zhang, Y., & Huang, Z. (2020). Uncovering inconspicuous places using social media check-ins and street view images. *Computers, Environment and Urban Systems, 81*(101478), 1–14.

Zhu, D., Zhang, F., Wang, S., Wang, Y., Cheng, X., Huang, Z., & Liu, Y. (2020). Understanding place characteristics in geographic contexts through graph convolutional neural networks. *Annals of the American Association of Geographers, 110*(2), 408–420.

Chapter 7
The Bottom-Up Approach and De-mapping Direction of GIS

Xun Shi, Meifang Li, and Xia Li

Abstract We see that GIS is under a major expansion of incorporating more bottom-up methods. The bottom-up approach does not seek to build general/global and therefore likely complicated and delicate models or problem solvers. Instead, it employs local and simple operations, and resorts to intensive computation to achieve the global solution. The burgeoning and adoption of the bottom-up approach are motivated by the contemporary application problems dealt with by GIS, featuring complex systems and high uncertainty, and facilitated by the explosive advancement of modern computing capacity. We use problems of classification, assessment, estimation, and prediction to illustrate the distinction between the top-down and bottom-up approaches. We also point out that an outcome of this new expansion of GIS is that mapping is receding from its center-stage position in GIS.

Keywords Top-down · Bottom-up · Computation · Uncertainty · De-mapping

7.1 Introduction

As in many sciences, the top-down approach used to be dominant in GI Science. The top-down approach features generality and determinism. First, seeing each local space or individual as a substantiation of the global situation or general pattern, the top-down approach seeks to establish global/general models or laws (covering the entire geographic area, entire temporal period, and entire population), and then applies such models or laws to local areas or individuals to achieve classification, characterization, assessment, and/or prediction. The global/general models or laws

X. Shi (✉) · M. Li
Geography Department, Dartmouth College, 6017 Fairchild, Hanover, NH 03755, USA
e-mail: xun.shi@dartmouth.edu

M. Li
e-mail: meifang.li@dartmouth.edu

X. Li
School of Geographic Sciences, East China Normal University, Shanghai 200241, China
e-mail: lixia@geo.ecnu.edu.cn

B. Li et al. (eds.), *New Thinking in GIScience*,
https://doi.org/10.1007/978-981-19-3816-0_7

can be in the form of analytical procedures, differential equations, regression models, artificial neural networks, decision trees, and so on. Second, the top-down approach is often deterministic, assuming accurate and precise data, representative samples, constant parameter values, fixed processes, and rational behaviors. Within the context of GIS, some early, basic, and most commonly used analytical procedures can be considered top-down, including simple buffer, simple overlay, certain network analyses, and some interpolations. The parameter setting used in those analyses is typically deterministic, represented by the clear-cut boundary, fixed buffer distance, even distribution, centroid representation, constant driving speed, crisp classification, and modeled semivariance.

The success of Newton's laws in physics encouraged and eventually defined the top-down methodology in sciences such as physics, chemistry, and biology. However, the limitations of the top-down approach are also obvious: it needs or assumes controlled experiments; it tends to ignore uncertainty in the data and process; and, while it is likely to oversimplify the complicated and complex reality, ironically the model itself tends to be complicated and delicate, as it tries to be general to cover all different situations.

On the other hand, in a sense, GIS was originally developed to automate the mapping process, for both visualization and analysis. The map has been the primary means for a human to handle spatial information. This is because such information is essentially about geometric features and relationships, and to human eyes and brains, nothing is more intuitive and effective to interpret, understand, and analyze such geometric information than seeing it represented and presented as graphics. Besides the cartographic functionality, some early and basic GIS analytical tools, such as buffer and overlay, can be considered as an emulation of manual processes with hardcopy maps.

The map-based operations of GIS are likely to be top-down. This is because what they are emulating, the manual process with hardcopy maps, is limited by calculation capability, manpower, and time. Due to such limitations, for the manual process people have to develop clever methods to summarize the data and generalize the methods, as well as to rely on assumptions, so as to make the process feasible and useful.

We see that GIS is currently under a major expansion of incorporating bottom-up methods and relatedly, exploring the de-mapping direction. This is due to that on the one hand, the GI Science is challenged by the size and complexity of contemporary application problems, such as climate change, globalization, urbanization, public health, land use dynamics, just to name a few, and on the other hand, the GI technology is facilitated by the explosive advancement of data handling capability and computing power supported by the modern computer technology. Section 7.2 will elaborate on these motivation and facilitation. Section 7.3 uses problems of classification, assessment, estimation, and prediction as examples to illustrate the distinction between the top-down and bottom-up approaches. Section 7.4 gives some concluding remarks.

7.2 Motivation and Facilitation for GIS to Incorporate Bottom-Up Methods

Geographic problems or phenomena are often too complicated to be characterized by global/general models (Batty & Longley, 1994). As Brown et al. (2005) summarized, "Geographical processes, such as diffusion of disease, wildfire spread, ecological evolution, transport and residential development, urban dynamics, and land-use changes, are usually highly complex and often include non-linear and emergence phenomena, stochastic components, feedback loops, and multiple equilibriums over various spatial and temporal scales. When characterizing or simulating these phenomena, small randomness may lead to a great deviation in the resulting pattern because of feedback effects". Unlike certain fields such as physics, chemistry, and biology, it is almost impossible to conduct condition-controlled experiments for analyzing and understanding geographic problems. The output from the effort of creating a general model or problem solver for geographic problems, especially those aforementioned contemporary regional or global problems, can be too complicated and too delicate, containing much more incompleteness, simplification, and uncertainties than its counterparts in other areas.

Alternatively, it seems that the bottom-up approach is often better than the top-down approach in representing, simulating, and understanding complex geographical phenomena. This notion was originated from Wolfram's argument that most, if not all, complex phenomena are eventually outcomes of local simple phenomena and their interactions, and a complex phenomenon at the global/general scale can be modeled by applying simple rules to local spaces or individuals (Wolfram, 2002). The bottom-up approach does not seek to build a global/general model beforehand. Instead, it builds relatively simple rules that characterize local situations and individual properties and behaviors, as well as their interactions. It then applies these rules to local spaces and/or individuals. This local and simple application may run repeatedly to mimic the real-world progression. It is expected that global/general patterns or solutions *emerge* from this evolving process. Typical implementations of the bottom-up approach include case-based reasoning (CBR), kernel density estimation (KDE), cellular automata (CA), and agent-based modeling (ABM). The nature of the bottom-up approach determines its characteristics or advantages, which can be summarized as follows:

- The bottom-up approach features simple rules for local modeling, leading to simple and local inference procedures, which are relatively easy to build, adjust, run, and calibrate, and as a result, are more likely to be robust.
- The bottom-up approach favors empirical simulation through intensive computation, and thereby has less reliance on statistical and other assumptions.
- The bottom-up approach is good at modeling stochastic processes and in turn, the uncertainty inherent to such processes, as it is easy for the approach to incorporate a certain amount of randomness.

- The bottom-up approach is good at representing interactions/feedback between localities and individuals, and thereby can capture subtle variation in dynamics and long-term evolution.

Methodologically and technically, the bottom-up approach is rather about computation than about deduction and/or statistical modeling. Therefore, it only became feasible when modern computing technologies became available. The quality of the output from a bottom-up method is closely related to the computing capacity the method can exploit. The current explosive advancement of computing capacity is undoubtedly a determining factor for the rapid popularization of the bottom-up approach.

Another facilitator of the bottom-up approach is the increasing availability of the data featuring vast extensivity and great details. By vast extensivity we mean large spatial extent, long temporal period, and big population. By great details we mean high spatial resolution, high temporal resolution, and less-aggregation. This type of data, on the one hand, impose demands for novel methodologies, as it became increasingly clear that conventional methodologies (typically being top-down) are not able to take full advantage of such data due to their size and detail level; on the other hand, the extensivity of and the detailed information in the data are necessary to support the bottom-up approach, which by nature is data-hungry.

Spatial analysis and spatial modeling that take the bottom-up approach represent a paradigm-shifting change, as they are no longer emulating map-based manual processes. As a result, they do not require input data to be in the conventional (graphic) map form and do not necessarily generate conventional maps as the output. With the bottom-up approach, the function of maps retreats to the secondary, only serving the purpose of human visualization. Even human visualization starts to lose its critical status, especially at the beginning of the analytical process (the widely regarded *exploratory data analysis* will become less critical). The function of the visualization seems to become primarily just for the human to interpret, understand, and utilize the output from the analysis.

7.3 Examples of Bottom-Up Methods

In this section, we use four generic problems to specifically compare top-down and bottom-up methods.

Classification: Widely used top-down classifiers include regression, maximum likelihood, artificial neural network, decision tree, genetic algorithm, Bayesian inference, support vector machine, and all their variants and descendants. They are top-down, because they all seek to build a global/general classifier through manipulating the available information, typically samples. They then classify new subjects using the constructed global classifier. On the contrary, the k-NN (k nearest neighbors) method and its alike, e.g., CBR (case-based reasoning), can be considered bottom-up, because they do not build a general classifier beforehand. Instead, when given a new subject, they use the samples (or cases) that are most similar to the new subject to achieve the classification. The similarity can be considered as the distance in the feature space and/or the spatial (or spatiotemporal) space, and thus the solution is considered local. In the field of knowledge systems, the top-down methods are referred to as *eager learning*, and the bottom-up methods are referred to as *lazy learning* (Shi et al., 2004).

Assessment: An application example of assessment is cluster (hotspot) detection in spatial epidemiology. The detection is to find out if the disease intensity at a location is abnormally high. A top-down method will build a general model, typically in the form of a polynomial, to characterize the spatial variation of the disease intensity in the study area, and then evaluates if the intensity at a location statistically significantly deviates from the general trend of the area, according to the general model (Shi & Wang, 2015). A bottom-up method, on the other hand, only measures the intensity locally and evaluates it locally (typically through Monte Carlo simulations), without trying to summarize the local values into a general representation of spatial variation (Shi & Wang, 2015). In a sense, the bottom-up methods in such detections correspond to the non-parametric methods in statistics.

Estimation: Individualized epidemic modeling for communicable diseases can be an example to illustrate the distinction between the top-down and bottom-up approaches to estimation. Conventional epidemic modeling is top-down, which works at the population level and treats the population spatially as an entirety. It represents the epidemic process with a series of differential equations (i.e., the general model) and estimates the parameter values in the equations (i.e., estimates characteristics of the epidemic, e.g., the reproduction number, R_t), by fitting actual data to the equations. Alternatively, the Epidemic Forest is a bottom-up modeling method. This method constructs tree structures to represent individual-level transmission relationships and their spatiotemporal features. It can fully utilize individual-level information about a disease case in all aspects, from genetic, biomedical, and epidemiological, to demographic, socioeconomic, geographic, and temporal. Epidemic characteristics and their spatiotemporal patterns can then be empirically derived from the tree structures (i.e., the forest) (Li et al., 2019, 2020).

Prediction: The distinction between the top-down and bottom-up approaches might be most obvious on the prediction problem. A typical example is modeling the land-use change. While a top-down method directly generates an overall predicted scenario, essentially an extrapolation with a general model derived from the historical data, the cellular automata (CA), a typical bottom-up method, simply lets the local situation evolve through interactions among nearby locations, each having particular properties determined by various human and physical factors, resulting in the *emergence* of the overall scenario from the local evolution (Chen et al., 2020; Li et al., 2017; Liu et al., 2017).

7.4 Concluding Remarks

The earliest GIS emulated map-based manual analytical processes. Ever since then, GIS has been expanding to incorporate analyses that do not rely on conventional map (graphic) representation. Many such analyses only use local information to generate local results, primarily for revealing or characterizing spatial variation. Besides those longstanding focal raster processes for terrain analysis or image processing, the relatively new geographically weight regression (GWR) is another example of such methods, and they can be generally called *kernel-based* methods. They are precursors of more formalized bottom-up methods exemplified by those discussed in Sect. 7.3. We see that GIS is currently under a major growth spurt of incorporating formalized modern bottom-up methods, due to the motivation and facilitation discussed in Sect. 7.2.

As an outcome of this type of expansion, mapping may give its center-stage position to computation in dealing with applications featuring complex systems and high uncertainty. In such applications, the importance of mapping in data representation is diminishing, and the map will be mainly for information presentation. GIS with visualization as auxiliary rather than essential functionality may become common.

References

Batty, M., & Longley, P. A. (1994). *Fractal cities: A geometry of form and function*. Academic Press Professional Inc.

Brown, D. G., Page, S., Riolo, R., Zellner, M., & Rand, W. (2005). Path dependence and the validation of agent-based spatial models of land use. *International Journal of Geographical Information Science, 19*(2), 153–174.

Chen, G., Li, X., Liu, X., Chen, Y., Liang, X., Leng, J., Xu, X., Liao, W., Qiu, Y. A., & Wu, Q. (2020). Global projections of future urban land expansion under shared socioeconomic pathways. *Nature Communications, 11*(1), 1–12.

Li, M., Shi, X., & Li, X. (2020). Integration of spatialization and individualization: The future of epidemic modelling for communicable diseases. *Annals of GIS, 26*(3), 219–226.

Li, M., Shi, X., Li, X., Ma, W., He, J., & Liu, T. (2019). Epidemic forest: A spatiotemporal model for communicable diseases. *Annals of the American Association of Geographers, 109*(3), 812–836.

Li, X., Chen, G., Liu, X., Liang, X., Wang, S., Chen, Y., Pei, F., & Xu, X. (2017). A new global land-use and land-cover change product at a 1-km resolution for 2010 to 2100 based on human–environment interactions. *Annals of the American Association of Geographers, 107*(5), 1040–1059.

Liu, X., Liang, X., Li, X., Xu, X., Ou, J., Chen, Y., Li, S., Wang, S., & Pei, F. (2017). A future land use simulation model (FLUS) for simulating multiple land use scenarios by coupling human and natural effects. *Landscape and Urban Planning, 168*, 94–116.

Shi, X., & Wang, S. (2015). Computational and data sciences for health-GIS. *Annals of GIS, 21*(2), 111–118.

Shi, X., Zhu, A.-X., Burt, E. J., Qi, F., & Simonson, D. (2004). A case-based reasoning approach to fuzzy soil mapping. *Soil Science Society of America Journal, 68*, 885–894.
Wolfram, S. (2002). *A new kind of science*. Wolfram Media Champaign.

Chapter 8
The Geography of Geography

Weihe Wendy Guan

Abstract There are many definitions for geography, most contain the word space or place. In order to foresee the future of geography, let us first examine the presence of the discipline, in particular, its variation in space. This chapter illustrates the distribution of global leading higher education institutions and compare that with the distribution of those leading the study of geography. Are they mostly overlapping? Or in some countries, do they deviate from each other? Among the leading institutions for the study of geography, are they focusing on physical geography, human geography, geographic information science, or all sub-disciplines? Among the leading institutions that are not strong in the study of geography, what are the related disciplines they choose to focus on? Is there a geographic variation in the composition of geographic education? If yes, how to describe it, and how to explain it? Do these patterns reveal any insight to the future of the discipline?

Keywords Geography · Higher education · Future

8.1 The Questions

Let us begin from a basic question—What is geography? We set off to search for the answer and found many. Below are a few examples.

- "Geography is the study of places and the relationships between people and their environment." (National Geographic Society, 2021) This definition focuses on people—places as defined by people, and environment as related to people.
- "Geography is a science that deals with the description, distribution, and interaction of the diverse physical, biological, and cultural features of the earth's surface." (Merriam Webster, 2021) This definition focuses on the earth's surface—inclusive to natural and cultural features.

W. W. Guan (✉)
Center for Geographic Analysis, Harvard University, 1737 Cambridge St., Suite 350, Cambridge, MA 02138, USA
e-mail: wguan@cga.harvard.edu

B. Li et al. (eds.), *New Thinking in GIScience*,
https://doi.org/10.1007/978-981-19-3816-0_8

- "Geography is the science of place and space. Geographers ask where things are located on the surface of the earth, why they are located where they are, how places differ from one another, and how people interact with the environment." (American Association of Geographers, 2021) This definition focuses on the inquiry—where, why, and how.

Instead of listing more definitions, let us try to summarize the key concepts we found in them: Geography studies the earth's surface; it is about space, place, regional difference, change over time, movement, and human–environment interaction; it explores natural and cultural features, phenomena, inhabitants, events, and processes; it is descriptive and analytical; qualitative and quantitative…

A better summary might be illustrated by the 5-sphere diagram—geography studies the earth's lithosphere, hydrosphere, atmosphere, pedosphere, and biosphere, plus the rapidly expanding human sphere (Fig. 8.1).

But there is a problem—all these spheres are already claimed by other sciences: lithosphere is the domain for geology and geomorphology, hydrosphere for hydrology and oceanography, atmosphere for climatology and meteorology, pedosphere for pedology, biosphere for biology and ecology, and human sphere for humanities, social sciences, as well as the relatively new field of environmental science (Fig. 8.2).

The new question becomes: Is Geography studying everything except objects in outer space? Or nothing unique to it at all?

Fig. 8.1 The five spheres of earth science

Fig. 8.2 Disciplines and their domains in earth science

Perhaps what matters more is not in what it studies, but how it studies. Jack Dangermond of ESRI in 2017 promoted a term "The Science of Where" (Vardhan, 2017). He was talking about a new image for the Geographic Information Science and Systems (GIS). "The Science of Where is applying a data-driven approach that uses geography to unlock the understanding (of the world)." Perhaps we could say Geography is The Science of Where in itself? The uniqueness of geography is at the "where", and the future of geography might lie in the "how"—spatial data science (or geographic data science, geodata science) represents a new approach for geographers to conduct their studies, a new trend comes with the rising tide of data science.

But wait, the definitions of geography are not so clear on whether geography is a study, a science, or an applied science. In general, a Study deals with observation, description, classification (what, who, where, when); a Science aims at explanation and prediction (why, how, how many); and an Applied Science, which leads to planning or engineering, focuses on prescription (what if). They are not exclusive to one another, rather stages in an evolution of human understanding. The future of geography might lie in this evolution.

Can we find any evidence to verify these bold predictions? Can we put the promise of geographic data science to the test, and search for evidence of geography's evolution from data?

8.2 The Exploration

Since the late 1940s, the identity and integrity of geography as an academic discipline has been in question (Smith, 1987). Many geography departments changed their names, and the trend has been accelerating in recent decades (Frazier, 2017). How is geography faring in the global high education landscape? We set off to find data that can present us an answer.

8.2.1 The Data

As a rudimentary attempt, we looked for data online on how universities are teaching geography. In particular:

- Where are the global leading higher education institutions, and where are universities leading the study of geography? Is there a regional difference?
- Among the leading institutions for geography, what sub-disciplines do they teach? Is there a regional difference?
- What topics are in the geography courses? Is there a regional difference?

There are many sources reporting the ranking of universities worldwide, but most of them do not rank universities by their geography standing. We found two sources useful, one is the QS Top Universities website (QS Quacquarelli Symonds Limited, 2019), the other is a database behind the 2019 AAG Guide to Geography Programs in the Americas (American Association of Geographers, 2021), in addition to individual university websites.

The QS Top Universities website provides worldwide university ranks by overall standing and geography standing. We visited websites of the top 100 universities leading in geography and populated a database with the type of degrees each of them offers in physical geography, human geography, regional geography, GI science, environmental and urban studies, historical geography, and general geography. We also georeferenced all the universities by location.

The AAG guide provided a list of the geography sub-disciplines of 311 universities in the Americas. We could not find similar data for other regions in the world, thus arbitrarily picked just the top three highly ranked universities leading geography in Europe and Asia respectively and extracted their geography course titles from the university websites, to form the corpse of sub-disciplines for comparison. The results are mapped and charted, and the findings are summarized below.

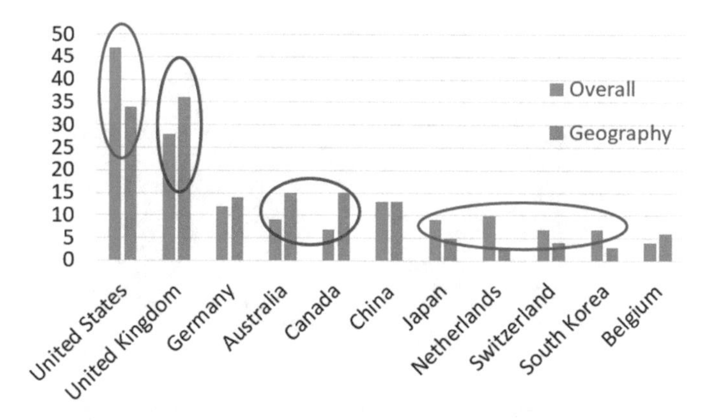

Fig. 8.3 Number of universities per country among the Global Top 200. Blue is by QS comprehensive ranking, and brown is by QS ranking of the geography subject

8.2.2 The Findings

Among the top 200 universities globally, if by overall standing (QS World University Rankings based on six key metrics), the US leads the count; if by geography standing (QS World University Rankings by subject), the UK leads. Australia and Canada also have more universities included in the top 200 by geography than that by overall standing. At a glance, the British Commonwealth is stronger in geography education than the rest of the world (Fig. 8.3).

From the websites of the top 100 universities leading in geography, we recorded the type of degrees each of them offers in (1) human-social geography (including human geography, historical geography, and regional geography), (2) physical and environmental geography (including environmental science, urban studies, and physical geography), and (3) Geographic Information Science, as well as general geography. The general pattern from the data shows that in UK and EU, the first category (human, regional and historical geography) seems more prevalent; in North America, the third category (GIS) seems more dominating; while in Asia, the second category (physical, environmental, GIS) seem more common. But the differences are not easy to present visually, and the data was manually gathered, most likely incomplete and lack of a thorough quality control.

To verify this pattern, we explored data from the AAG (for the American Universities) and individual university websites (Oxford, Cambridge and UCL for UK; NUS, HKU and PKU for Asia). In the course names word cloud, "Human" stands out in the UK universities (Fig. 8.4); "GIS" stands out in the American universities (Fig. 8.5); while "Urban" and "Planning" stand out in the Asian universities (Fig. 8.6). Geography and environmental are high frequency words in all three regions.

33	Geography	4	Political	3	Ecology
9	Environmental	4	Physical	2	Understanding
7	Human	4	Earth	2	Globali sation
6	Change	4	Urban	2	Biogeography
5	Environment	3	Fieldclass	2	Cartography
5	Development	3	Historical	2	Geopolitics
5	Processes	3	Overseas	2	Quaternary
5	Society	3	Cultural	2	Urbanism
5	Global	3	Economic	2	Politics
5	Field	3	Methods	2	Thinking

Fig. 8.4 Word cloud with geography course names in 3 UK universities (Oxford, Cambridge, UCL)

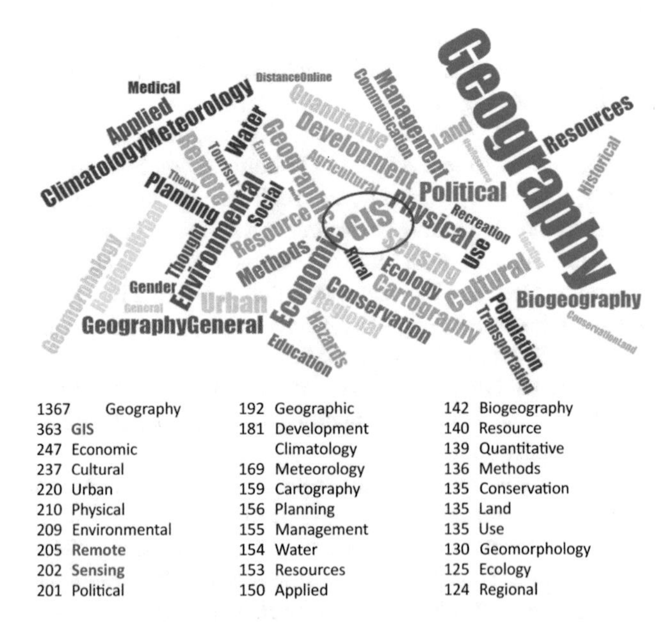

1367	Geography	192	Geographic	142	Biogeography
363	GIS	181	Development	140	Resource
247	Economic		Climatology	139	Quantitative
237	Cultural	169	Meteorology	136	Methods
220	Urban	159	Cartography	135	Conservation
210	Physical	156	Planning	135	Land
209	Environmental	155	Management	135	Use
205	Remote	154	Water	130	Geomorphology
202	Sensing	153	Resources	125	Ecology
201	Political	150	Applied	124	Regional

Fig. 8.5 Word cloud with geography discipline names in 311 American universities

37	Geography	9	Social	6	Human
25	Environmental	9	Global	6	Space
21	Urban	8	Geographies	5	Management
21	Planning	8	Environment	5	Resources
16	Introduction	8	Design	5	Research
16	Ecology	7	Methods	5	Sciences
14	Practice	7	Change	5	Tourism
11	Science	6	Development	5	Applied
10	Physical	6	Cultural	5	Remote
10	Field	6	World	5	Sensing

Fig. 8.6 Word cloud with geography course names in 3 Asian universities (NUS, HKU, PKU)

8.3 The Future

The preliminary data exploration shows that the UK is leading the world in geography study, particularly in the study of human geography; the Americas are embracing the GI science; while Asia is pushing forward on the applied front, not only in urban planning but also in resource management, development, tourism, etc.

Perhaps geography is evolving from a study to a science and continues towards an applied science. As a study, geography has a long history in fields such as Human Geography, Regional Geography, or Historical Geography, led by European universities. As a science, geography is thriving in fields such as Earth Science, Geographic Information Science, or Geospatial Data Science, led by North American universities. As an applied science, geography is branching out to relatively new fields such as GeoDesign, GeoPlanning, or GeoEngineering, led by Eastern Asian universities.

Nonetheless, the future of geography will evolve along multiple fronts. As Peng Liang wrote in his essay of "Geography: Un-diminishable Value" (in Chinse): "Geography reveals the laws of nature and demonstrates the essence of humanity." "The charm of geography lies at three levels: it serves the needs for human survival; it presents the beauty of nature; and it frames the background for civilization." (Liang, 2014).

Acknowledgements The author thanks David Liu, a high school student intern at Harvard's Center for Geographic Analysis in the summer of 2019 and now a student at Brandeis University, who collected most of the data used in this study. This chapter is revised from a position paper presented in the Future of Geography Conference in 2019.

References

American Association of Geographers. (2021, December 21). *Guide to Geography Programs in the Americas*. Retrieved from American Association of Geographers. http://www3.aag.org/guide

American Association of Geographers. (2021, December 21). *What Geographers Do*. Retrieved from American Association of Geographers. http://www3.aag.org/cs/jobs_and_careers/what_geographers_do/overview

Frazier, A. E. (2017). Renaming and Rebranding within U.S. and Canadian Geography Departments, 1990–2014. *The Professional Geographer, 69*(1), 12–21.

Liang, P. (2014). Geography: Un-diminishable Value. *Newsletters of the Geographical Society of China, 131*(4–5).

Merriam Webster. (2021, December 21). *Geography*. Retrieved from Merriam Webster Dictionary. https://www.merriam-webster.com/dictionary/geography

National Geographic Society. (2021, December 21). *What is Geography?* Retrieved from National Geographic Society. https://www.nationalgeographic.org/education/what-is-geography/

QS Quacquarelli Symonds Limited. (2019, 10 1). *QS World University Rankings by Subject*. Retrieved from QS Top Universities. https://www.topuniversities.com/subject-rankings/2019?utm_source=topnav

Smith, N. (1987). Academic war over the field of geography: The elimination of geography at Harvard, 1947–1951. *Annals of the Association of American Geographers, 77*(2), 155–172.

Vardhan, H. (2017, July 11). *Applying the Science of Where Through GIS*. Retrieved from Geospatial World. https://www.geospatialworld.net/blogs/the-science-of-where-gis/

Chapter 9
Classification and Description of Geographic Information: A Comprehensive Expression Framework

Guonian Lv, Zhaoyuan Yu, Linwang Yuan, Mingguang Wu, Liangchen Zhou, Wen Luo, and Xueying Zhang

Abstract Geography is a comprehensive discipline that studies spatial–temporal patterns, evolution processes, and interaction mechanisms of geographic objects and phenomena. With the evolution of the world from a binary space to a ternary space, it is urgent to deepen and expand the understanding, expression, and mining of geographic information. Most current GIS models use the geometry + combination to express geographic information. Geographic processes, including interplay among features, cannot be directly modeled under the above notion. Geography analyzes spatial and temporal structure of macroscopic patterns as a whole and studies evolutionary processes from the perspective of comprehensive integration, and reveals system structures from the perspective of the integrated role of

G. Lv (✉) · Z. Yu · L. Yuan · M. Wu · L. Zhou · W. Luo · X. Zhang
School of Geography, Nanjing Normal University, Nanjing 210023, Jiangsu, China
e-mail: gnlu@njnu.edu.cn

Z. Yu
e-mail: yuzhaoyuan@njnu.edu.cn

L. Yuan
e-mail: yuanlinwang@njnu.edu.cn

M. Wu
e-mail: Wumingguang@njnu.edu.cn

L. Zhou
e-mail: Zhouliangchen@njnu.edu.cn

W. Luo
e-mail: Luowen@njnu.edu.cn

X. Zhang
e-mail: Zhangxueying@njnu.edu.cn

Jiangsu Center for Collaborative Innovation in Geographical Information Resource Development and Application, Nanjing 210023, Jiangsu, China

Key Laboratory of Virtual Geographic Environment, Ministry of Education, Nanjing Normal University, Nanjing 210023, Jiangsu, China

B. Li et al. (eds.), *New Thinking in GIScience*,
https://doi.org/10.1007/978-981-19-3816-0_9

multiple elements. Based on the concept of ternary space, we identify seven dimensions of geographic information elements, which include semantics, spatial location, geometric structure, attribute, interrelationship, evolution process, and interplay mechanism. We also discussed how such representation framework can be employed under a geometric algebra approach to represent geographic scenes and to achieve a unified representation of the seven dimensions.

Keywords Geographic information · Ternary space · Elements of information · Dimensions of geographical description · Geographic information classification

9.1 Introduction

With the rapid development of ICT (Information, Communication and Technology), such as cloud computing, internet of things, virtual reality and other technologies, great progress and improvement have been made on the acquisition, analysis, and display of geographic information. Geographic information has become interlink, dynamic and ubiquitous. With the continuous increase of the heterogeneity of geographical information, the traditional geographic information description schema is prone problems, such as incomplete attribute, inaccurate semantics, and unclear evolutionary processes. How to build a comprehensive description framework of geographic information for the era of ICT is a frontier scientific problem.

Geographical information contains a variety of physical, human, social, and many other aspects. Geographic information can capture comprehensive properties of earth processes and phenomena that are multidimensional, and dynamic. In recent years, the description of geographic space has changed from a binary space (physical, and social-human world) to a ternary world (physical, social-human, and information world). In this context, geographic information description should include the interpretation of time, place, process, law, mechanism, and other elements. Therefore, we suggest a description framework that integrally expresses the seven dimensions: semantics, spatial location, geometric structure, attribute, interrelationship, evolution process, and interplay mechanism (Lü et al., 2017).

9.2 The Connotation of Geographic Information

9.2.1 Overall Framework

Based on the concept of a ternary world, we aim to develop an abstract mapping mechanism of the ternary world of physical, human, and information to the computer from the perspective of geography. In this chapter, a geographic information classification system is analyzed systematically, and a framework consisting of seven-dimensional descriptors of geographic information is given (Fig. 9.1). The physical,

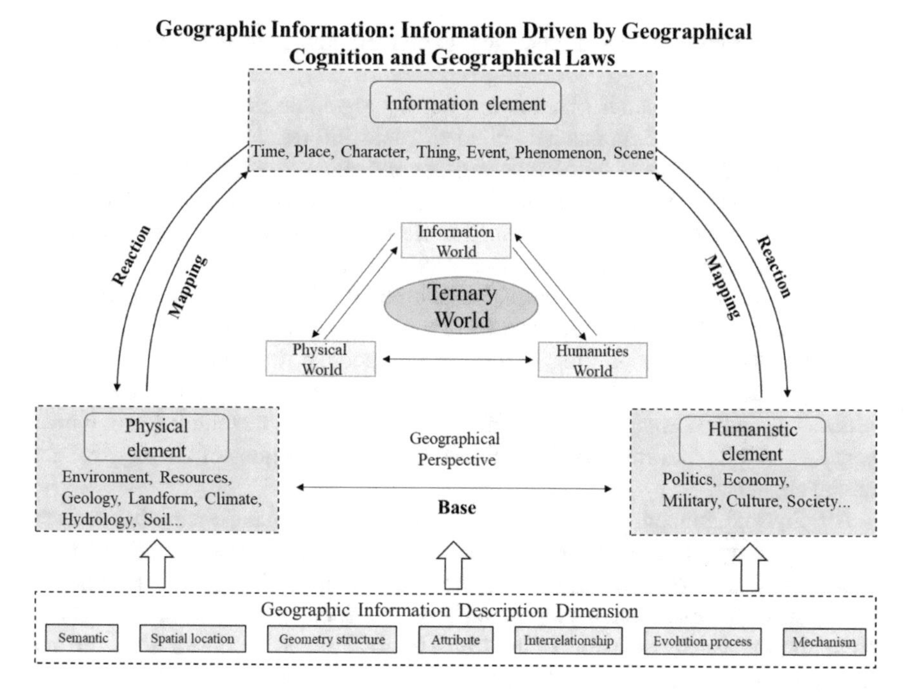

Fig. 9.1 Geographic information description framework

human, and information worlds are composed of various geographical elements. These geographical elements, which are the key concepts and expression units of geography, are interconnected and can be transformed as geographic information. The comprehensive expression of geographic information can then be organized with a unified and complete expression model. This unified expression model, as an organization of information, can provide a theoretical foundation that enables geographic information to incorporate complex geographic objects and geographic processes into geographic information systems.

9.2.2 Information Elements for Ternary Space

The traditional binary space is suited for the physical world which only perceived to be consisting of two parts: physical and human. However, with the rapid development of ICT, the word has evolved into a new generation that covers the physical world, the human world, and the information worlds (Zhou, 2015). The physical world refers to the physical environment in which humans live and the material systems contained therein; the human world refers to the sum of human behavior and social activities; and the information world is a virtual world built on the physical space and social space where the physical and human geographic information transformed from the real world is captured, stored, managed, expressed, and analyzed in the form of a digital twin.

Based on the perspective of a ternary world, geographic elements can be divided into physical elements, humanistic elements, and information elements. Physical elements include geology, landforms, weather and climate, hydrology, soil, biology, and other elements. Human elements include political, economic, military, cultural, social, and historical elements. Information elements include time, place, character, object, event, phenomenon, scene. The information world can carry and map the physical world and the human world. Meanwhile, it can further influence and even reconstruct the physical and human worlds employing multi-situation simulation and digital twins (Guo & Ying, 2017). In this context, information organization in geographic space needs a comprehensive description of physical, human, and information in the ternary world. The physical elements are the foundation of human elements, and the information elements are further abstractions of the physical and human elements.

All physical, human, and information elements complement each other and can be expressed by information elements. The development of GIS has accelerated the mutual integration of physical, human, and information worlds. The introduction of ternary space has expanded the representation dimensions of geographic information to capture the ternary elements of the world (Guo et al., 2018). The abstraction and description of the physical and human worlds from the perspective of information elements can be summarized into seven elements, namely time, place, character, object, event, phenomenon, and scene. The seven elements of information are the top-level abstraction of the ternary world, and each element can be further divided.

9.2.3 Seven Dimensions for Geographical Information Description

Geography is a science that studies spatial patterns, temporal evolutions of the human living environment, and the interaction between humans and their environment in Earth surface systems (Chen et al., 2019). Geographical information is fundamentally different from other information, with characteristics that are often regional, multidimensional, and changing over time and space. Traditionally, geographic information description follows the schema of "location + geometry + attribute" using discrete geometric objects such as points, lines, polygons, and volumes to approximate the complex real geographic world. This schema shows weakness in organizing complex and continuous geographic objects and in modeling geographic processes.

Recently, the concept of geographic scene has been applied to the expression of geographic information. A geographic scene is a specific synthetic region comprising physical, human, and information factors and their mutual relationships and interactions. A geographic scene has a specific structure and functions and is characterized by comprehensiveness. In our previous research, we proposed a representation model consisting of six elements of geographic scene (Lü et al., 2018): semantics, location, the shapes and attributes of geographic elements, the relationships among elements, and evolutionary processes. The ternary space model can be adopted to extend the concept of geographic scenes by refining the dimensions of shape and relationship as geometric structure and interrelationship, and by adding interaction mechanism as an additional dimension. These seven dimensions, namely semantics, spatial location, geometric structure, attributes, interrelationships, evolution processes, and interplay mechanisms, will be discussed as follows.

9.2.3.1 Semantics

Semantics refers to the geographic characteristics of information elements in geographic scenes that have been processed and recognized by humans. It is mostly indirect and obtained through reasoning. The semantic description of information elements may include definitions, classification systems, and schematic diagrams (schematics). Among them, definition is the connotation, including time, place, and scene. Classification system refers to the hierarchy of differentiating and associating diverse geographic elements in a given context. Schematic diagram is a graphical (often abstracted) representation of the connotation; it usually embodies a graphical description and decomposition of its essential characteristics.

9.2.3.2 Spatial Location

Spatial location is a description of the location of geographic entities, identifying where a geographic element is located, or where an event/process/phenomenon occurs. Coordinates are widely used to encode spatial location. More broadly, landmarks, placenames, and addresses, or even spatial relationships (e.g., near, opposite), can also provide useful references about spatial location as spatial identifiers.

9.2.3.3 Geometric Structure

Geometric structure is a description of the geometric forms of various geographic entities, including shapes, orientations, and reflections There are many types of geometric forms such as points, lines, polygons, and bodies in the existing geographic information system. In addition to traditional object descriptions, geometric descriptions and expressions can also be carried out through pixels and voxels. Pixel is a basic and atomic component of the two-dimensional grid, and a voxel can be considered as a three-dimensional volume pixel, which represents a basic and atomic component of the regular grid in a three-dimensional space.

9.2.3.4 Attribute

Attribute components are mainly dedicated to the non-spatial characteristics of spatial entities, such as land cover, temperature, and population. Attribute components can be recorded from multiple perspectives, such as geometric, physical, chemical, biological, cultural, social, and economic, among others. Note that attribute components can also evolve with temporal and/or spatial components of geographic entities.

9.2.3.5 Interrelationship

Interrelationship is used to describe the spatial and/or temporal relationships between information elements. The traditional geographic information representation model highlights spatial relationships (e.g., distance relationships, topological relationships, directional relationships), and often overlooks the interplay relationships among features. The relationships that expressed by differential equations, chemical equations, and information diagrams are also overlooked by current GIS data models. More interrelationships that focus on geology and landform, climate and hydrology, landform and vegetation and other physical elements, regions and economic development, social relations, as well as communities and other human factors could be included to form a comprehensive description of the relationships among physical, human, and information worlds.

9.2.3.6 Evolution Process

Evolution process refers to the change of information elements over time. The description of the evolution process should first include the description of the time information, including the point in time (t_i), the time snapshot ($\Delta t, dt$), or the process (∂t) description of the full life cycle. Descriptions of the state and behavior of the element at the time could also be attached.

9.2.3.7 Interplay Mechanism

Behind visual appearance, there are interplay mechanisms among geographic elements. These mechanisms mainly reveal the possible cause of material migration, energy conversion, and information transmission. The mechanism of actions among elements in the real world can be described, perceived, and analyzed through various functions such as scenes, maps, networks, and models. On the other hand, the mechanism of action can also be used as constraints and rules for the expression of geographic information.

9.3 Example of the New Geographic Information Description

For the seven-dimensional expression of geographic information, it is necessary to establish an innovative data model. Unlike traditional GIS data models that mainly focus on the organization of spatiotemporal information with related attributes, a new data model should be developed with consideration of all the seven dimensions. Thus, semantics, spatial location, geometric structure, attributes, interrelationships, evolution processes, and interplay mechanisms must be expressed structurally and unified in the data model. In our previous works, geometric algebra (GA), which is a high dimensional algebra system, is used to develop such data models. GA integrates geometric and algebraic expressions organically, and realizes the unified description of time and space, continuous and discrete, as well as unified measurement and operation in high-dimensional space through unique reversible geometric product operations and rich geometric algebra operators. The multi-dimensional unity and the coordinate independence of GA are used to construct a multi-dimensional unified expression and calculation model of geographic information, which provides high-dimensional expression and calculation support for various information elements (Fig. 9.2).

GA provides a blade and multivector structure for organizing and representing the complex structure of the ternary world and facilitates object expression and measurement through geometric product. The multivector structure provides the fundamental containers that can connect different types of elements with different dimensions. In GA, spatial location is represented as the basis of GA and the geometric structures can be developed based on the Grassmann structure and multivector structures (Yuan et al., 2012). Evolution process can be represented using versors, which can be used as calculation operators to construct differential equations or discrete dynamical

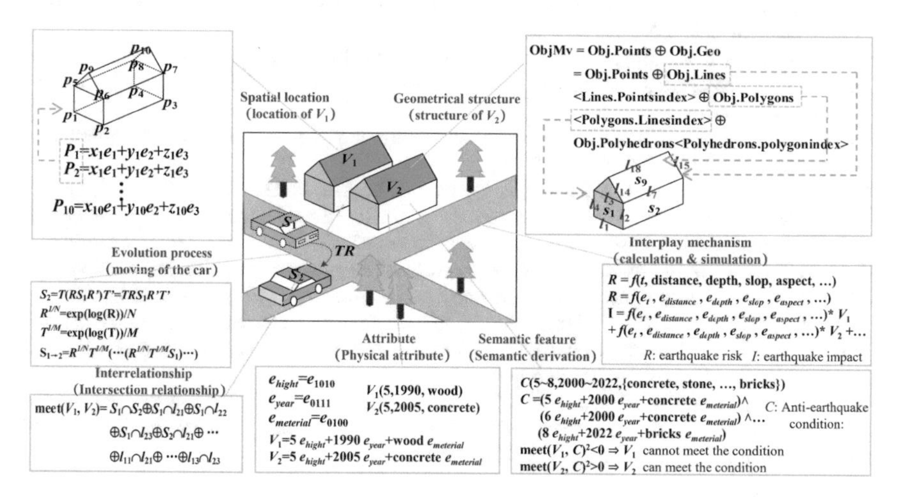

Fig. 9.2 Geographic information description with geometric algebra

representations (Yu et al., 2015, 2016). Semantics, attribute, interrelationship, and interplay mechanism are usually unstructured and cannot be represented as geometries or simple algebraic equations. To integrate these four components in the model, a GA-based distributed knowledge representation mechanism (Patyk, 2010) is used for coding them. Such integrated representation provides the mathematical and theoretical basis for integrated information modeling of geographic multi-factors (Yuan et al., 2019).

9.4 Conclusion

Based on the concept of ternary space, we proposed a seven dimensions framework to represent geographic information, which includes, semantics, spatial location, geometric structure, attribute, interrelationship, evolution process, and interplay mechanism. This representation model should enable the systematic integration of geographic information with multiple scales, dimensions, attributes, and spatial properties, which solves the problems of incomplete expression of information elements and lack of geographic characteristics. On this basis, the proposed framework expands the scope and domain of geographic information science and can serve as the basis for the development of a new generation of the geographic information system.

References

Chen, F. H., Fu, B. J., Xia, J., Wu, D., & Wu, S. H. (2019). Major advances in studies of the physical geography and living environment of China during the past 70 years and future prospects. *Science China Earth Sciences, 62*, 1665–1701.

Guo, R. Z., Chen, Y. B., Ying, S., Lü, G. N., & Li, Z. L. (2018). Geographic visualization of pan-map with the context of ternary spaces. *Geomatics and Information Science of Wuhan University, 43*(11), 1603–1610.

Guo, R. Z., & Ying, S. (2017). The rejuvenation of cartography in ICT era. *Acta Geodaetica Et Cartographica Sinica, 46*(10), 1274–1283.

Lü, G. N., Chen, M., Yuan, L. W., Zhou, L. C., Wen, Y. L., Wu, M. G., Hu, B., Yu, Z. Y., Yue, S. S., & Sheng, Y. H. (2018). Geographic scenario: A possible foundation for further development of virtual geographic environments. *International Journal of Digital Earth, 11*(4), 356–368.

Lü, G. N., Yuan, L. W., & Yu, Z. Y. (2017). Surveying and mapping geographical information from the perspective of geography. *Acta Geodaetica Et Cartographica Sinica, 46*(10), 1549–1556.

Patyk, A. (2010). Geometric Algebra Model of Distributed Representations. In E. Bayro-Corrochano & G. Scheuermann (Eds.), *Geometric Algebra computing in engineering and computer science* (pp. 401–430). Springer.

Yu, Z. Y., Luo, W., Hu, Y., Yuan, L. W., Zhu, A. X., & Lü, G. N. (2015). Change detection for 3D vector data: A CGA-based Delaunay–TIN intersection approach. *International Journal of Geographical Information Science, 29*, 2328–2347.

Yu, Z. Y., Wang, J. J., Luo, W., Hu, Y., Yuan, L. W., & Lü, G. N. (2016). A dynamic evacuation simulation framework based on geometric algebra. *Computers, Environment and Urban Systems, 59*, 208–219.
Yuan, L. W., Lü, G. N., Luo, W., Yu, Z. Y., Yi, L., & Sheng, Y. H. (2012). Geometric algebra method for multidimensionally-unified GIS computation. *Chinese Science Bulletin, 57*(7), 802–811.
Yuan, L. W., Yu, Z. Y., & Luo, W. (2019). Towards the next generation GIS: A geometric algebra approach. *Annals of GIS, 25*(3), 195–206. https://doi.org/10.1080/19475683.2019.1612945
Zhou, C. H. (2015). Prospects on pan-spatial information system. *Progress in Geography, 34*(2), 129–131.

Chapter 10
On the Third Law of Geography

A-Xing Zhu

Abstract Laws are statements of relation of phenomena that currently hold under given conditions and powerful ways for people to communicate and even advance human understanding about the world around us. Currently, three general principles in geography have been named as Law of Geography. The first is the spatial autocorrelation principle, which describes the relation among the attribute values of a given geographic variable over distance. The second is the spatial heterogeneity principle, describing the uncontrolled variance of geographic variables. The third is the geographic similarity principle which describes the resemblance of geographic phenomena under similar geographic configurations (geographic contexts). The Third Law of Geography is different from the first two in that it emphasizes the individual representation of single samples using the similarity in geographic configuration. This focus on individual representation offers a completely new perspective on geographic analyses and knowledge discovery. There are three key issues to be addressed for the Third Law of Geography to fully manifest itself in this capacity: the characterization of geographic configuration; the integration of the individual representation-based techniques with the average model-based techniques; and application of the Third Law in the broader range of geographic subfields and related disciplines.

Keywords First Law of Geography · Spatial autocorrelation · Second Law of Geography · Spatial heterogeneity · Third Law of Geography · Geographic similarity

A.-X. Zhu (✉)
Department of Geography, University of Wisconsin-Madison, Madison, WI 53706, USA
e-mail: azhu@wisc.edu

School of Geography, Nanjing Normal University, Nanjing 210023, Jiangsu, China

State Key Laboratory of Resources and Environmental Information System, Institute of Geographic Sciences and Natural Resource Research, Chinese Academy of Sciences, Beijing 100101, China

B. Li et al. (eds.), *New Thinking in GIScience*,
https://doi.org/10.1007/978-981-19-3816-0_10

10.1 About Laws of Geography

The Merriam-Webster Dictionary has two definitions for law in the context discussed here (Entry 6a and 6b) (Merriam-Webster, online). 6a defines it as "a statement of an order or relation of phenomena that so far as is known is invariable under the given conditions" and 6b defines it as "a general relation proved or assumed to hold between mathematical or logical expressions". In both definitions one can clearly see that laws are not universally true and should not be taken as such. They are true "under the given conditions", can even be "assumed" to hold. For example, even Newton's three Laws of Motion are not universal. They only hold for objects which are idealized as single point masses (Truesdell et al., 2003) and they do not hold for the motion of deformable bodies (Lubliner, 2008), not for reference frames which are in acceleration (Chabay & Sherwood, 2015), not for objects that are at the small scale of atom (Peskin & Schroeder, 1995). Even with these conditions (or constraints), these laws advanced our understanding about the world around us in two ways. First, they provided the basis for understanding the motion of objects and used as a unified quantitative explanation of a wide range of physical phenomena. Second, they, through their conditions under which they hold, directed efforts in furthering our understanding. For example, the conditions posed on objects for which these laws hold inspired researchers to study the motions about other objects, which led the discoveries synthesized as Euler's laws of motion (for rigid body) (McGill & King, 1995), Cauchy's continuum mechanics (fluid body) (Kurrer, 2018), quantum mechanics (at the scale of atomic particles) (Feynman et al., 1964).

Currently, there are three general principles in geography which have been named as Laws of Geography. The first is about the spatial continuity (spatial autocorrelation) exhibited by geographic phenomena (The First Law of Geography, "near things are more related than distant things") (Tobler, 1970). The second is about the spatial heterogeneity of geographic phenomena (The Second Law of Geography, "uncontrolled variance") (Anselin, 1989; Goodchild, 2004). The third is about the geographic similarity, describing the resemblance of geographic configurations between two locations (The Third Law of Geography, ""the more similar the geographic configurations between two points, the more similar the values of the target geographic variable at these two locations") (Zhu & Turner, 2022; Zhu et al., 2018). If one uses the above definitions of law to assess the validity of these principles as law, one might find that they do fit the definition. They each do capture the "order or relation" of geographic phenomena that generally hold but with conditions. There are exceptions to these orders or relations. For examples, a cliff exhibits discontinuity in elevation; the surface of temperature over a calm lake surface is quite uniform; and a human may not have similar thoughts at the similar geographic scenes. Nevertheless, the spatial continuity described in the First Law, the spatial heterogeneity in the Second Law, and the geographic similarity in the Third Law are all well-known norms (order) and relations exhibited by geographic phenomena and hold true at the vast majority of times. In this regard, the scale independence (Phillips, 2022) may be another Law of Geography (maybe the Fourth in the order of being named as law, not in its importance). These properties guided us well in understanding geographic phenomena and in our interaction with the geographic environment.

Laws of Geography have two properties: synopsized and descriptive (Zhu et al., 2020). By "synopsized" I mean that the scholars who named these principles as laws were not the persons who initially discovered these principles. In fact, what expressed in these principles exists in the form of general knowledge (or common knowledge) which has been accumulated by researchers in the field over a long time. The naming scholars of these respective laws synthesized and distilled the understanding, and then abstracted the understanding to a law like statement. For example, the knowledge on spatial continuity of geographic variable is well-known and even widely applied before Tobler named it as the First Law of Geography. The same can be said about the other two Laws of Geography and the law of scale independence. By "descriptive" I mean these laws are expressed in a qualitative form, not in a quantitative explanation as we are accustomed to the laws in physics. For example, the Second Law describes the nature of spatial heterogeneity as "uncontrolled variance" and the Third Law describes the effects of geographic resemblance on a target geographic variable in terms of "the more similar…, the more similar". This way of expressing geographic principles exhibits the nature of geographic knowledge and suits the custom so far geographers use to describe our understanding about geographic phenomena (Zhu et al., 2020).

10.2 The Third Law of Geography

In searching for an alternative to the average model approach to spatial prediction, Zhu et al. (2015) discovered the importance of individual representation of single samples through geographic configuration in capturing local variations in spatial prediction. Based on this line of work, Zhu et al. (2018) abstracted the commonly known geographic similarity to a law like statement, "The more similar the geographic configurations of two points (areas), the more similar the values (processes) of the target variable at these two points (areas)", referred to as the geographic similarity principle, which can be simplified to as "the more similar the geographic configurations between two points, the more similar the status of the target variable". The essence of what captured in the Third Law of Geography is the use of individual representation of single samples, which argues that a sample can be used to represent locations or areas which are similar to itself in geographic configuration (Fig. 10.1). If one of the two locations is a sample (where status of the target geographic variable is known), then the status of the target variable at the other location can be assessed by the similarity in geographic configuration between these two points (Fig. 10.2) (Zhu, 2022; Zhu & Turner, 2022).

There are three unique aspects about what was stated in the Third Law of Geography (Zhu & Turner, 2022). The first is the use of individual representation of single samples. Under the Third Law, status of a target geographic variable at an unvisited location is assessed by the status of the target variable at one and only one

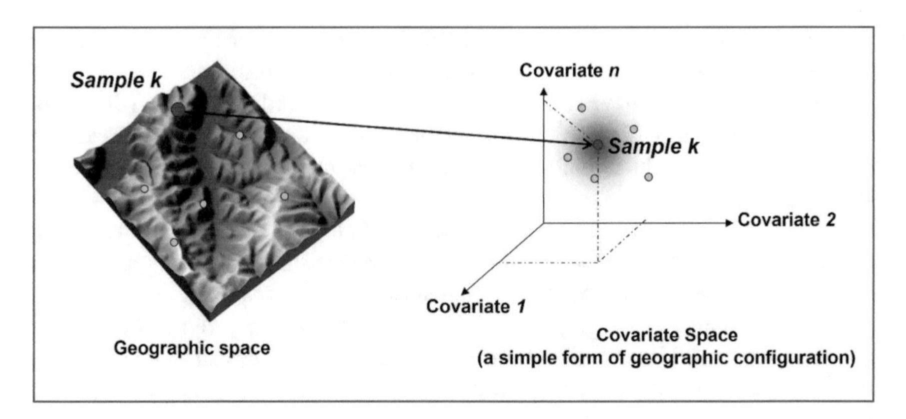

Fig. 10.1 Explicit use of the individual presentation of a single sample. Sample k can be used to represent the points which are similar (or close) to itself in geographic configuration [revised from (Zhu, 2022)]

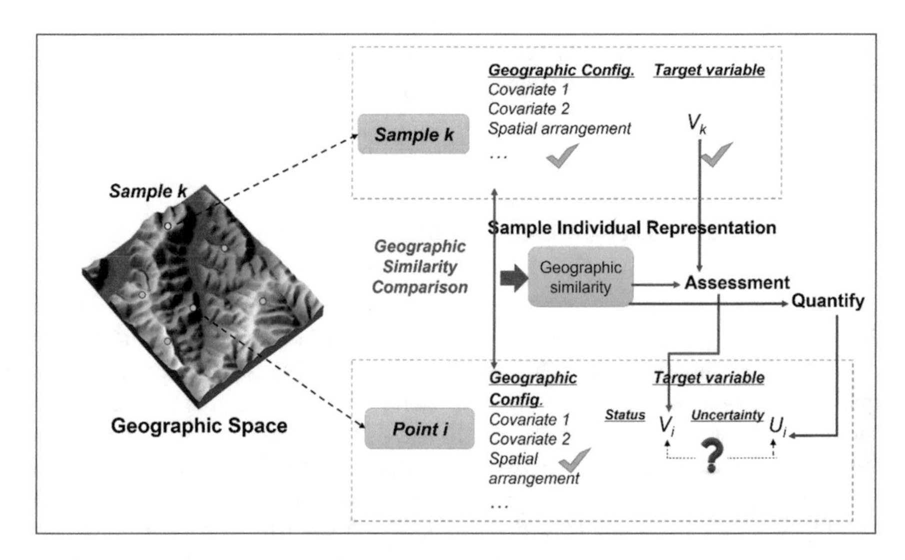

Fig. 10.2 Assessment based on individual representation of single samples [revised from (Zhu, 2022)]

sample location at a time (Zhu et al., 2015), not by the average model developed from the entire sample set. The second aspect is that the assessment is done through the comparison of the locations, not through the construction of a relationship (an average model) derived from all samples, which (constructing the average model) is the common practice adopted in most geographic learning and knowledge discovery (Jiang, 2015). The third aspect is that the comparison is measured using the similarity in geographic configuration between the two locations, not by a linkage of geographic configuration to the status of the target variable directly.

These unique aspects or characteristics of the Third Law of Geography provides a significant opportunity to conduct geographic analyses and knowledge discovery from a completely different angle than the widely adopted average model approach in current geographic knowledge discovery (Zhu, 2022). This significance comes from the benefits using the individual representation of single samples and can be summarized in four areas: *sample requirement*, *relationship requirement*, *uncertainty provision*, and *knowledge retention*.

No sample requirement: One of the key requirements for the average model to work well (applicable to the entire geographic area under study) is that the samples used to develop this model are representative of the area under concern, which is often achieved through the requirements on the sample size and sample distribution (de Gruijter et al., 2006; Wheeler et al., 2013). Achieving a good sample representation of the study area is a non-trivial task due to budget limitation and accessibility constraints (Meyer et al., 2015). More importantly, geographers are now relying more on volunteered data from non-specialists (volunteered geographic information, VGI) (Goodchild, 2007; Heipke, 2010) and becoming an important source of data for geographic knowledge discovery (Sui et al., 2013). However, these data sources have their limitations or biases on the level of representation due to the ad-hoc nature in their collection (Mullen et al., 2015). These limitations or biases would lead to biased, even wrong, conclusions, just as one would be completely misled if one relies on the selection process (sampling process) from a social media app, which is based on the preference of the audience, not on the representation of what is happening in the world. The use of the individual representation of single samples coupled with uncertainty provision allows the removal of the requirement for good representation of the study area in the sample set, which significantly reduces the risk of drawing biased or wrong conclusions from biased samples.

No relationship requirement: The average model approach also imposes a stationarity requirement on the relationship captured in the model so that the developed model can be applied to the entire study area under concern. The applicability of the model (the likelihood of the stationarity requirement being met) depends on two aspects: the representation of the study area in the samples as discussed above and the variation of the actual relationship over the study area. Even if the samples are representative of the study area, the developed average model may still not be appropriate for some parts of the study area due to the complex nature of geographic phenomena which leads to a great variation of their relationships to other geographic factors, just as stipulated in the Second Law of Geography. For example, when examining the relationship between vegetation coverage and soil depth (here defined as the depth to the bottom of the B horizon) for an area with three different parent materials (assumed to be granite, limestone, and mudstone). Clearly, the relationships are significantly different among the three areas. Even though the samples used are representative of the entire area, the average model developed from the samples would still not work well for the entire area due to the vast differences in how vegetation coverage is related to the soil depth among the three regions of different parent materials. This is because what is captured in the samples, the different relationships for the three sub regions, is not represented in the model which is the average condition in the samples. One may argue that there are more sophisticated learning methods (such as machine learning techniques) which can be used to address this issue. This is true but there are two caveats with these types of learning techniques. One is that they are so data dependent that generalization is a major issue for these techniques. The other is that these techniques are often unable to work with relationships that vary continuously over space. The Third Law does not require a specific relationship to be built, thus there is no requirement of stationarity (Zhu & Turner, 2022). The differences in relationships among geographic variables can be easily accounted for as long as they are captured in the samples (Zhu et al., 2015).

Uncertainty provision: Uncertainty about the assessment of status of a geographic variable at a location is an important part of the information that should be provided because it can provide quality information about the assessment at that location and can be used to assess the quality of the resulting decision and decision alternatives (Zhang & Goodchild, 2002). This information is more than often not provided for each location from current geographic analytical techniques. Through the emphasis on the individual representation of single samples through the similarity in geographic configuration, the Third Law of Geography offers an effective way to quantify the uncertainty associated with the assessment of status of the target variable at each location (Liu et al., 2020; Zhu et al., 2015). This is achieved through the employment of similarity in geographic configuration between the sample and the location of assessment. If the geographic configuration at the location of assessment is similar to that at a sample, then there is a high confidence to use that sample to assess the status of the target variable at that location. If none of the samples available is similar in geographic configuration with the location of assessment, using any of these samples to assess the status of the target variable at the location would lead to high uncertainty. Thus, similarity in geographic configuration between the most similar sample and the location of assessment can be used to quantify the uncertainty associated with the assessment at each location using these samples. It has been shown that the so quantified uncertainty is a good surrogate to the quality of the assessment (Zhu et al., 2015). Due to the fact that uncertainty is available for every location, the spatial variation of assessment quality can be provided as well (Zhu et al., 2015). This spatial variation of uncertainty (assessment quality) can be used to effectively allocate additional sampling efforts (Li et al., 2016; Zhang et al., 2016) and to mitigate sampling bias (Zhang & Zhu, 2019).

Knowledge retention and representation: Existing knowledge in geography, like in many other disciplines, exists in the form of averages (means) or generalized rules. This form of knowledge retention and representation is certainly important due to its simplicity and clarity, but this form alone is insufficient (Taleb, 2017). The negative consequences of this insufficiency or tolerance of difficult-to-predict events are best exemplified in Taleb's Black Swan Theory (Taleb, 2010). One of the elements in this insufficiency is the neglect of the individuality captured in samples (or the individual representation of single samples). The Third Law focuses on the usefulness of the individual contributions from single samples, which effectively complements what is missing in current forms of geographic knowledge retention and representation.

In summary, these four elements of significance suggest that the Third Law of Geography can serve as a new theoretical basis for fundamentally transforming geographic analyses. By adding the individual representation-based approach, the Third Law effectively broadens geographic analyses from a single average model doctrine to duo model doctrine, or even a doctrine integrating the average model approach with the individual representation-based approach. This will significantly enhance the analytical capacity of geographers and related practitioners.

10.3 Issues to Address

There are three key issues which need to be addressed to maximize the benefits brought forward by the Third Law of Geography. The first is the characterization of geographic configuration. The term, geographic configuration, means the structure of geographic conditions around a point, including the conditions of the covariates (other geographic factors that covary with the target variable), the ranking (weights) of these covariates, the spatial arrangement of the conditions of these covariates, as well as the scale at which the configuration is characterized. Given these, geographic configuration is a spatial statement, sometimes people would treat it as "spatial context". However, I prefer geographic configuration or geographic context over spatial context because the latter is likely to be interpreted as tied to spatial distance alone. The following questions need to be answered about geographic configuration first before applying the Third Law. What covariates should be included for a given target variable? How to weight these covariates? How to quantify spatial arrangement of the conditions from the covariates around the location of interest? What would be the geographic configuration for an area or an area at different scales?

The second key issue is the development of spatial analytical methods/tools based on the Third Law of Geography. Efforts are underway in spatial prediction, such as new prediction techniques (Zhu et al., 2015), uncertainty assessment methods (Liu et al., 2020), sample selection (Zhu et al., 2019), and sample-bias mitigation (Zhang & Zhu, 2019). These are far too few if the Law is going to be applicable to the discipline of geography as whole. New methods/techniques need to be developed for other spatial analysis as well as for the integration of individual representation-based approach with the average model approach.

The third key issue is more applications in other sub-fields of geography and related disciplines. Currently researches on the Third Law of Geography are mostly in physical science side of geography (such as soil mapping, landslide susceptibility mapping, wildlife habitat suitability mapping). There is a great need for examples in other subfields, particularly from human geography perspective (such as analysis of crimes, assessment of ideation of suicide). The applications in human geography will present new issues and new challenges to solve, which will definitely enrich the concepts behind the Third Law.

10.4 Summary

The Third Law of Geography, "The more similar the geographic configurations between two points, the more similar the status of the target variable", synopsizes an important principle in Geography which focuses on the individual representation of single samples. It has been shown that this principle can serve as a theoretical basis for addressing the challenges facing the analytical ability based on the average model approach, such as specific sample requirement, relationship requirement (including stationarity), uncertainty deficiency, and over-generalization in knowledge retention and representation. By addressing these key challenges in geographic analyses, the Third Law provides a brand-new perspective and opportunity for geographic analyses, which could potentially change the landscape for geographic analyses and knowledge discovery.

Geographic analyses are experiencing an important change, from small area focused studies to large spatial extent research, from coarse spatial granularity (resolution) to fine grained modeling requiring high level of spatial details with uncertainty evaluation, which calls for the use of the geographic similarity principle as advocated by the Third Law. At the same time, the development of location-enabled personal devices and the increasing ability for capturing information on geographic environment have changed the landscape of data provision which greatly enhance our ability to characterize geographic configuration, in turn makes the geographic similarity principle viable for its wider application in different sub-fields of geography and related fields.

Acknowledgements The work reported here was supported by grants from National Natural Science Foundation of China (Project No.: 41871300 and Project No.: 41431177) and from the 111 Program of China (Approved Number: D19002). Supports through the Vilas Associate Award, the Hammel Faculty Fellow Award and the Manasse Chair Professorship from the University of Wisconsin-Madison are greatly appreciated.

References

Anselin, L. (1989). What is special about spatial data? Alternative perspectives on spatial data analysis, Technical Report 89-4. National Center for Geographic Information and Analysis.

Chabay, R., & Sherwood, B. (2015). *Matter & interactions Vol 1: Modern mechanics* (pp. 34–35). Wiley.

de Gruijter, J., Brus, D., Bierkens, M., & Knotters, M. (2006). *Sampling for natural resource monitoring* (p. 332). Springer.

Feynman, R., Leighton, R., & Sands, M. (1964). *The Feynman Lectures on Physics. 3*. California Institute of Technology. ISBN 978-0201500646.

Goodchild, M. F. (2004). The validity and usefulness of laws in geographic information science and geography. *Annals of the Association of American Geographers, 94*, 300–303. https://doi.org/10.1111/j.1467-8306.2004.09402008.x

Goodchild, M. F. (2007). Citizens as sensors: The world of volunteered geography. *GeoJournal, 69*, 211–221. https://doi.org/10.1007/s10708-007-9111-y

Heipke, C. (2010). Crowdsourcing geospatial data. *ISPRS Journal of Photogrammetry and Remote Sensing, 65*(6), 550–557. https://doi.org/10.1016/j.isprsjprs.2010.06.005

Jiang, B. (2015). Geospatial analysis requires a different way of thinking: The problem of spatial heterogeneity. *GeoJournal, 80*, 1–13.

Li, Y., Zhu, A. X., Shi, Z., Liu, J., & Du, F. (2016). Supplemental sampling for digital soil mapping based on prediction uncertainty from both the feature domain and the spatial domain. *Geoderma, 284*, 73–84. https://doi.org/10.1016/j.geoderma.2016.08.013

Liu, J., Zhu, A. X., Rossiter, D., Du, F., & Burt, J. (2020). A trustworthiness indicator to select sample points for the individual predictive soil mapping method (iPSM). *Geoderma, 373*, 114440.

Lubliner, J. (2008). *Plasticity theory*. Dover Publications. ISBN 978-0-486-46290-5.

Kurrer, K.-E. (2018). The history of the theory of structures. *Searching for equilibrium* (pp. 978–979). Wiley. ISBN 978-3-433-03229-9.

McGill, D. J., & King, W. W. (1995). *Engineering mechanics, an introduction to dynamics* (3rd ed.). PWS Publishing Company. ISBN 0-534-93399-8.

Meyer, C., Kreft, H., Guralnick, R., & Jetz, W. (2015). Global priorities for an effective information basis of biodiversity distributions. *Nature Communications, 6*, 8221. https://doi.org/10.1038/ncomms9221

Merriam-Webster, Online https://www.merriam-webster.com/dictionary/law

Mullen, W. F., Jackson, S., Croitoru, A., Crooks, A., Stehanidis, A., & Agouris, P. (2015). Assessing the impact of demographic characteristics on spatial error in volunteered geographic information features. *GeoJournal, 80*(4), 587–605. https://doi.org/10.1007/s10708-014-9564-8

Peskin, M., & Schroeder, D. (1995). *An introduction to quantum field theory*. Westview Press. ISBN 978-0-201-50397-5.

Phillips, J. (2022). The law of scale independence. *Annals of GIS*, in press.

Sui, D., Elwood, S., & Goodhild, M. (Eds.). (2013). *Crowdsourced geographic knowledge: Volunteered geographic information in theory and practice*. Springer.

Taleb, N. N. (2010). The Black Swan. *The impact of the highly improbable: With a new section: "On robustness and fragility"* (2nd ed., p. 444). Random House Trade Paperbacks.

Taleb, N. N. (2017). Probability, risk, and extremes. In D. Needham, & J. Weitzdorfer (Eds.), *Extremes* (Darwin College Lectures) (p. 10). Cambridge University Press. https://fooledbyrandomness.com/DarwinCollege.pdf

Tobler, W. R. (1970). A computer movie simulating urban growth in the Detroit Region. *Economic Geography, 46*(sup1), 234–240. https://doi.org/10.2307/143141

Truesdell, C. A., Becchi, A., & Benvenuto, E. (2003). *Essays on the history of mechanics: In memory of Clifford Ambrose Truesdell and Edoardo Benvenuto* (p. 207). Birkhäuser. ISBN 978-3-7643-1476-7.

Wheeler, D., Shaw, G., & Barr, S. (2013). *Statistical techniques in geographic analysis* (p. 354). Taylor & Francis.

Zhang, G., & Zhu, A. X. (2019). A representativeness-directed approach to mitigate spatial bias in VGI for the predictive mapping of geographic phenomena. *International Journal of Geographical Information Science, 33*(9), 1873–1893.

Zhang, J.-X., & Goodchild, M. F. (2002). *Uncertainty in geographical information*. Taylor & Francis.

Zhang, S. J., Zhu, A. X., Liu, J., Qin, C. Z., & An, Y. M. (2016). A heuristic uncertainty directed field sampling design for digital soil mapping. *Geoderma, 267*, 123–136.

Zhu, A. X. (2022). Geographic similarity: What is the big deal?". *Annals of American Association of Geographers* (under review).

Zhu, A. X., Liu, J., Du, F., Zhang, S. J., Qin, C. Z., Burt, J., Behrens, T., & Scholten, T. (2015). Predictive soil mapping with limited sample data. *European Journal of Soil Sciences, 66*, 535–547.

Zhu, A. X., Lv, G., Liu, J., Qin, C. Z., & Zhou, C. (2018). Spatial prediction based on Third Law of Geography. *Annals of GIS, 24*(4), 225–240. https://doi.org/10.1080/19475683.2018.1534890

Zhu, A. X, Lv, G. N., Zhou, C. H., & Qin, C. Z. (2020). Geographic similarity: Third Law of Geography?. *Journal of Geo-information Science (in Chinese) 22*(4), 673–679. https://doi.org/10.12082/dqxxkx.2020.200069.

Zhu, A. X., Miao, Y., Liu, J. Z., Bai, S. B., Zeng, C., Ma, T. W., & Hong, H. Y. (2019). A similarity-based approach to sampling absence data for landslide susceptibility mapping using data-driven methods. *Catena, 183*, 104188.

Zhu, A. X., & Turner, M. (2022). How is the Third Law of Geography Different?. *Annals of GIS*, in press.

Chapter 11
Human Mobility and the Neighborhood Effect Averaging Problem (NEAP)

Mei-Po Kwan

Abstract The neighborhood effect averaging problem (NEAP) was discovered in 2018. It arises when human mobility is ignored when assessing individual exposures to environmental factors (e.g., noise and air pollution). Neighborhood effect averaging occurs because most people move around in their daily life, and as a result, their mobility-based exposures would tend toward the average of the population or participants of the study area. Assessments of individual exposures or their health impacts based only on residential neighborhoods do no capture people's exposures in non-residential neighborhoods and thus may lead to erroneous findings (because people's daily mobility may amplify or attenuate the exposures they experienced in their residential neighborhoods). To date, there has been limited research on the NEAP and its effects on research findings. This chapter provides a succinct overview of the NEAP and relevant recent studies on the problem. It also highlights the need to mitigate the NEAP in research and its policy implications, especially concerning the situations of socially disadvantaged groups.

Keywords Neighborhood effect · Health geography · Environmental health · Air pollution · Ethnic segregation

11.1 Introduction

A major focus of geographic research is to discover the relationships between different phenomena over space (e.g., how geographic contexts and environmental factors like land use or air pollution affect people's health). To derive the relevant variables to examine these relationships, much of past research is based on an area-based or zone-based framework. For example, the relationships between toxic environmental substances and health are examined by first deriving environmental and health variables from fixed areal or contextual units like census tracts and then analyzing the relationships between the variables using appropriate methods (e.g.,

M.-P. Kwan (✉)
Department of Geography and Resource Management and Institute of Space and Earth Information Science, The Chinese University of Hong Kong, Shatin, Hong Kong, China
e-mail: mpk654@gmail.com

B. Li et al. (eds.), *New Thinking in GIScience*,
https://doi.org/10.1007/978-981-19-3816-0_11

spatial regression). As a result of using an area-based framework, all such past studies faced the well-known modifiable areal unit problem (MAUP) because using different zonal schemes or spatial scales for deriving area-based variables may yield different results (Fotheringham & Wong, 1991). Much research has been conducted to deepen our understanding of the MAUP, its effects on research findings, and methods for mitigating it (e.g., identifying and using the best areal division, neighborhood size, or geographic scale) (e.g., Mu & Wang, 2008).

In recent years, a new methodological problem in geographic analysis was discovered. It is the uncertain geographic context problem (UGCoP) discovered by Kwan (2012a, 2012b). It refers to the problem that findings on the effects of area-based attributes (e.g., land-use mix) on *individual* behaviors or outcomes (e.g., physical activity) could be affected by how contextual units or neighborhoods for deriving the environmental variables are geographically delineated. Because no researcher has complete and perfect knowledge of the precise geographic context that influences *individual* behaviors or outcomes being examined, no study that uses area-based environmental variables can fully overcome the problem. As discussed in Kwan (2012a), the UGCoP poses serious inferential challenges and is a fundamental methodological problem. Note that the UGCoP is not relevant to area-based outcomes such as cancer or crime rates of census tracts because it is not clear how a true causally relevant geographic context may be meaningfully conceptualized or delineated for an area-based group, such as all the individuals who live in the same census tract.

While the UGCoP seems similar to the modifiable areal unit problem (MAUP), it is a different kind of problem because it is not due to the use of different zonal schemes or spatial scales for area-based variables (Kwan, 2018a). Instead, the UGCoP is due to the use of arbitrary areal units (e.g., buffer areas of different types and sizes around individuals' homes) for deriving area-based variables because of the lack of knowledge about the precise spatial and temporal configurations of the contextual or environmental factors that exert influence on the individual behavior or experience under study. As a result, methods for addressing the MAUP do not necessarily solve the UGCoP. This poses a major challenge to all studies that derive contextual or environmental variables based on arbitrarily delineated areal units (e.g., different types and sizes of home buffers). Unlike addressing the MAUP, addressing the UGCoP calls for more accurate delineations of the "true causally relevant" geographic context and more accurate measurements of the pertinent geographic or environmental variables.

A major source of the UGCoP is a result of ignoring people's daily mobility and real-time microenvironments (e.g., in a restaurant) over space and time. For instance, most studies to date use static administrative areas (e.g., census tracts) and people's home locations to derive measures of their exposures to environmental factors (i.e., exposure measures). However, people move around in their daily life and thus are exposed to environmental contexts not only in their residential neighborhoods but also in many other areas (Kwan, 2009, 2013). The UGCoP thus arises from our lack of knowledge of people's precise exposures as their daily lives unfold in space and time and the use of static and fixed delineations of neighborhood or geographic context that ignore human mobility. The static residence-based approach of past research ignores individuals' daily mobility and may lead to erroneous exposure assessments and health effect estimates (Chen & Kwan, 2015; Kwan et al., 2019; Park & Kwan, 2017; Yoo et al., 2015).

In 2018, one particular manifestation of the UGCoP, which is also a result of ignoring people's daily mobility, was discovered: the neighborhood effect averaging problem (NEAP) (Kwan, 2018b). While there have been considerable studies on the UGCoP to date, there is limited research on the NEAP and its effects on research findings. This chapter provides a succinct overview of the NEAP and relevant recent studies on the problem. It also highlights the need to mitigate the NEAP in research and its policy implications, especially concerning the situations of socially disadvantaged groups.

11.2 The Neighborhood Effect Averaging Problem

In many social science disciplines, the neighborhood effect has played an important role in understanding the influences of neighborhood environments on various social and behavioral outcomes (e.g., health or crime). Past studies tend to assess the neighborhood effect based on people's residential neighborhood (i.e., the neighborhoods where people live). However, people move around in their daily lives to conduct different activities (e.g., work, attend school, or shop) outside of their residential neighborhoods. They are thus also exposed to environmental factors in those non-residential neighborhoods. Adopting a residence-based approach and ignoring people's non-residential exposures associated with their daily mobility (i.e., mobility-based exposures) may introduce errors in the assessments of people's exposures to environmental factors and their health impacts.

One specific kind of error due to ignoring people's daily mobility and their non-residential exposures arises from the phenomenon of neighborhood effect averaging. It was first observed by Kwan (2018b) based on several recent studies (Dewulf et al., 2016; Shafran-Nathan et al., 2017; Yu et al., 2018). Neighborhood effect averaging occurs because when an individual travels in his/her daily life, the values of the environmental factors (e.g., noise or air pollution) the person is exposed to would tend to have a wider range than the value of the environmental factors in the person's residential neighborhood. Take air pollution as an example, this means that air pollution levels are higher in some non-residential neighborhoods and lower in others when a person moves around outside his/her residential neighborhood. Now, consider the high and low end of the pollution levels in the study area. For people who live in residential neighborhoods with a high air pollution level, it is more likely for them to be exposed to neighborhoods with lower levels of air pollution when they move around. For people who live in residential neighborhoods with a low air pollution level, it is more likely for them to be exposed to neighborhoods with higher levels of neighborhoods when they move around. This means that individual exposures to air pollution would tend toward the average value of air pollution in the study area when people's daily mobility and their exposures to non-residential are taken into account. This is the phenomenon of neighborhood effect averaging.

As a result of neighborhood effect averaging, assessing individual exposures to environmental factors based on people's residential neighborhood may lead to erroneous results in studies of mobility-dependent exposures (e.g., noise and air pollution) and their health impacts. Kwan (2018b) first discovered this problem and called it the neighborhood effect averaging problem (NEAP). She suggested that "for health outcomes that are also affected by exposures to environmental factors in people's nonresidential neighborhoods as they move around in their daily life (mobility-dependent exposures), using residence-based neighborhoods to estimate individual exposures to and the health impact of environmental factors will tend to overestimate the statistical significance and effect size of the neighborhood effect because it ignores the confounding effect of neighborhood effect averaging that arises from human daily mobility" (Kwan, 2018b).

11.3 Recent Studies on the NEAP

The NEAP was first discovered by Kwan (2018b) based on several studies on people's exposures to air pollution (e.g., Dewulf et al., 2016; Yu et al., 2018). To provide further evidence on the existence of the NEAP, Kwan and her collaborators conducted several studies that examined individual exposures to air pollution, traffic congestion, and other ethnic groups in different studies areas (e.g., Los Angeles in the U.S. and Xining in China). These studies showed that the NEAP exists in other kinds of individual exposures and that they are closely associated with the patterns of daily mobility of different social groups.

For example, in a study that investigated whether ignoring people's daily mobility and activity patterns would affect assessments of people's exposures to traffic congestion, Kim and Kwan (2019) found that people's exposures to traffic congestion are significantly different between two assessments: when only commuting trips are considered and when non-community trips are also considered. They observed the existence of the NEAP: people's exposures to traffic congestion tend toward the average value of the participants in the study area when their daily activity patterns are considered. The results indicate that ignoring even part of people's daily mobility and activity patterns (i.e., non-commuting trips) can lead to erroneous findings.

Subsequently, Kim and Kwan (2021a, 2021b) published two papers that provide further evidence that the NEAP exists when examining individual exposures to environmental factors. One of these papers investigated people's exposures to ozone in Los Angles using an activity-travel diary dataset. The study concluded that the NEAP is observed and that older, non-working, and low-income people, and women (when compared to younger, employed, and high-income people, and men) experienced less neighborhood effect averaging because they have lower levels of daily mobility. In the second paper that used the same dataset, Kim and Kwan (2021a) found that certain groups of people (e.g., non-workers like students and homemakers) experience very little downward averaging (i.e., their mobility-based exposures do not significantly decrease even when their daily mobility is taken into account). In other words, because of the limited daily mobility of these social groups, it is difficult for them to lower the high levels of exposure in their residential neighborhoods by traveling and exposing to lower levels of ozone in areas outside of their residential neighborhoods.

While these studies examined whether the NEAP exists and its potential impacts on research findings in the assessments of people's exposures to traffic congestion and air pollution, research has also been conducted to investigate the effects of the NEAP on research findings in other kinds of exposures. For instance, a study by Tan et al. (2020) used people's limited exposures to other ethnic groups as a measure of ethnic segregation based on an environmental exposure conceptual framework. According to this exposure-based conceptualization, lower exposures to people of other ethnic groups mean higher levels of ethnic segregation and vice versa. Based on a comparison of the activity-travel patterns of the Han majorities and the Hui ethnic minorities in Xining, China, the study concluded that the NEAP exists when investigating people's exposures to other ethnic groups. Participants tend to have higher ethnic exposures in their non-residential activity locations if they have low exposures to the other ethnic group in their residential neighborhoods (i.e., if they live in highly segregated neighborhoods). On the other hand, participants tend to have lower ethnic exposures in their non-residential activity locations if they have higher exposures to the other ethnic group in their residential neighborhoods (e.g., if they live in highly mixed neighborhoods).

Extending research on the NEAP to other realms, Huang and Kwan (2021) examined how assessments of individual exposure to COVID-19 risk are affected by the NEAP and the UGCoP. Using the COVID-19 data on an open-access government website and the individual-level activity data of 60 confirmed COVID-19 cases (infected persons) in Hong Kong, the study represented COVID-19 risk environments using case-based and venues-based high-risk locations. The COVID-19 risk of each of the 60 selected cases is then evaluated by three approaches based on their exposure to the case-based or venue-based risk environments: the mobility-based approach, the residence-based approach, and the activity-space-based approach. The results of this study indicate that the UGCoP and the NEAP exist in the assessment of COVID-19 risk. It concluded that COVID-19 studies need to address the uncertainties due to the UGCoP and the NEAP by considering people's daily mobility. Ignoring peoples' daily mobility and its interactions with the complex and dynamic COVID-19 risk environments may lead to misleading results and misinform government non-pharmaceutical intervention measures.

11.4 Implications of the NEAP

These studies indicate that research findings on people's mobility-dependent exposures such as traffic congestion, air pollution, and infectious disease are likely to be affected by the NEAP. Researchers in geographic and environmental health research thus need to pay special attention to the pertinent methodological and policy implications.

For instance, without taking into account people's daily mobility, assessments of individual mobility-dependent exposures like air pollution may be erroneous because of the NEAP. Further, since neighborhood effect averaging may have different impacts on the mobility-dependent exposures of different social groups, policy-makers need to recognize such differential impacts when developing policy measures to address the needs of different social groups (especially the socially disadvantaged groups). On the positive side, people with high daily mobility may experience more neighborhood effect averaging that mitigates their high exposures in their residential neighborhoods. However, people with low daily mobility (e.g., low-income people and older adults) may experience little neighborhood effect averaging, and as a result, may not be able to mitigate the impacts of the high exposures found in their residential neighborhoods (Huang & Kwan, 2021; Kim & Kwan, 2019; Ma et al., 2020). Policymakers should thus pay special attention to the experiences of the socially disadvantaged groups and develop intervention measures to address their needs (e.g., increasing the mobility of those who live in disadvantaged neighborhoods).

Although we now have a better understanding of the NEAP thanks to these recent studies, we need to be aware of how the NEAP happens and recognize the situations in which it does not apply. For example, the NEAP may not influence individual exposures to factors that occur mainly in or around people's residential neighborhoods (e.g., social capital) because it was largely observed in mobility-dependent exposures. Further, for people with very low mobility (e.g., older adults), the NEAP seems less relevant because it is largely due to people's daily mobility. Also, the environmental factors under examination need to vary considerably in space and time in the study area for the NEAP to have its effect. Otherwise, the NEAP will not be observed because people's exposures would not be significantly different even if they move around or visit non-residential neighborhoods. Further, while our understanding of the NEAP has advanced considerably in recent years, the relevant studies to date are largely on individual exposures to traffic congestion and air pollution. Therefore, future studies also need to investigate whether the NEAP exists in other kinds of mobility-dependent exposures (e.g., green spaces and healthy food environment) because different environmental factors have different space–time dynamics and interactions with human movement patterns, leading to different manifestations of the NEAP.

References

Chen, X., & Kwan, M.-P. (2015). Contextual uncertainties, human mobility, and perceived food environment: The uncertain geographic context problem in food access research. *American Journal of Public Health, 105*(9), 1734–1737.

Dewulf, B., Neutens, T., Lefebvre, W., Seynaeve, G., Vanpoucke, C., Beckx, C., & van de Weghe, N. (2016). Dynamic assessment of exposure to air pollution using mobile phone data. *International Journal of Health Geographics, 15*, 14.

Fotheringham, A. S., & Wong, D. W. S. (1991). The modifiable areal unit problem in multivariate statistical analysis. *Environment and Planning A, 23*(7), 1025–1044.

Huang, J., & Kwan, M.-P. (2021). Uncertainties in the assessment of COVID-19 risk: A study of people's exposure to high-risk environments using individual-level activity data. *Annals of the American Association of Geographers* (online).

Kim, J., & Kwan, M.-P. (2019). Beyond commuting: Ignoring individuals' activity-travel patterns may lead to inaccurate assessments of their exposure to traffic congestion. *International Journal of Environmental Research and Public Health, 16*(1), 89.

Kim, J., & Kwan, M.-P. (2021a). Assessment of sociodemographic disparities in environmental exposure might be erroneous due to neighborhood effect averaging: Implications for environmental inequality research. *Environment Research, 195*, 110519.

Kim, J., & Kwan, M.-P. (2021b). How neighborhood effect averaging may affect assessment of individual exposures to air pollution: A study of ozone exposures in Los Angeles. *Annals of the American Association of Geographers, 111*(1), 121–140.

Kwan, M.-P. (2009). From place-based to people-based exposure measures. *Social Science and Medicine, 69*(9), 1311–1313.

Kwan, M.-P. (2012a). How GIS can help address the uncertain geographic context problem in social science research. *Annals of GIS, 18*(4), 245–255.

Kwan, M.-P. (2012b). The uncertain geographic context problem. *Annals of the Association of American Geographers, 102*(5), 958–968.

Kwan, M.-P. (2013). Beyond space (as we knew it): Toward temporally integrated geographies of segregation, health, and accessibility. *Annals of the Association of American Geographers, 103*(5), 1078–1086.

Kwan, M.-P. (2018a). The limits of the neighborhood effect: Contextual uncertainties in geographic, environmental health, and social science research. *Annals of the American Association of Geographers, 108*(6), 1482–1490.

Kwan, M.-P. (2018b). The neighborhood effect averaging problem (NEAP): An elusive confounder of the neighborhood effect. *International Journal of Environmental Research and Public Health, 15*, 1841.

Kwan, M.-P., Wan, J., Tyburski, M., Epstein, D. H., Kowalczyk, W. J., & Preston, K. L. (2019). Uncertainties in the geographic context of health behaviors: A study of substance users' exposure to psychosocial stress using GPS data. *International Journal of Geographical Information Science, 33*(6), 1176–1195.

Ma, X., Li, X., Kwan, M.-P., & Chai, Y. (2020). Who could not avoid exposure to high levels of residence-based pollution by daily mobility? Evidence of air pollution exposure from the perspective of the neighborhood effect averaging problem (NEAP). *International Journal of Environmental Research and Public Health, 17*(4), 1223.

Mu, L., & Wang, F. (2008). A scale-space clustering method: Mitigating the effect of Scale in the analysis of zone-based data. *Annals of the Association of American Geographers, 98*(1), 85–101.

Park, Y. M., & Kwan, M.-P. (2017). Individual exposure estimates may be erroneous when spatiotemporal variability of air pollution and human mobility are ignored. *Health & Place, 43*, 85–94.

Shafran-Nathan, R., Yuval, Levy, I., & Broday, D. M. (2017). Exposure estimation errors to nitrogen oxides on a population scale due to daytime activity away from home. *Science of the Total Environment, 580*, 1401–1409.

Tan, Y., Kwan, M.-P., & Chen, Z. (2020). Examining ethnic exposure through the perspective of the neighborhood effect averaging problem: A case study of Xining, China. *International Journal of Environmental Research and Public Health, 17*, 2872.

Yoo, E., Rudra, C., Glasgow, M., & Mu, L. (2015). Geospatial estimation of individual exposure to air pollutants: Moving from static monitoring to activity-based dynamic exposure assessment. *Annals of the Association of American Geographers, 105*(5), 915–926.

Yu, H., Russell, A., Mulholland, J., & Huang, Z. (2018). Using cell phone location to assess misclassification errors in air pollution exposure estimation. *Environmental Pollution, 233*, 261–266.

Chapter 12
How to Form and Answer the *So What* Question in GIScience

Lan Mu

Abstract *So What* is about justifying contribution to knowledge. It is often presented as the relevance, significance, and broader value of the research. Building upon some successful formats from education, medicine, and geography, we present a style to form and answer the *So What* question in GIScience. The format is *Where? Why? How?* and *SO What?* (WWHO). It is also referred to as the "Gazing on the Peak" format, inspired by the famous poem by Du Fu.

Keywords So What · GIScience · Relevance · Significance

12.1 Introduction

Thinkers and researchers constantly seek good questions, from "asking the right question is half the answer" by Aristotle (384–322 BC) to "the formulation of a problem is far more essential than its solution" By Albert Einstein (1879–1955). Those walking down the academic career path have undoubtedly encountered the famous *So What* question, a politer version of *Who Cares*. A presenter spends 90% of the time elaborating on how advanced, complicated, and efficient a geospatial algorithm is and throws tons of equations on the screen. The audience is lost, and the famous question appears, *So What*?

In preparing this chapter, I realized that I have been thinking about this question for GIScience research for years but never dedicated time to writing it. We constantly ask the *So What* question during students' thesis and dissertation proposals and defenses, corresponding to the initial *Overarching* question. In reviewing manuscripts, we have also seen many articles, including our own, suffering from a rudimentary lack of clarity regarding explaining their significance and relevance.

So What is about justifying contribution to knowledge. It is often presented as the relevance, significance, and broader value of the research (Selwyn, 2014). In

L. Mu (✉)
Department of Geography, University of Georgia, Athens, GA 30602, USA
e-mail: mulan@uga.edu

B. Li et al. (eds.), *New Thinking in GIScience*,
https://doi.org/10.1007/978-981-19-3816-0_12

medicine, the *So What* question leads to the difference between academic and practice. In clinical trial research, it is estimated that one-third of the time between the initial idea and publication is spent fighting about the research question and answering the *So What* (Riva et al., 2012).

GIScience is an applied and practical discipline. Like clinicians and researchers in the medical field, much effort has been made to close the academic-practitioner gap in GIScience. This chapter reviews some successful formats from education, medicine, and geography and searches for a style to form and answer the *So What* question in GIScience.

12.2 The "So What" Question in Education, Medical Research and Geography

12.2.1 The Relevance in Technology Education

In an editorial, Selwyn (2014) explained why some articles cannot be published due to the missing answer to the *So What* question. He used the examples of two submissions, and both were about using computer games in classrooms. One described using a game in Grade 7 history classroom, and the other was about another game in Garde 9 science classroom. While both pieces of research were exciting, neither article explained why their project might be of interest or relevance beyond the teachers and students in those two classrooms. The editors decided that neither translated automatically into scholarly, academic writing due to the inadequate "wow" factor. According to Selwyn (2014), the *So What* question in this kind of research should demonstrate how such teaching methods "add to the understanding of the social complexities of digital technology and media use in education."

From the perspective of education research, the *So What* question is briefed as four types of relevance—the relevance to educational practice, the relevance to policy, the relevance to other academic research and writing, and the relevance to theory (Selwyn, 2014).

12.2.2 The PICOT Format in Medical Research

In medical research, the format of forming research questions was first introduced in 1995 as PICO (P—Population, I—Intervention, C—Comparison, O—Outcome), and later expanded to PICOT (T—Time) (Richardson et al., 1995). To close the gap between clinical practice and research and to facilitate raising research questions and answering the *So What* question, the PICOT format is advocated in medical trial research. The authors form the research question by following the PICOT format. For examples,

- "In adults with chronic neck pain (P), what is the minimum dose of manipulation necessary (I) to produce a clinically important improvement in neck pain (O) compared to supervised exercise (C) at six weeks (T)?" (Riva et al., 2012)
- "In patients without preoperative anemia (P) undergoing cardiac or orthopedic surgery (I), does treatment with: (1) intravenous iron alone; or (2) intravenous iron with recombinant erythropoietin; compared with (3) placebo (C), administered a day after surgery, increase hemoglobin concentration (O) 7 days after surgery (T)?" (Thabane et al., 2009).

The PICOT format is easy to follow. It provides clinicians and researchers an initial basis for mutual understanding and communication, and helps frame and answer the relevance question. The PICOT format generally works well in comparative studies or association studies, such as the association between exposure and outcome(s). The format has been routinely advocated in evidence-based medicine (Thabane et al., 2009).

Additionally, other formats or styles have been proposed, such as PICO (stands for Population, Intervention, Comparison, and Outcomes), PICOS (stands for Population, Intervention, Comparison, Outcomes, Setting), SPIDER (stands for Sample, Phenomenon of Interest, Design, Evaluation, and Research type) (Methley et al., 2014) and PESICO (stands for Person, Environments, Stakeholders, Intervention, Comparison, and Outcome) (Schlosser et al., 2007).

After the research question is set up using the PICOT or other format, its answer, especially the I—intervention and O—outcome parts, can address the *So What* question. Asking the right question is INDEED half the answer.

12.2.3 The WWO Format in Geography

Geographer Shannon McCune wrote "Geography: *W*here? *W*hy? *SO* What?" (WWO) in 1970 (McCune, 1970). He emphasized that "the geography taught in America must be realistic—it must deal with this current, changing world; in order to fit students for their lives, moreover, it must deal with the world of tomorrow." He went on decomposing the geography study into three questions.

The first question is *Where*? McCune considered it the beginning or the first steppingstone of geography inquiry. He pointed out, amazingly, more than 50 years ago, that "rapid transit and fast communications are forcing the reading of maps in the light of new geographical relationships." He also mentioned that "geographers are fortunate to be aided in their task of seeking the answer to their question *where* by the development of new tools such as the satellite photograph, the remote sensor images, and the computer-drawn map."

The second question *Why?* rises when just answering the question *where* does not satisfy geographers. "Why are observable things, including man and all his activities, *where* they are on this earth?" In order to answer the *Why*, geographers bring a deep concern to the uneven distribution of resources, both physical and socioeconomic, in the world and the geographic variation. The answer to this question really distinguishes and highlights geography from others and, therefore, the discipline's core. McCune predicted the future *Why* question-solving for geographers, that "Machines can gather and analyze all this new data, but then these machines have to be asked the right questions. Wisdom, foresight, and concern—these must be used as the vast array of geographical facts are interpreted and theorized."

The third question *So What* is inevitable following *Where* and *Why*. The *So What* question will help students "see the relevance of what they study to their lives today and for the years to come." McCune used examples of the Vietnam War, landing on the moon, unequal human and natural resource distributions and more, to illustrate the *So What* by utilizing phrases such as global responsibility, and relevance to the world's needs.

For example, the following statement has the WWO format: In Alaska (*Where*), will a newly established oil field change the measures of the environment and life (*Why*), and therefore deteriorate or enhance the environment of today and tomorrow (*So What*).

12.3 The WWHO or the "Gazing on the Peak" Format in GIScience

Formats of relevance, PICOT, and WWO from education, medical research, and geography set excellent and easy-to-follow examples. When thinking of using them for GIScience research, I realized that we could not use those directly since GIScience has its unique element. GIScience is rooted in geography and deals with all kinds of phenomena on the earth's surface. Therefore, P—Population and C—Comparison are not necessarily present in all GIScience research. Meanwhile, GIScience is a data and application-driven field, so concrete methods or procedures are required in almost all research projects. Thus, the *How* question. When framing and answering research questions in GIScience, we can build upon McCune's WWO, add the *How*, and create a format: ***W**here? **W**hy? **H**ow?* and ***SO** What?* (WWHO).

A well-known Chinese poem came to my mind when I took a moment to write down this thought: "Gazing on the Peak" by Du Fu, one of the greatest Chinese poets in the Tang Dynasty. It has all the elements we are looking for in WWHO (Table 12.1) format. The first two lines answer the *Where* question. The following four lines address the *Why* question for Mountain Tai (Daizong)'s breathtaking beauty. The seventh line identifies and answers the *How* question (i.e., how to enjoy the scenic landscape)—climb up to its highest summit. The last line summarizes the *So What*—with one sweeping view see how small all other mountains are. For this reason, we can also refer to WWHO as the "Gazing on the Peak" format.

To further illustrate the format, I searched on the Web of Science Core Collection using the keywords of "GIS" and "health" and retrieved 5519 articles. The top two original research articles with the highest number of citations (excluding review

Table 12.1 Examples following the *WWHO* format

	《望岳》杜甫 Gazing on the peak by Du Fu[a]	Luo and Wang (2003)	Eeftens et al. (2012)
Where	岱宗夫如何 齐鲁青未了 And what then is Daizong like? over Qi and Lu, green unending	Chicago region	20 European study areas at 20 sites per area
Why	造化钟神秀 阴阳割昏晓 荡胸生层云 决眦入归鸟 Creation compacted spirit splendors here Dark and light, riving dusk and dawn Exhilirating the breast, it produces layers of cloud; splitting eye-pupils, it has homing birds entering	Examine spatial accessibility to primary care	Estimate individual exposure if air pollution
How	会当凌绝顶 Someday may I climb up to its highest summit	The two-step floating catchment area (FCA) method	Land use regression (LUR) models with GIS-derived variables
So What	一览众山小 with one sweeping view see how small all other mountains are	Improve the designation of health professional shortage areas	Air pollution concentrations at small scale, i.e., the home addresses of participants, can be estimated

[a] Translated by Stephen Owen (2015)
Luo and Wang (2003) and Eeftens et al. (2012)

articles and software/patent articles) were authored by Luo and Wang (2003) and Eeftens et al. (2012). I analyzed these two articles using the WWHO format (Table 12.1).

12.4 Conclusion

Forming questions is as important as answering them. In order to answer the *So What* question in GIScience, we reviewed formats of relevance, PICOT, and WWO from education, medical research, and geography. Building upon those excellent examples, I propose a WWHO format to form and answer the *So What* question in GIScience: ***W****here?* ***W****hy?* ***H****ow?* and ***SO*** *What?* It is also referred to as the "Gazing on the Peak" format, inspired by the famous poem by Du Fu. Use the format to frame research questions in GIScience, then answer them. Presenting the format clearly in both the abstract and conclusion sections increases the readability of articles, justifies contribution to knowledge and articulates the research's relevance, significance, and broader value.

References

Eeftens, M., Beelen, R., de Hoogh, K., Bellander, T., Cesaroni, G., Cirach, M., Declercq, C., Dedele, A., Dons, E., de Nazelle, A., Dimakopoulou, K., Eriksen, K., Falq, G., Fischer, P., Galassi, C., Grazuleviciene, R., Heinrich, J., Hoffmann, B., Jerrett, M., ... Hoek, G. (2012). Development of land use regression models for PM2.5, PM2.5 absorbance, PM10 and PMcoarse in 20 European study areas; Results of the ESCAPE project. *Environmental Science & Technology, 46*(20), 11195–11205.

Luo, W., & Wang, F. H. (2003). Measures of spatial accessibility to health care in a GIS environment: Synthesis and a case study in the Chicago region. *Environment and Planning B: Planning & Design, 30*(6), 865–884.

McCune, S. (1970). Geography: Where? Why? So What? *Journal of Geography, 69*(8), 454–457.

Methley, A. M., Campbell, S., Chew-Graham, C., McNally, R., & Cheraghi-Sohi, S. (2014). PICO, PICOS and SPIDER: A comparison study of specificity and sensitivity in three search tools for qualitative systematic reviews. *BMC Health Services Research, 14*(1), 579.

Owen, S., Warner, D. X., & Kroll, P. W. (2015). *The Poetry of Du Fu*. De Gruyter.

Richardson, W. S., Wilson, M. C., Nishikawa, J., & Hayward, R. S. (1995). The well-built clinical question: A key to evidence-based decisions. *ACP Journal Club, 123*(3), A12–A13.

Riva, J. J., Malik, K. M., Burnie, S. J., Endicott, A. R., & Busse, J. W. (2012). What is your research question? An introduction to the PICOT format for clinicians. *The Journal of the Canadian Chiropractic Association, 56*(3), 167.

Schlosser, R. W., Koul, R., & Costello, J. (2007). Asking well-built questions for evidence-based practice in augmentative and alternative communication. *Journal of Communication Disorders, 40*(3), 225–238.

Selwyn, N. (2014). 'So What?'... a question that every journal article needs to answer. *Learning, Media and Technology, 39*(1), 1–5.

Thabane, L., Thomas, T., Ye, C., & Paul, J. (2009). Posing the research question: Not so simple. *Canadian Journal of Anesthesia/journal Canadien D'anesthésie, 56*(1), 71–79.

Chapter 13
Prospects on Causal Inferences in GIS

Bin Li

Abstract Although causal reasoning is a tradition in geographic inquiry, adaptations of statistical and computational causal inference frameworks developed in the past decades have been limited in GIS. To facilitate spatial causal analysis, GIS should develop new data models and software tools for the discovery of causal structures as well as the identification and estimation of causal effects. Event-based and scenario-based spatio-temporal models are promising concepts. Spatially explicit causal models can be developed by integrating spatial statistical models with existing computational and statistical models for causal analysis. There are limits to quantitative approaches to causal inferences; a comprehensive causal analysis should include qualitative analysis.

Keywords Causal inference · Potential outcome · Causal diagram · Spatial statistics · GIS

13.1 Introduction

Causal inference is considered one of the most important ideas in statistics in the past 50 years (Gelman & Vehtari, 2021). The 2021 Nobel Prize in Economic Science was awarded to Joshua Angrist and Guido Imbens "for their methodological contributions to the analysis of causal relationships."[1] Prior to that, the 2011 Turing Award in computer science was given to Judea Pearl, for his "fundamental contributions to artificial intelligence through the development of a calculus for probabilistic and causal reasoning."[2] However, both significant advances in causal reasoning have had limited impacts on the theoretical and methodological development in geography. The advances in statistical causal inferences have yet to be implemented in

[1] https://www.nobelprize.org/prizes/economic-sciences/2021/press-release/.

[2] https://amturing.acm.org/award_winners/pearl_2658896.cfm.

B. Li (✉)
Department of Geography & Environmental Studies, Central Michigan University, Mount Pleasant, MI 48858, USA
e-mail: li1b@cmich.edu

B. Li et al. (eds.), *New Thinking in GIScience*,
https://doi.org/10.1007/978-981-19-3816-0_13

GIScience despite the ubiquitous use of GIS in social governance and management, where rigorous causal inferences are in high demand. Integration of spatial effects in current causal inference frameworks presents opportunities for geography and GIScience. This essay speculates on some of the niche areas for research and development in GIS. I will show that while causal inference is not completely new to geography and GIScience, spatial causality analysis with statistical and computational approaches remains scarce. Methodological and technical elements, however, do exist for developing spatially explicit causal inference models.

13.2 Causal Inference Is Not New

Humans have always been trying to find the cause of the phenomena occurring around them and to themselves, pursuing the laws and logic that relate events and outcomes to each other. Some attribute these occurrences to the supreme power of Gods or deities who created the rules for the world's operation. Scientists believe that a phenomenon is the outcome of processes governed by some laws. To find the cause of an event or an outcome is either to uncover the laws and logics that chain them together, or to derive explanations and conclusions based on experiences and observations. The tradition of deductive and inductive reasoning has perpetuated modern day scientific inquiries.

Although statistical learning through regression is a contemporary inductive mode of reasoning, it was historically believed that its use was limited to revealing association but not causality between outcomes and causes. Regression modeling often concludes at establishing associations, quantifying estimation uncertainty, and sometimes achieving predictions at certain levels of accuracy, while falling short of causal reasoning. Such a state of inquiry is far from satisfactory, particularly for economists who aspire to intervene with economic operations through policies, for biomedical scientists who need to evaluate the efficacy of a new drug, or for geographers who pride themselves on their mission to acquire knowledge about human–environment relationship. Causal inference, or identifying the cause of a phenomenon and determining what would happen to the phenomenon when the cause changes, has attracted major research efforts in the science community in recent decades; in particular, several influential theoretical frameworks have been established, represented by the potential outcome models (Angrist et al., 1996; Fisher, 1936; Holland, 1986; Imbens & Rubin, 2010, 2015; Neyman, 1923; Rosenbaum & Rubin, 1983; Rubin, 1997) and the causal diagram model (Pearl & Mackenzie, 2018; Pearl, 2009a). The presentation of the 2011 Tuning Award to Pearl and the 2021 Nobel Prize in Economic Science award to Angrist and Imbens for their contributions to causal inferences are recognitions of the significant achievement in research and applications of causal inferences. Among statistical scientists, development in causal inferences is considered as one of the most important ideas in the past 50 in statistics (Gelman & Vehtari, 2021).

While causal inference has yet to become mainstream in geography and GIScience, it is by no means neglected. Mercer conducted regression modeling of causal relationships between housing quality and social economic factors in the 70s, which was one of the earlier attempts to tackle statistical causal inferences in geography (Mercer, 1975). As a rudimentary look into the academic landscape of causal inference in geography and GIScience, we selected a group of journals and conducted a keyword search of two phrases, "difference in differences" and "propensity score," which are signature methods for causal analysis. We regard a paper containing either of these two phrases as a paper that involves causal inference. As Table 13.1 shows, after a gap of several decades since Mercer's paper, there is a surge of causal analysis with the potential outcome framework in economic geography, regional science, and epidemiology. Among the selected journals, those centered at economics and epidemiology and with strong geographic associations, i.e., *Urban Economics*, *Regional and Urban Economics*, and *Epidemiology*, have experienced a surge of papers involving causal inferences after 2000. Among the geography journals, economic geography stands out in the group due to its strong ties with economics. The *Annals of AAG*, a flagship journal in geography, started to publish papers that explored causal relationships with statistical/computational methods. These publications dealt with a range of theoretical, technical, and applied topics, mostly within the potential outcome framework. Meanwhile, research on and

Table 13.1 Frequencies of key phrases for statistical causal inferences with select journals in geography and related disciplines

Journals	"Difference in differences"	"Propensity score"
Economic Geography	33	23
Journal of Economic Geography	26	21
Urban Geography	5	3
Annals of AAG	7	3
Professional Geography	1	2
International Regional Science Review	7	14
Urban Economics	125	50
Regional and Urban Economics	200	73
Epidemiology	895	115
Geographical Analysis	5	2
Spatial Statistics	1	2
Geographical Systems	2	1
Computer Environment and Urban Systems	0	0
International Journal of GISci	0	0
Cartography and Geographic Information Science	0	1
Transactions in GIS	0	0
Remote Sensing of Environment	0	2

with causal diagrams is scarce, except for epidemiology. Without diving into comprehensive literature reviews (which can be found in Akbari et al., 2021; Baum-Snow & Ferreira, 2015; Reich et al., 2020) these numbers give us a rough picture of the consideration of causal inference within geography and related disciplines. Though geography has been slow to integrate causal inference into its methodological frameworks compared to its sister disciplines, there are clear indications that geographers have taken notice of research opportunities in causal inferences, signified by the emergence of spatially explicit causal models (Kolak & Anselin, 2020; Zhang et al., 2019).

In the realm of GIS, scenario modeling, which has been practiced through geoprocessing and simulation and applied to themes such as urban growth, land use change, and environmental management (Batty & Longley, 1986; Chen et al., 2020; Li & Yeh, 2000, 2002; Swetnam et al., 2011; Torrens & Torrens, 2004), can also be considered a type of causal inference. It tries to answer the question of "what if" the state of one or more factors changes over time, how specific aspects or the entire system would respond (Batty, 2007). Although statistical methods for causal inferences have yet to enter GIS' mainstream, causal reasoning is nevertheless one of its theoretical and technical traditions.

13.3 Spatial Statistical Causal Inference Is New

Our keyword search revealed an obvious absence of research on causal inferences in GIScience; most existing causality papers in geography journals were published in recent years. Research on spatially explicit causal inferences is in its infancy. Software tools are hard to come by, therefore conducting spatial causality studies requires integrating different software packages with programming. Causal inferences present a new frontier for geographic research in general, and for GIScience in particular (Carré & Hamdani, 2021).

Geography and GIScience are new to all three sub-areas of statistical causal inferences, i.e., causal discovery, causal identification, and estimation of causal effects (Tikka & Karvanen, 2018). Discovering causal rules and structures from data based has been an active area of research in general data mining (Bhoopathi & Rama, 2017; Silverstein et al., 2000) and geospatial data mining (Bleisch et al., 2014; Galton et al., 2015; Yuan, 1996). They often were based on computational approaches that are not designed to adjust for confounding effects and do not necessarily produce representations readily adoptable to statistical models for identification and estimation of causal effects. Causal structure discovery algorithms based on the causal graph model (Glymour et al., 2019; Pearl, 2009b) have yet to be widely adopted to geospatial research and applications of existing tools are rarely seen (Ramsey et al., 2018), with a few exceptions in epidemiology and public health (Fleischer & Roux, 2008; Textor et al., 2011, 2016).

Given a causal structure, the causal effect is not necessarily identifiable from observations. Pearl proposed the do-calculus rules and accompanying algorithms to help solve the identifiability problem (Pearl, 1995; Shpitser & Pearl, 2006), which has been implemented in the R package "causaleffect" (Tikka & Karvanen, 2018), supplementing "DAGitty", another tool for graph-based causal inferences that can be used to determined instrumental variables (Textor et al., 2016). Applications of causal discovery tools such as TETRAD and MIM (Haughton et al., 2006) as well as causal effect identification tools have been scarce in geography and GIScience. In addition, spatial effects—whether direct or indirect—have not been explicitly incorporated in these algorithms, leaving room for refinement.

Estimation of causal effects is the area that has made noticeable progress in geographic research in recent years. A niche area identified by geographers is the Stable Unit Treatment Value, or the SUTVA assumption, which is key to the estimation of treatment effects in the counterfactual framework (Imbens & Rubin, 2015) and often violated by spatial dependence and spatial heterogeneity (Kolak & Anselin, 2020). Kolak and Anselin (2020) recently showed that spatial panel models (Elhorst, 2014) can be extended to relax the SUTVA assumption when performing difference-in-differences analysis while noting that "estimation of treatment effects is in its infancy in regional science," which mirrors the situation across the spectrum of methodologies in estimating causal effects.

13.4 Relevance to GIS

GIS as a computer software system can contribute to all three aspects of computational and statistical causal reasoning, i.e., discovery of causal structure, identification and estimation of causal effects. I will focus on exploring the computational and software perspectives: data models and database models, data processing functions, core functions for causal discovery, identification, and estimation.

Data models are digital representations of real-world systems, which determine how computational algorithms can be designed, implemented, and executed efficiently (Fox et al., 1988). The original vector-raster data models and their extensions play crucial roles in solving geo-spatial problems with GIS (Zeiler, 1999). It is hard to imagine how to calculate slopes for a region without the raster or the TIN data models, or to find the shortest routes without the network data model. Implementing tools for conducting statistical causal inferences in GIS will require new data models that can represent geographic phenomena and processes that are multi-scale, multi-dimensional, and multi-representational. The event-based data model advocated by Yuan (2020), scenario-based data model outlined by Lü et al. (2019), legacy system in use (netCDF) (Rew & Davis, 1990), along with the numerous spatio-temporal data models proposed in the past decades (Gong, 1997; Yuan, 1996; Yuan et al., 2014), provide the theoretical and technical foundations for developing data models and database models that can accommodate spatial statistical causal reasoning.

For geo-spatial objects, their relationships in geographic space and through time also require formal representations. Spatial weights matrix and semi-variograms are two common approaches to representing spatial relationships, which can be extended to the time dimension via spatio-temporal weights matrix in the former case and spatio-temporal covariance functions (spatio-temporal Kriging) in the latter (Cressie & Wikle, 2015), supplementing the traditional panel data model commonly used in spatial econometrics (Elhorst, 2014). These computational representations of spatio-temporal relationships are fundamental for subsequent implementation of spatial causal inference algorithms. For example, a synthetic variable can be generated through the eigenvector spatial filtering (ESF) methodology, serving as a proxy of spatio-temporal relationship (Griffith, 2012), and potentially solving critical issues in spatial causal inferences. We will elaborate on this point shortly.

Data manipulation and processing ensures the data sets from various sources and different models to be integrated and conformed to the requirements of computational and statistical causal models. Areal interpolation, for example, is commonly used to downscale or upscale data from different spatial units where data were collected to a common unit. While GIS functions for areal interpolation have evolved from simple weighted average methods (Goodchild & Lam, 1980) to the more rigorous geostatistical approach that can accommodate both Gaussian and count data (Krivoruchko et al., 2011), there are many issues with non-Gaussian data remained to be resolved (Comber & Zeng, 2019; De Oliveira, 2014). Lastly, all the areal interpolation algorithms must address the Modifiable Areal Unit Problem (MAUP) (Openshaw & Taylor, 1979), and its sibling Modifiable Temporal Unit Problem (MTUP). The latter is particularly relevant when longitudinal data need to be combined to perform causal analysis (Cheng & Adepeju, 2014).

As discussed in the previous section, efforts in integrating spatial components with the causal diagram framework put forward by Pearl have been scarce despite active research in discovering causal rules and relationships in spatial data sets. Because geographic configuration can affect both causal factors and the outcome, controlling geographic confounding effects may be necessary. Instead of using absolute geographic locations (spherical or plane coordinates), a synthetic variable formed by a linear combination of eigenvectors from the spatial weights matrix or spatial temporal weights matrix can enter the causal diagram, participating in the causal structure discovery and identification processes.

Traditional methods for conducting causal inferences with the potential outcome framework, such as propensity score, instrumental variables, and difference in differences have assumed away spatial effects. There have been efforts in filling such gaps. Propensity score matching reduces geographic confounding by including proximity-based variables in logistic regression or employing more rigorous spatial probit models that account for neighborhood effects (De Castris & Pellegrini, 2015). Instrumental variable analysis uses geographic region or ESF generated synthetic variable as an instrument to mitigate confounding effects (Le Gallo & Páez, 2013; Vertosick et al., 2017). Difference in differences analysis can be spatially extended using the spatial panel model (Kolak & Anselin, 2020). ESF-based methods have the advantage of maintaining the same model specification and estimation of existing non-spatial models, particularly in modeling non-Gaussian outcomes (Griffith et al., 2019). Software implementations of these spatially explicit methods and models as well as mainstream causal inference models are available mostly as R packages (Bivand et al., 2013; Millo et al., 2012; Murakami, 2017; Rosseel, 2012; Textor et al., 2016; Tikka & Karvanen, 2018). Integrating them with commercial GIS platforms can be done through bridging APIs as demonstrated by the ESF Tool, which links R functions and ArcGIS tools to perform advanced spatial regression modeling (Koo et al., 2018).

13.5 Conclusions

As a mode of academic inquiry, causal reasoning is a tradition in geography and GIScience. As theories and methods, statistical and computational causal inferences have yet to enter the mainstream of research and development in the discipline. This is particularly true in graph-based causal inferences. Today, the opportunity is ripe to bring geography to causal inferences and to elevate the functionalities of GIS beyond geometric/cartographic modeling and regression modeling.

Meanwhile, we should be mindful of the limitations of quantitative approaches, particularly the quasi-experimental frameworks discussed in this essay. Social systems are complex, and many causal factors may not be quantifiable and measurable. When resorting to surrogate and composite variables to represent them, we may obtain causal effects that are inaccurate or even misleading (Berrie, 2019). In addition, causal structures for social, economic, and political phenomena could be too complex to be quantified. Most critical is the assumption of uni-directional causal relationship in current causal inference frameworks, which does not allow for two-way interactions between causal variables. Lastly, there are substantial critiques on quantitative causal reasoning from the perspectives of political economy and social theory (Poore & Chrisman, 2006; Walters & Vayda, 2009), which calls for an integration of quantitative and qualitative approaches to causal reasoning.

Acknowledgements The author would like to thank Dr. A-Xing Zhu for reviewing the earlier draft of the chapter and providing insightful comments.

References

Akbari, K., Winter, S., & Tomko, M. (2021). Spatial causality: A systematic review on spatial causal inference. *Geographical Analysis*. (Online Version of Record before inclusion in a issue) https://doi.org/10.1111/gean.12312.

Angrist, J. D., Imbens, G. W., & Rubin, D. B. (1996). Identification of causal effects using instrumental variables. *Journal of the American Statistical Association, 91*(434), 444–455.

Batty, M. (2007). *Cities and complexity: Understanding cities with cellular automata, agent-based models, and fractals*. The MIT Press.

Batty, M., & Longley, P. A. (1986). The fractal simulation of urban structure. *Environment and Planning A, 18*(9), 1143–1179.

Baum-Snow, N., & Ferreira, F. (2015). Causal inference in urban and regional economics. In *Handbook of regional and urban economics* (Vol. 5, pp. 3–68). Elsevier.
Berrie, L. (2019). *Causal inference methods and simulation approaches in observational health research within a geographical framework* (PhD Thesis). University of Leeds.
Bhoopathi, H., & Rama, B. (2017). Causal rule mining for knowledge discovery from databases. In *2017 International Conference on Intelligent Computing and Control Systems (ICICCS)* (pp. 978–984).
Bivand, R. S., Pebesma, E., & Gómez-Rubio, V. (2013). *Applied spatial data analysis with R* (2nd ed., 2013 edition). Springer.
Bleisch, S., Duckham, M., Galton, A., Laube, P., & Lyon, J. (2014). Mining candidate causal relationships in movement patterns. *International Journal of Geographical Information Science, 28*(2), 363–382.
Carré, C., & Hamdani, Y. (2021). Pyramidal framework: guidance for the next generation of GIS spatial-temporal models. *ISPRS International Journal of Geo-Information, 10*(3), 188.
Chen, G., Li, X., Liu, X., Chen, Y., Liang, X., Leng, J., Xu, X., Liao, W., Qiu, Y., & Wu, Q. (2020). Global projections of future urban land expansion under shared socioeconomic pathways. *Nature Communications, 11*(1), 1–12.
Cheng, T., & Adepeju, M. (2014). Modifiable temporal unit problem (MTUP) and its effect on space-time cluster detection. *PLoS ONE, 9*(6), e100465.
Comber, A., & Zeng, W. (2019). Spatial interpolation using areal features: A review of methods and opportunities using new forms of data with coded illustrations. *Geography Compass, 13*(10), e12465.
Cressie, N., & Wikle, C. K. (2015). *Statistics for spatio-temporal data.* Wiley.
De Castris, M., & Pellegrini, G. (2015). *Neighborhood effects on the propensity score matching.*
De Oliveira, V. (2014). Poisson kriging: A closer investigation. *Spatial Statistics, 7*, 1–20.
Elhorst, J. P. (2014). Spatial panel data models. In *Spatial econometrics* (pp. 37–93). Springer. https://doi.org/10.1007/978-3-642-40340-8_3
Fisher, R. A. (1936). Design of experiments. *British Medical Journal, 1*(3923), 554–554.
Fleischer, N. L., & Roux, A. D. (2008). Using directed acyclic graphs to guide analyses of neighbourhood health effects: An introduction. *Journal of Epidemiology & Community Health, 62*(9), 842–846.
Fox, G. C., Johnson, M. A., Lyzenga, G. A., Otto, S. W., Salmon, J. K., & Walker, D. W. (1988). *Solving problems on concurrent processors* (Vol. 1). Prentice Hall.
Galton, A., Duckham, M., & Both, A. (2015). Extracting causal rules from spatio-temporal data. In *International Conference on Spatial Information Theory*, 23–43.
Gelman, A., & Vehtari, A. (2021). What are the most important statistical ideas of the past 50 years? *Journal of the American Statistical Association, 116*(536), 2087–2097.
Glymour, C., Zhang, K., & Spirtes, P. (2019). Review of causal discovery methods based on graphical models. *Frontiers in Genetics, 10*, 524.
Gong, J. (1997). An object oriented spatio temporal data model in GIS. *Acta Geodaetica et Cartographic Sinica, 4.*
Goodchild, M. F., & Lam, N.S.-N. (1980). Areal interpolation: A variant of the traditional spatial problem. *Geo-Processing, 1*(3), 297–312.
Griffith, D. (2012). Space, time, and space-time eigenvector filter specifications that account for autocorrelation. *Estadística Española, 54*(177), 7–34.
Griffith, D., Chun, Y., & Li, B. (2019). *Spatial regression analysis using eigenvector spatial filtering.* Academic.
Haughton, D., Kamis, A., & Scholten, P. (2006). A review of three directed acyclic graphs software packages: MIM, Tetrad, and WinMine. *The American Statistician, 60*(3), 272–286.
Holland, P. W. (1986). Statistics and causal inference. *Journal of the American Statistical Association, 81*(396), 945–960.
Imbens, G. W., & Rubin, D. B. (2010). Rubin causal model. In *Microeconometrics* (pp. 229–241). Springer.

Imbens, G. W., & Rubin, D. B. (2015). *Causal inference in statistics, social, and biomedical sciences*. Cambridge University Press.

Kolak, M., & Anselin, L. (2020). A spatial perspective on the econometrics of program evaluation. *International Regional Science Review, 43*(1–2), 128–153.

Koo, H., Chun, Y., & Griffith, D. A. (2018). Integrating spatial data analysis functionalities in a GIS environment: Spatial analysis using ArcGIS Engine and R (SAAR). *Transactions in GIS, 22*(3), 721–736.

Krivoruchko, K., Gribov, A., & Krause, E. (2011). Multivariate areal interpolation for continuous and count data. *Procedia Environmental Sciences, 3*, 14–19.

Le Gallo, J., & Páez, A. (2013). Using synthetic variables in instrumental variable estimation of spatial series models. *Environment and Planning A, 45*(9), 2227–2242.

Li, X., & Yeh, A.G.-O. (2000). Modelling sustainable urban development by the integration of constrained cellular automata and GIS. *International Journal of Geographical Information Science, 14*(2), 131–152.

Li, X., & Yeh, A.G.-O. (2002). Neural-network-based cellular automata for simulating multiple land use changes using GIS. *International Journal of Geographical Information Science, 16*(4), 323–343.

Lü, G., Batty, M., Strobl, J., Lin, H., Zhu, A.-X., & Chen, M. (2019). Reflections and speculations on the progress in Geographic Information Systems (GIS): A geographic perspective. *International Journal of Geographical Information Science, 33*(2), 346–367.

Mercer, J. (1975). Metropolitan housing quality and an application of causal modeling. *Geographical Analysis, 7*(3), 295–302.

Millo, G., Piras, G., et al. (2012). splm: Spatial panel data models in R. *Journal of Statistical Software, 47*(1), 1–38.

Murakami, D. (2017). spmoran: An R package for Moran's eigenvector-based spatial regression analysis. ArXiv Preprint ArXiv:1703.04467.

Neyman, J. S. (1923). On the application of probability theory to agricultural experiments. Essay on principles. Section 9. *Annals of Agricultural Sciences, 10*, 1–51.

Openshaw, S., & Taylor, P. J. (1979). A million or so correlation coefficients: Three experiments on the modifiable areal unit problem. *Statistical Applications in the Spatial Sciences, 21*, 127–144.

Pearl, J. (1995). Causal diagrams for empirical research. *Biometrika, 82*(4), 669–688.

Pearl, J. (2009a). Causal inference in statistics: An overview. *Statistics Surveys, 3*, 96–146.

Pearl, J. (2009b). *Causality: Models, reasoning and inference* (2nd ed.). Cambridge University Press.

Pearl, J., & Mackenzie, D. (2018). *The book of why: The new science of cause and effect*. Basic Books.

Poore, B. S., & Chrisman, N. R. (2006). Order from noise: Toward a social theory of geographic information. *Annals of the Association of American Geographers, 96*(3), 508–523.

Ramsey, J. D., Zhang, K., Glymour, M., Romero, R. S., Huang, B., Ebert-Uphoff, I., Samarasinghe, S., Barnes, E. A., & Glymour, C. (2018). TETRAD—A toolbox for causal discovery. In *8th International Workshop on Climate Informatics*.

Reich, B. J., Yang, S., Guan, Y., Giffin, A. B., Miller, M. J., & Rappold, A. G. (2020). A review of spatial causal inference methods for environmental and epidemiological applications. ArXiv Preprint ArXiv:2007.02714.

Rew, R., & Davis, G. (1990). NetCDF: An interface for scientific data access. *IEEE Computer Graphics and Applications, 10*(4), 76–82.

Rosenbaum, P. R., & Rubin, D. B. (1983). The central role of the propensity score in observational studies for causal effects. *Biometrika, 70*(1), 41–55.

Rosseel, Y. (2012). Lavaan: An R package for structural equation modeling and more. Version 0.5-12 (BETA). *Journal of Statistical Software, 48*(2), 1–36.

Rubin, D. B. (1997). Estimating causal effects from large data sets using propensity scores. *Annals of Internal Medicine, 127*(8_Part_2), 757–763.

Shpitser, I., & Pearl, J. (2006). Identification of joint interventional distributions in recursive semi-Markovian causal models. *Proceedings of the National Conference on Artificial Intelligence, 21*(2), 1219.

Silverstein, C., Brin, S., Motwani, R., & Ullman, J. (2000). Scalable techniques for mining causal structures. *Data Mining and Knowledge Discovery, 4*(2), 163–192.

Swetnam, R. D., Fisher, B., Mbilinyi, B. P., Munishi, P. K., Willcock, S., Ricketts, T., Mwakalila, S., Balmford, A., Burgess, N. D., & Marshall, A. R. (2011). Mapping socio-economic scenarios of land cover change: A GIS method to enable ecosystem service modelling. *Journal of Environmental Management, 92*(3), 563–574.

Textor, J., Hardt, J., & Knüppel, S. (2011). DAGitty: A graphical tool for analyzing causal diagrams. *Epidemiology, 22*(5), 745.

Textor, J., van der Zander, B., Gilthorpe, M. S., Liśkiewicz, M., & Ellison, G. T. (2016). Robust causal inference using directed acyclic graphs: The R package 'dagitty.' *International Journal of Epidemiology, 45*(6), 1887–1894.

Tikka, S., & Karvanen, J. (2018). Identifying causal effects with the R package causaleffect. ArXiv Preprint ArXiv:1806.07161.

Torrens, P., & Torrens, P. (2004). *Geosimulation*. Wiley.

Vertosick, E. A., Assel, M., & Vickers, A. J. (2017). A systematic review of instrumental variable analyses using geographic region as an instrument. *Cancer Epidemiology, 51*, 49–55.

Walters, B. B., & Vayda, A. P. (2009). Event ecology, causal historical analysis, and human–environment research. *Annals of the Association of American Geographers, 99*(3), 534–553.

Yuan, M. (2020). Why are events important and how to compute them in geospatial research? *Journal of Spatial Information Science, 2020*(21), 47–61.

Yuan, M. (1996). Temporal GIS and spatio-temporal modeling. In *Proceedings of Third International Conference Workshop on Integrating GIS and Environment Modeling*, Santa Fe, NM (p. 33).

Yuan, M., Nara, A., & Bothwell, J. (2014). Space–time representation and analytics. *Annals of GIS, 20*(1), 1–9.

Zeiler, M. (1999). *Modeling our world: The ESRI guide to geodatabase design*. ESRI, Inc.

Zhang, H., Li, X., Liu, X., Chen, Y., Ou, J., Niu, N., Jin, Y., & Shi, H. (2019). Will the development of a high-speed railway have impacts on land use patterns in China? *Annals of the American Association of Geographers, 109*(3), 979–1005.

Chapter 14
Bayesian Methods for Geospatial Data Analysis

Wei Tu and Lili Yu

Abstract This chapter provides an applied introduction to model two types of point-based geospatial data using Bayesian methods. Unlike frequentist inference, Bayesian inference describes unknown statistical parameters with a prior distribution. With this foundation, Bayesian approach provides a valuable alternative to analyze geospatial data. We begin the chapter by introducing the basic concepts and benefits of Bayesian inference and survey four selected Bayesian models and methods, including Bayesian spatial interpolation, spatial epidemiology/disease mapping, Bayesian hierarchical models, and Bayesian spatial autoregressive models, for their applications in geospatial data analysis. Then we discuss some popular software packages to perform Bayesian analysis. We conclude the chapter by encouraging geospatial researchers and practitioners to add Bayesian methods in their toolboxes.

Keywords Geospatial data analysis · Case and count data · Bayesian inference · Markov Chain Monte Carlo (MCMC) · Bayesian spatial interpolation · Spatial epidemiology/disease mapping · Bayesian hierarchical models · And Bayesian spatial autoregressive (SAR) models

14.1 Introduction

Geospatial data analysis has always been deeply rooted in two main inference paradigms in statistics, the classical frequentist inference and the younger but fasting-growing Bayesian inference (Haining, 2014). The two paradigms represent two ontologically (model-building) and epistemologically (integrating knowledge from other sources) different statistical reasonings (Withers, 2002). Geographers have for decades recognized the great potential of Bayesian inference. For instances,

W. Tu (✉)
Department Geology and Geography, Georgia Southern University, Statesboro, GA 30460, USA
e-mail: wtu@georgiasouthern.edu

L. Yu
Department of Biostatistics, Epidemiology and Environmental Health Sciences, Jiann-Ping Hsu College of Public Health, Georgia Southern University, Statesboro, GA 30460, USA
e-mail: lyu@georgiasouthern.edu

B. Li et al. (eds.), *New Thinking in GIScience*,
https://doi.org/10.1007/978-981-19-3816-0_14

Bennett (1985) suggested that Bayesian approaches held the greatest potential in advancing spatial analysis; Hepple (1995) introduced the Bayesian analysis in spatial and network econometrics; Fotheringham et al. (2000) discussed Bayesian inference along with classical inference in their quantitative geography textbook; Withers (2002) provided a comprehensive review on the methodological and substantive benefits of Bayesian methods in human geography research and encouraged geographers to try the new approach; and Lawson and Banerjee (2009) conducted a comprehensive technical review of Bayesian spatial analysis, which covered a broad range of topics including spatial data types, basic concepts and algorithms of Bayesian inference, Bayesian models and examples for point processes and disease mapping, and software packages for Bayesian modeling.

This chapter is organized into five sections: after a brief introduction, Sect. 14.2 reviews the basics of Bayesian inference and its potentials in geospatial research; Sect. 14.3 discusses four applications and models; Sect. 14.4 provides a short discussion on the implementation of Bayesian models; and the last section offers some concluding remarks.

14.2 Bayesian Inference

Between the two major inference frameworks in statistics, the frequentist inference is the conventional approach. It interprets the probability as the long-run frequency or repeatable experiments. Therefore, it can estimate the parameters, which are considered as unknown but fixed quantities, based on the sample data. On the other hand, the Bayesian inference is based on Bayes' theorem, named after Thomas Bayes. In addition to the long-run frequency which can be obtained from sample data (Greenland, 2000), this approach includes subjective experience of uncertainty (De Finetti, 1974) to interpret the probability. The subjective experience relies on previous knowledge on uncertainty to describe the distributions of the parameters, which are considered as unknown but random variables and we call them as prior distributions of the parameters. Bayesian methods combine the information from both sample data and the prior distributions to produce posterior distribution. The model fitting and the implications of the resulting posterior distribution can also be evaluated (Gelman, 2014).

To be specific, let y be a random variable with distribution $f(y|\theta)$. A sample is collected for the random variable y with independent observations $y_1, \ldots, y_n$, then the likelihood is defined as: $L(\theta|y_1, \ldots, y_n) = \prod_{i=1}^{n} f(y_i|\theta)$.

The likelihood summarizes the information about θ based on the observations (Tanner, 1998) and is used by the frequentist inference to estimate the parameter θ. Therefore, the frequentist inference uses information only from sample data. In comparison, Bayesian inference is based on information from the likelihood as well as the prior distribution. Let $p(\theta)$ be the prior distribution of θ which represents the subjective knowledge of θ, then the Bayes' theory calculates the posterior distribution (Kreft & de Leeuw, 2007) as:

$$
\begin{aligned}
p(\theta|y_1,\ldots,y_n) &= \frac{p(\theta)L(\theta|y_1,\ldots,y_n)}{p(y_1,\ldots,y_n)} \\
&= \frac{p(\theta)L(\theta|y_1,\ldots,y_n)}{\int p(y_1,\ldots,y_n|\theta)p(\theta)\mathrm{d}\theta}
\end{aligned}
\tag{14.1}
$$

The posterior distribution is a conditional distribution, i.e., the distribution of θ given the sample data. It updates the knowledge about θ from the prior distribution using the information from the sample data. Then it can be used to make inference about θ, such as mean, credibility interval which corresponds to confidence interval of the frequentist approach.

As the Bayes' theory incorporates the prior information, it adds complexity to the computation. The integration of denominator in Eq. (14.1) may not have a closed form. Because the analytical and numerical integration are often not intractable, especially for high dimension integration, methods were proposed to approximate the posterior distribution. The most popular method is Markov Chain Monte Carlo (MCMC) (Berger, 2000; Cappe & Robert, 2000), which includes Gibbs sampler and Metropolis Hastings algorithm.

The MCMC method is based on the theory of Law of Large Numbers which states that an expectation can be efficiently approximated by a Monte Carlo estimator. Therefore, the basic idea of the MCMC method is to make inference based on samples drawn from posterior distribution. Specifically, it first generates sequences of dependent observations which is called Markov chains, then inference is done using these samples, such as estimating expectation of the parameter using the sample mean. It has been proved that, although the samples are dependent, the observations in these samples can be considered as independent and identical from the true posterior distribution when the Markov Chain is long enough (i.e., to infinity) and is under certain conditions (the chain must be finite, aperiodic, irreducible, and ergodic). To meet those conditions, some iterations at the beginning of the MCMC run need to be discarded and the process is called Burn-in samples. The number of Burn-in samples that need to be discarded can be determined by diagnostics, such as the Geweke Diagnostic, the Heidelberg and Welch Diagnostic, the Raftery and Lewis Diagnostic, and the Gelman and Rubin Multiple Sequence Diagnostic. In addition, the Gibbs sampler is the simplest MCMC algorithm, and it is a special case of Metropolis Hastings algorithm.

Bayesian inference are detailed in numerous textbooks (Congdon, 2014; Gelman, 2014). Several journal articles also provide extensive discussions of Bayesian methods that are directly dealing with geospatial data (Berger, 2000; Cappe & Robert, 2000; Hepple, 1995; Lawson & Banerjee, 2009). The following discussion addresses only selected applications of Bayesian inference on geospatial data analysis.

14.3 Applications of Bayesian Models in Geospatial Problems

We focus our discussions on models that analyze two types of geospatial data, point data with attributes and count data aggregated to areal units. The following discussions are confined to four selected methods and models, namely Bayesian spatial interpolation, spatial epidemiology/disease mapping, Bayesian hierarchical models, and Bayesian spatial autoregressive models.

14.3.1 Bayesian Spatial Interpolation

The classical interpolation method such as Kriging relies on the BLUP (Best Linear Unbiased Predictor) and substitutes maximum likelihood estimates for the model parameters (Lam, 1983). The Bayesian approach, on the other hand, first computes a posterior distribution for model parameters and then computes the posterior predictive distribution by marginalizing over (averaging over) the posterior distribution. The major advantage of the latter solution is that the inference is supported by proper and moderately informative priors on the weakly identified correlation function parameters (Lawson & Banerjee, 2009; Mugglin et al., 1999). Bayesian approach fuses information from multiple sources in the development of models, so it can better handle uncertainty in the interpolation results. Bayesian spatial interpolation methods are applied most commonly in environmental studies (Brown et al., 1994; Cooley et al., 2007; Fuentes & Raftery, 2005). More recently, Bayesian-based spatiotemporal methods have been developed to analyze rapidly increasing collections of and access to spatiotemporal data (Christakos, 2000; Cressie & Wikle, 2011; Esmaeilbeigi et al., 2020; Haining & Li, 2021; Li & Revesz, 2004; Sahu et al., 2010, 2015; Susanto et al., 2016), because of the above-mentioned advantages.

14.3.2 Bayesian Models for Disease Mapping, Risk Estimate, and Prediction

There are two common types of disease data. The first type is case event data, where the locations of cases (points) are usually known residential addresses of patients. These data form a spatial point process. But such data are usually unavailable, particularly for large study areas. The second type is aggregated counts of cases (events), which are more common and accessible. The boundaries of aggregation that form the basic spatial units of the study region are typically subjective with respect to the disease process (such as zip code areas). Bayesian models have long been applied on disease mapping, risk estimate, and prediction (Besag & Newell, 1991; Greenland, 2006, 2007, 2009; Lawson, 2018; Wakefield & Morris, 2001; Waller et al., 1997). A wise choice of the prior distribution can inform the models by bringing in epidemiological domain knowledge and other information from the study area, which can lead to more reliable model results. In addition, the Bayesian approach is more flexible and effective in dealing with sparse data or rare events. Lawson and Banerjee (2009) illustrated the technical details of specifying Bayesian models for analyzing the two types of data and highlighted the applications of count data.

14.3.3 Bayesian Hierarchical Models

Hierarchical (multilevel) regression models have long been used in geospatial research to explicitly incorporate data collected at various spatial scales of observations, for instance, individuals nested in neighborhoods and neighborhoods in cities. Hierarchical models are naturally Bayesian because the distributions of regression coefficients across various clusters (groups, geographic regions etc.) can be treated as a special type of prior distribution. The "empirical Bayes" method estimates regression coefficients as weighted average of the coefficients obtained from sample data from all clusters. In this case, sample data are used to form the prior population distribution, so there are no prior distributions for the hyperparameters. The "Pure Bayes" method generates prior distributions for the hyperparameters from a population. Though the two approaches commonly yield similar results, the latter approach explicitly takes account of prior uncertainty, so it usually generates larger posterior variance (Western, 1999). Moreover, datasets used in hierarchical models could be complex due to problems such as measurement error, censored or missing observations, complex multilevel correlation structures, and multiple endpoints. Comparing to frequentist procedures, Bayesian procedures are not only more flexible in handling the above data issues, but also easier to justify the theoretical properties in the model (Congdon, 2021; Dunson, 2001; McGlothlin & Viele, 2018).

14.3.4 Bayesian Spatial Autoregressive Models

Spatial autoregressive (SAR) models differ from standard regression models in that they account for spatial autocorrelation in the sample data (Griffith, 2009). Bayesian methods have been used to estimate SAR models for several decades (Hepple, 1979; LeSage, 1997, 2000) and the motivation was driven by several advantages of the approach: it can accommodate the presence of an unknown form of heteroskedasticity in the disturbance term in SAR models; it can produce posterior distribution of spatial lag parameters; it can help choose between a logit or probit model; and it by nature can allow prior knowledge to be introduced in the model when available. Doğan and Taşpınar (2014) compared the robust method of moments (GMM) estimator and the estimators based on the Bayesian MCMC approach for SAR models with heteroskedasticity of an unknown form. Their results indicate that the Maximum

Likelihood Estimation (MLE) and the Bayesian estimators impose relatively greater bias on the SAR parameter estimation when there is a negative spatial dependence in the model, they also found that the Bayesian estimators perform better than the robust GMM estimator in terms of finite sample efficiency. LeSage and Chih (2018) developed a Bayesian heterogeneous coefficients SAR panel model to estimate spill-in and spill-out effects for wage in the contiguous US states.

Litterman (1986) proposed the Bayesian vector autoregressive model (BVAR) to overcome the collinearity and overparameterization that are typically found in unrestricted vector autoregressive models (VAR). The Bayesian approach can specify coefficients with varying weights and the estimated coefficients are therefore a combination of prior knowledge and the information from sample data. Like other Bayesian SAR models, BVAR models have mostly been applied in economic and regional forecasting research (Cuaresma et al., 2016; Puri & Soydemir, 2000).

14.4 Bayesian Implementation

Bayesian models are often fit using MCMC techniques. Many software packages can perform MCMC estimation with varying degrees of difficulty and different sampling procedures. The most popular one is WinBUGS (Bayesian inference Using Gibbs Sampling). This free software package employs both Gibbs sampling and Metropolis–Hastings updating methods for a wide range of models. It allows specifying models, sampling from the posterior distribution of parameters, diagnosing model convergence, and creating graphical and analytic output (Lunn et al., 2009). GeoBUGS, a GIS add-on module of WinBUGS, can be used to fit spatial models and to produce a range of map products from model output.

JAGS (Just Another Gibbs Sampler) is an open-source and cross-platform Bayesian analysis program that uses the same model description language as WinBUGS. It can specify Bayesian models and generate samples from the posterior distribution (Plummer, 2003). JAGS users usually rely on R packages such as "coda" and "mcmcplot" to test model convergence, analyze model output, and generate graphics of model results. The "rjags" package of the R software provides an interface to access the JAGS library.

STAN is another specialized software package for Bayesian analysis. Different from JAGS and WinBUGS samplers, it uses a Hamiltonian Monte Carlo and No-U Turn sampling procedure due to their abilities to handle nonconjugate priors and high posterior correlations (Stan Development Team, 2021). Like JAGS, the R package "rstan" is commonly used to access the STAN library from R and the same R packages for JAGS can be used to analyze model output and produce result graphics.

Bayesian analysis is also facilitated by a growing number of R packages. For instances, "brms" is for fitting Bayesian generalized (non-)linear multivariate multilevel models using STAN for full Bayesian inference; "geoR" is for geostatistical analysis including variogram-based, likelihood-based, and Bayesian methods; "spBayes" is for spatially varying short-length time series data; "spTimer" is for fitting large hierarchical Bayesian spatiotemporal models; and "CARBayes" is for fitting a class of univariate and multivariate spatial generalized linear mixed models for areal unit data, with inference in a Bayesian setting using MCMC simulation.

A growing number of researchers have been adopted INLA (the integrated nested Laplace approximation) as an alternative method for approximating Bayesian inference over the past 10 or so years (Blangiardo & Cameletti, 2015; Rue et al., 2009). The INLA methodology focuses on models that can be expressed as latent Gaussian Markov random fields (GMRF), therefore it works for a large family of models. It also enjoys significant computational advantages over classic methods such as MCMC in dealing with complex models. The method can be implemented using the R-INLA package (R-INLA Project).

14.5 Some Concluding Thoughts

For more than four decades, the Bayesian inference has been proposed as an alternative inference to overcome some intrinsic issues in the classical statistical inference in geospatial inquires. For instances, Summerfield (1983) questioned the validity and relevance of applying the classical statistical inference to population data in geography research. Bennett (1985) argued Bayesian approaches had offered powerful alternative theory and techniques to advance statistical inference in spatial science. Haining (2014) highlighted three areas of development in spatial data analysis in the coming years after summarizing the major progress in spatial statistics in the first decade of the twenty-first century: spatial data mining; the "new" geostatistics; and the Bayesian spatial hierarchical modeling. Although the word "Bayesian" appears only in the last area, Bayesian methods have also been applied in the other two areas (Diggle & Lophaven, 2006; Gelfand & Banerjee, 2017; Zhang et al., 2019).

The Bayesian approach has provided geospatial researchers a versatile alternative to fit a wide range of models. In addition to the benefit of including prior knowledge to models, the Bayesian approach is much more flexible because almost any model assumption can be treated as a priori. In addition, the fact that geospatial datasets are not always samples and they could be populations or apparent populations (irreplicable observations) has made it difficult to do classical inference. The Bayesian approach provides a debatable solution to this unique inference challenge in geospatial research (Berk et al., 1995; Mendoza et al., 2021). Moreover, the Bayesian approach can accommodate the needs in the rapidly growing space–time modeling (Faghmous & Kumar, 2014; Holmström et al., 2015).

The opponents of the Bayesian approach hold two fundamental objections to the method. One is that the approach might be abused as an automatic inference engine, and the other is the subjectivity in the choice of prior distribution (Gelman et al., 2013). Like many other authors, we view Bayesian inference as an ontologically and epistemologically different approach with appealing statistical properties that the classical inference lacks. The true value of the approach, however, will need to be assessed by whether it can advance geospatial reasoning in the long run (Withers,

2002). We encourage you to explore the great potential offered by this compelling alterative approach to tackle the problems in the fascinating geospatial world.

Acknowledgements The authors thank Professors Xun Shi and Bin Li for their constructive comments and suggestions to improve the chapter. We are also grateful for the research assistance provided by Harrison Currin. The contents of this publication are solely the responsibility of the authors and do not necessarily represent views of the book editors.

References

Bennett, R. J. (1985). A reappraisal of the role of spatial science and statistical inference in geography in Britain. *Espace Geographique, 14*(1), 23–28.
Berk, R. A., Western, B., & Weiss, R. E. (1995). Statistical inference for apparent populations. *Sociological Methodology, 25*, 421–458.
Berger, J. O. (2000). Bayesian analysis: A look at today and thoughts of tomorrow. *American Statistical Association, 95*, 1269–1276.
Blangiardo, M., & Cameletti, M. (2015). *Spatial and spatio-temporal bayesian models with R-INLA*. Wiley.
Brown, P. J., Le, N. D., & Zidek, J. V. (1994). Multivariate spatial interpolation and exposure to air pollutants. *Canadian Journal of Statistics, 22*, 489–509.
Cappe, O., & Robert, C. P. (2000). Markov chain monte carlo: 10 years and still running. *Journal of the American Statistical Association, 95*, 1282–1285.
Christakos, G. (2000). *Modern spatiotemporal geostatistics*. Oxford University Press.
Congdon, P. (2014). *Applied Bayesian modelling*. Wiley.
Congdon, P. (2021). *Bayesian hierarchical models: With applications using R* (2nd ed.). Chapman & Hall/CRC.
Cooley, D., Nychka, D., & Naveau, P. (2007). Bayesian spatial modeling of extreme precipitation return levels. *Journal of the American Statistical Association, 102*(479), 824–840.
Cressie, N. A., & Wikle, C. K. (2011). *Statistics for spatio-temporal data*. Wiley.
Cuaresma, J. C., Feldkircher, M., & Huber, F. (2016). Forecasting with global vector autoregressive models: A Bayesian approach. *Journal of Applied Econometrics, 31*(7), 1371–1391.
De Finetti, B. (1974). *Theory of probability: A critical introductory treatment* (Vol. 1). Wiley.
Doğan, O., & Taşpınar, S. (2014). Spatial autoregressive models with unknown heteroskedasticity: A comparison of Bayesian and robust GMM approach. *Regional Science and Urban Economics, 45*, 1–21.
Diggle, P., & Lophaven, S. (2006). Bayesian geostatistical design. *Scandinavian Journal of Statistics, 33*(1), 53–64.
Dunson, D. B. (2001). Commentary: Practical advantages of Bayesian analysis of epidemiologic data. *American Journal of Epidemiology, 153*(12), 1222–1226.
Esmaeilbeigi, M., Chatrabgoun, O., Hosseinian-Far, A., Montasari, R., & Daneshkhah, A. (2020). A low cost and highly accurate technique for big data spatial-temporal interpolation. *Applied Numerical Mathematics, 153*, 492–502.
Faghmous, J. H., & Kumar, V. (2014). Spatio-temporal data mining for climate data: Advances, challenges, and opportunities. *Studies in Big Data, 1*, 83–116.
Fotheringham, A. S., Brunsdon, C., & Charlton, M. (2000). *Quantitative geography: Perspectives on spatial data analysis* (1st ed.). Sage Publications.
Fuentes, M., & Raftery, A. E. (2005). Model evaluation and spatial interpolation by Bayesian combination of observations with outputs from numerical models. *Biometrics, 61*(1), 36–45.

Gelfand, A., & Banerjee, S. (2017). Bayesian modeling and analysis of geostatistical data. *Annual Review of Statistics and Its Application, 4*(1), 245–266.

Gelman, A. (2014). *Bayesian data analysis*. CRC Press: Boca Raton.

Gelman, A., Carlin, J. B., Stern, H. S., Dunson, D. B., Vehtari, A., & Rubin, D. B. (2013). *Bayesian data analysis* (3rd ed.). Chapman and Hall/CRC.

Greenland, S. (2000). Principles of multilevel modelling. *International Journal of Epidemiology, 29*(1), 158–167.

Greenland, S. (2006). Bayesian perspectives for epidemiological research: I. Foundations and basic methods. *International Journal of Epidemiology, 35*(3), 765–775.

Greenland, S. (2007). Bayesian perspectives for epidemiological research. II. Regression analysis. *International Journal of Epidemiology, 36*(1), 195–202.

Greenland, S. (2009). Bayesian perspectives for Epidemiologic research: III. Bias analysis via missing-data methods. *International Journal of Epidemiology, 38*(6), 1662–1673.

Griffith, D. A. (2009). Spatially autoregressive models. In R. Kitchin & N. Thrift (Eds.), *International encyclopedia of human geography* (pp. 396–402). Elsevier.

Haining, R. (2014). *Spatial data and statistical methods: A chronological overview* (pp. 1277–1294). Springer.

Haining, R., & Li, G. (2021). *Modelling spatial and spatial-temporal data: A Bayesian approach* (1st ed.). Chapman and Hall/CRC.

Hepple, L. W. (1979). Bayesian analysis of the linear model with spatial dependence. In C. Bartels & R. Ketellapper (Eds.), *Exploratory and explanatory statistical analysis of spatial data* (pp. 179–199). Springer.

Hepple, L. W. (1995). Bayesian techniques in spatial and network econometrics: 1. Model comparison and posterior odds. *Environment and Planning A: Economy and Space, 27*(3), 447–469.

Holmström, L., Ilvonen, L., Seppä, H., & Veski, S. (2015). A Bayesian spatiotemporal model for reconstructing climate from multiple pollen records. *The Annals of Applied Statistics, 9*(3), 1194–1225.

Kreft, G. G., & de Leeuw, J. (2007). *Introducing multilevel modeling*. SAGE Publications.

Lam, N. S. N. (1983). Spatial interpolation methods: A review. *The American Cartographer, 10*(2), 129–150.

Plummer, M. (2003). JAGS: A program for analysis of Bayesian graphical models using Gibbs sampling. In *Proceedings of the 3rd International Workshop on Distributed Statistical Computing*.

Lawson, A. B., & Banerjee, S. (2009). Bayesian spatial analysis. In *The SAGE Handbook of Spatial Analysis* (pp. 321–341). SAGE Publications.

Lawson, A. B. (2018). *Bayesian disease mapping: Hierarchical modeling in spatial epidemiology* (3rd ed.). Chapman and Hall/CRC.

Lesage, J. P. (1997). Bayesian estimation of spatial autoregressive models. *International Regional Science Review, 20*(1–2), 113–129.

LeSage, J. P. (2000). Bayesian estimation of limited dependent variable spatial autoregressive models. *Geographical Analysis, 32*(1), 19–35.

LeSage, J. P., & Chih, Y. Y. (2018). A Bayesian spatial panel model with heterogeneous coefficients. *Regional Science and Urban Economics, 72*, 58–73.

Li, L., & Revesz, P. (2004). Interpolation methods for spatio-temporal geographic data. *Computers, Environment and Urban Systems, 28*(3), 201–227.

Litterman, R. B. (1986). A statistical approach to economic forecasting. *Journal of Business and Economic Statistics, 4*, 1–4.

Lunn, D., Spiegelhalter, D., Thomas, A., & Best, N. (2009). The BUGS project: Evolution, critique and future directions. *Statistics in Medicine, 28*(25), 3049–3067.

McGlothlin, A. E., & Viele, K. (2018). Bayesian hierarchical models. *Journal of the American Medical Association, 320*(22), 2365–2366.

Mendoza, M., Contreras-Cristán, A., & Gutiérrez-Peña, E. (2021). Bayesian analysis of finite populations under simple random sampling. *Entropy, 23*(3), 318.

Mugglin, A. S., Carlin, B. P., Zhu, L., & Conlon, E. (1999). Bayesian areal interpolation, estimation, and smoothing: An inferential approach for geographic information systems. *Environment and Planning a: Economy and Space, 31*(8), 1337–1352.

Puri, A., & Soydemir, G. F. E. (2000). Forecasting industrial employment figures in Southern California: A Bayesian Vector Autoregressive model. *The Annals of Regional Science, 34*(4), 503–514.

Rue, H., Martino, S., & Chopin, N. (2009). Approximate bayesian inference for latent gaussian models by using integrated nested Laplace approximations. *Journal of the Royal Statistical Society: Series B (Statistical Methodology), 71*, 319–392.

Sahu, S. K., Gelfand, A. E., & Holland, D. M. (2010). Fusing point and areal level space-time data with application to wet deposition. *Journal of the Royal Statistical Society: Series C—Applied Statistics, 59*(1), 77–103.

Sahu, S. K., Shuvo Bakar, K., & Awang, N. (2015). Bayesian forecasting using spatiotemporal models with applications to ozone concentration levels in the Eastern United States. *Geometry Driven Statistics* (pp. 260–282). Wiley.

Stan Development Team. (2021). Stan user's guide. Stan. Retrieved December 11, 2021, from https://mc-stan.org/docs/2_18/stan-users-guide/index.html

Summerfield, M. A. (1983). Populations, samples and statistical inference in Geography. *The Professional Geographer, 35*(2), 143–149.

Susanto, F., de Souza, P., & He, J. (2016). Spatiotemporal interpolation for environmental modelling. *Sensors, 16*(8), 1245.

Tanner, M. A. (1998). *Tools for statistical inference: Methods for the exploration of posterior distributions and likelihood functions*. Springer.

Wakefield, J. C., & Morris, S. E. (2001). The Bayesian modeling of disease risk in relation to a point source. *Journal of the American Statistical Association, 96*(453), 77–91.

Waller, L. A., Carlin, B. P., Xia, H., & Gelfand, A. E. (1997). Hierarchical spatio-temporal mapping of disease rates. *Journal of the American Statistical Association, 92*(438), 607–617.

Western, B. (1999). Bayesian analysis for sociologists. *Sociological Methods & Research, 28*(1), 7–34.

Withers, S. D. (2002). Quantitative methods: Bayesian inference, Bayesian thinking. *Progress in Human Geography, 26*(4), 553–566.

Zhang, L., Datta, A., & Banerjee, S. (2019). Practical Bayesian modeling and inference for massive spatial data sets on modest computing environments. *Statistical Analysis and Data Mining: The ASA Data Science Journal, 12*(3), 197–209.

Chapter 15
GIS Software Product Development Challenges in the Era of Cloud Computing

Fuxiang Frank Xia

Abstract Over the past four decades, GIS software products have evolved from workstations to desktops to client–server systems, culminating in today's SaaS-based Cloud computing platforms (Platform as a Service—PaaS). Previous generations of GIS products were mostly self-contained with very limited interactions outside the system. The Geospatial database built on top of the traditional relational database (RDBMS) was usually part of the self-contained system as well. In the first decade of the twenty-first century, Web applications were often lightweight with limited GIS functionality. However, in this new era of cloud computing with SaaS, the whole paradigm has shifted to having full-blown GIS capabilities running in a browser or smart device. In the future, cloud computing will replace much of the desktop functionality we see today. This leads to a whole new way in which a GIS software product should be developed. The traditional waterfall approach to developing a product—from defining specifications, to prototyping, to reviewing with stakeholders—will not survive this rapidly changing era of SaaS. In this chapter, we will discuss the contemporary challenges of cloud computing, particularly with SaaS. We will focus on the various aspects of the development process including design, development, testing, and monitoring of a product system. We will also briefly discuss the critical importance of a team in the success of product development.

Keywords SaaS · Microservice · Big datastore · Spatial aggregation

15.1 Introduction

Starting in the early 1980s, the GIS software industry has become one of the most innovative and unique industries, to the point that even the biggest technology players have trouble competing with vendors specializing in the field.

One of the key reasons for this barrier to entry is that GIS software is spatial-oriented and requires special attention to its research and development. The industry

F. F. Xia (✉)
ESRI, 380 New York Street, Redlands, CA 92373, USA
e-mail: fxia@esri.com

B. Li et al. (eds.), *New Thinking in GIScience*,
https://doi.org/10.1007/978-981-19-3816-0_15

leaders must not only have a solid grasp of information technology trends, but also a deep understanding of geospatial users across many different fields.

GIS software development faces some common challenges across all platforms. In the early years, GIS functionality and geospatial databases were the biggest challenges for GIS vendors. In some cases, GIS vendors would have to innovate beyond what general information technology could offer (such as ESRI's geodatabase). In those days, GIS software was mostly a self-contained system, and the traditional waterfall development methodology worked great for that. This method starts with a defined product specification and prototype to prove viability to stakeholders, which leads directly into a robust development cycle. To ensure success in this process, there is quality assurance and testing, iterating, and product refining before release. In the first two decades, GIS software vendors set up a solid foundation for geospatial functionality and geodatabase capabilities.

In this new era of cloud computing and SaaS, articulated by industry visionaries like ESRI's Jack Dangermond, GIS capabilities should be made available to everyone via SaaS in any cloud platform. The evolution started over a decade ago through early Web GIS offerings such as ESRI's ArcWeb Service, a predecessor of ArcGIS Online, and continues to evolve as we face tremendous challenges as the information technology landscape changes. Starting in the next section, we will outline the specific challenges we are facing and explore possible solutions.

15.2 Challenges to Developing GIS Software as SaaS

Developing GIS software as SaaS is not as simple as moving all the GIS functionality to the Web and running a GIS server (or even a cluster of servers) on a cloud platform like Amazon AWS or Microsoft Azure. Here are a series of challenges one may face:

1. Adapting to the continuously evolving technology landscape: The rapid changes in information technology required to for developing a GIS product as SaaS pose huge challenges in approaching architecture, design, and development.
2. Security: In a self-contained client–server architecture, communication between the client and server is well-defined. Once the client is authenticated, it is assumed to be secure for the entire time when customers are using the system. Internet security breaches are inevitable, so it's paramount to design the system and all the services securely so customers can confidently adopt the technology. Security must be a core part of the development process and a key component of the system.
3. Service Interruption: The REST service architecture has been the standard choice for service-oriented architecture for a long time (Fielding, 2000). Now, microservice architecture is the default choice for implementing the REST services. It is assumed that any of these microservices could be offline due to various reasons such as resource scarcity and restarting of services.

4. Strategic adoption of open-source software: Traditionally, GIS software vendors use very limited open-source software. In the era of cloud computing, no single company can develop every component of its system from scratch. In fact, in order to offer software as SaaS, it is necessary to use open-source software ranging from platform to service orchestration to data storage. There is a delicate balance between in-house development and adopting open-source projects.
5. Verification and testing of SaaS products: Standard regression testing and manual testing of user interaction will not be sufficient for SaaS products. New ways must be adopted to verifying and testing the system and in particular, capability verification should be conducted from the start of the product development process.
6. Integrating with existing products: Most GIS software vendors do not develop a new product from scratch without carefully considering integration with existing products. Here, the challenges are both technical and organizational. In a technical sense, existing products may not support the APIs or protocols the new system requires. It may require significant development time to add the support of those API and protocols. In an organizational sense, collaboration with existing product teams is vital. It is crucial to have buy-in from both the leadership and team members of those existing products.
7. Production system monitoring: Once a SaaS system is deployed and used by customers, customers will almost always assume it will be available 24/7. A production monitoring system is a must to guarantee services to customers (in fact, product owners may have contractual obligations to said customers).
8. Old and new geospatial functionality/capability: Moving current geospatial functionality to the cloud will not be easy; some geospatial capabilities may require rewriting in a new language or divide a set of services into multiple microservices. In addition, new geospatial functions will also have to be developed to meet new requirements from users or new application fields that are only possible via SaaS.
9. Big Data and integration of GeoAI and Machine Learning capability: Recent advances in AI/Deep Learning have made it possible to develop many GeoAI capabilities. However, effectively integrating those capabilities in SaaS is still a big hurdle.
10. Team building: No software product development can succeed without an effective team—from product visionaries, managers, engineering leaders to product engineers and developers.

15.2.1 Agile Development Philosophy and Microservice Architecture

Traditionally, most software developers have used a waterfall-style approach to design software architecture, accounting for every detail in the development lifecycle, long before a single line of code is written. A slightly improved version of the waterfall style usually starts with product specification, then prototyping, stakeholder review and then full-scale development. However, in the era of SaaS, that model no longer works as it is far too slow for today's ever-changing technology landscape.

To overcome the limitation of the waterfall style approach, the industry has come to understand that the Agile methodology, popularized by the 2001 Manifesto for Agile Software Development (Beck et al., 2022), is the best approach. The product architecture must be flexible enough to accommodate many changes down the road—sometimes before the first version of the product is even released. One of the most important elements of the Agile methodology is called "Minimum Viable Architecture" (MVA). The MVA states that the architecture built for product release must be continuously improved over the life of the product. In other words, there is no need to over-design the product in the beginning, as the product may become obsolete by the time it reaches the market. Some platform technologies can change so much that the product may have to switch to a platform different from the original before it is ready for release. The real risk for a new GIS product is not becoming obsolete in geospatial capabilities, but the underlying architecture and platform it is built on. For example, the Distributed Cloud Operating System (DC/OS) was the most important player in the cloud container orchestration arena until Google introduced Kubernetes as an open-source version of its Borg system in 2014 (RisingStack, 2022). It quickly became the industry standard for container orchestration for cloud computing in 2016. If a SaaS GIS product started with DC/OS and didn't architect it correctly, it could take longer than expected to switch the system to a better technology. This leads us to another important aspect of SaaS: containerization of GIS services.

The traditional GIS products are mostly monolithic systems, where a heavyweight server provides all the services for a complicated workflow, which may involve tens and even hundreds of different services. Any of those services going down would bring the whole system down and interrupt every customer who uses the services. To overcome the limitation of traditional systems, a Microservices Architecture should be used to develop each service as self-contained with clear bounded context. Microservices are small, independent, and loosely coupled as shown in Fig. 15.1:

For example, a traditional map server may provide many different services:

- Feature services for delivering a set of geospatial features as GeoJSON to a browser client via a standard REST API
- Map services that render a set of features into a JPEG or PNG images via a similar REST API
- Tile services that deliver vector tiles to a client to be rendered in the requesting clients (browser, desktop, or smart devices)
- A geocoding service to return a geographic coordinate for a given street address

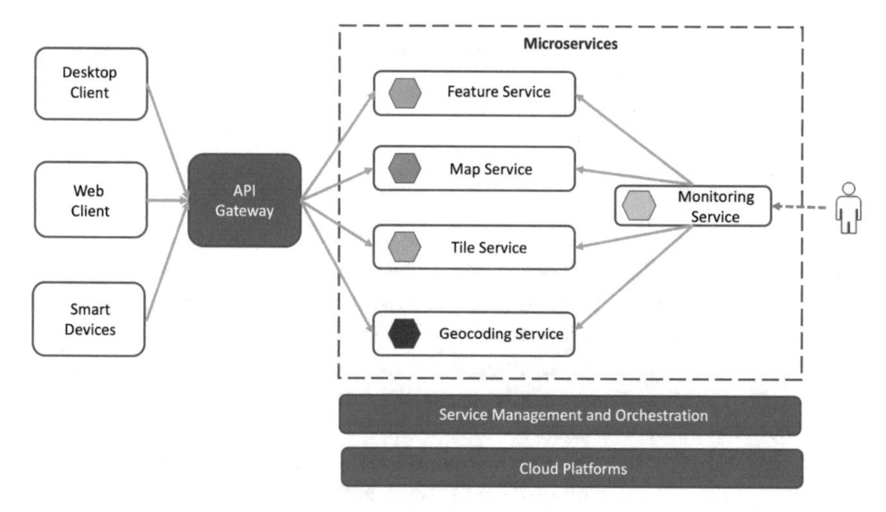

Fig. 15.1 A reference microservice architecture

- A routing service to return a GeoJSON feature for a route from a given starting and ending street addresses, or an JPEG/PNG image depicting a route on top of a basemap.

In a microservice architecture, each one of above services should be a self-contained "small" service. However, most geospatial services are computationally heavy compared to other Internet services. It still poses a great challenge to GIS architects to divide and design those services in a way that could be orchestrated to deliver the functionality and performance that meet GIS users' needs. Containerization of a well-designed microservice makes development and deployment of those services much easier than the traditional monolithic system. Any change or bug fixes could be isolated in a single container, deployed independently without affecting other services. Currently, the dominating container technology is Docker from docker.com.

The old style of managing a cluster of servers does not fit with the new era of SaaS. The Kubernetes play a critical role in the management/orchestration layer on top of the cloud platform providers. It manages all the service containers and can automatically scale up and down the system based on the system load. Getting it right is not easy and creating a set of tools to automatically deploy and update the service containers is even harder. The GIS product development team not only has to understand the characteristics of geospatial functionality running within each service container but also the way the Kubernetes orchestrates the containers. On top of that, it is not ideal to be locked into a single cloud provider. In other words, the system must be developed in such a way to separate the specific platform capability/functionality from the system itself so there is flexibility to switch to different cloud providers if needed. For any global GIS vendors, running the SaaS in multiple cloud platforms (such as Amazon AWS, Microsoft Azure, Google Cloud and Ali Yun) could be a required feature. In this case, there is no choice but to design its SaaS system as a cross cloud platform product.

15.2.2 Security

Security should be an integral part of system development from the very beginning. In a SaaS system, user authentication and authorization for using certain services may not even reside in the GIS SaaS system. The GIS SaaS system may have to connect to an outside service to authenticate connected users and the services they are allowed to access. In a SaaS system, to fully utilize the cloud resources to reduce overall cost, it becomes a common practice to have multiple customers use the same set of resources such as a cluster of VMs (Virtual Machines) or databases, which is called a multi-tenants system. In a multi-tenant system, any kind of security mix-up would not be tolerable since it could easily breach the service contractor with customers and greatly diminish their confidence in the product and related services. That is an internal security challenge. Externally, Internet security breaches happen so often, it's paramount to design the system and services securely so customers are confident while using the technology. The recent security vulnerability in Log4j/Log4Shell Java library (one of the most widely used Java libraries) underscores the seriousness of a security flaw in a system design that causes a real global crisis in the whole internet community (Newman, 2022; Nicholas, 2022).

15.2.3 Continuous Integration/Continuous Delivery (CI/CD)

Once a SaaS GIS system is deployed, it is often required to be available 24/7, and oftentimes users will expect this as standard. To meet this requirement, one must adopt a software development practice called Continuous Integration/Continuous Delivery (CI/CD). CI/CD combines development and operations teams with their day-to-day tasks. CI/CD creates organized code structures that support streamlined development and deployment processes that also allow for more frequent updates with fewer disruptions. However, the process alone does not guarantee a smooth deployment process. More tools are needed to harden the new functions and features before the SaaS system is delivered to the production system. This is where testing automation and chaos engineering can play a critical role.

15.2.4 *Shift-Left Testing, Testing Automation and Chaos Engineering*

In contrast to waterfall's linear development model, the "shift-left" methodology promotes early identification of issues by incorporating quality assurance at the outset of the software development process. By using some testing automation frameworks, the process can trigger automated responses and actions for the failed test cases. The testing automation process should contain a set of typical use cases that simulate the workflow users would create. And these test cases should be performed in an environment close to the real production system. The testing cases should use as much real-world data as possible, then supplement with simulated data if necessary. If the system is model driven, then it should have built-in functionality that will automatically validate the model before running it. A strong test automation strategy allows GIS vendors to bring products to the market faster, reduce operating costs, and deliver a positive experience to customers. In essence, shift left testing operates under this idea that prevention is the best medicine. Testing early and often means development teams and stakeholders always know where things stand at any given moment.

The chaos engineering methodology developed by Netflix brings another tool to harden the SaaS system. It experiments with the resilience of a SaaS system in a production or semi-production system by setting up a tool that would cause breakdowns in the system. These kinds of experiments can reveal where the most vulnerable parts of the whole system are and give developers an opportunity to fix them before they become a big problem for customers in the future.

15.2.5 *Integration with Existing Systems*

Most GIS software vendors do not develop a new product from scratch without thinking of integration with current existing products. Here the challenges are two-fold: one is technical where the existing products may not support the APIs or protocols the new system requires. This means the existing product team may require significantly more development time. The second challenge is the collaboration with the existing product team. It will be crucial to have buy-in from both the leadership and the team members of those existing products, especially if the new system needs the existing product(s) to add new functionalities or fix issues.

15.2.6 *Big Data Stores*

Due to the spatial characteristics of geospatial data, a product team must make a critical decision about how the data are stored. Traditionally, RDBMS dominated the field. But in the Big Data era, distributed data stores such as Elasticsearch, MongoDB, Cassandra, PostgreSQL-Citus, TimescaleDB and more have become strong competitors to traditional databases like Oracle and SQL Server. Interestingly, most of the distributed data store technologies are open source. More importantly, almost all of them have built-in geospatial data types. In a cloud environment, beyond the usual querying performance for all data stores, two additional criteria become very important to evaluate the data store technology: data partitioning and sharding. These are two different techniques for breaking up a large dataset into smaller subsets.

Data partitioning is the technique of distributing data across multiple tables, disks, or nodes to improve query processing performance or to increase database manageability. Data query performance can be improved in one of two ways. First, depending on how the data is partitioned, in some cases it can be determined a priori that a partition does not have to be accessed to process the query. For example, if a spatiotemporal dataset is partitioned with its temporal attribute, and the query contains a temporal constraint, it is easy to determine if a given partition contains the requested data from a client. Moreover, when data is partitioned or sharded across multiple data nodes, it can be accessed in parallel and I/O parallelism and query parallelism can be achieved.

Sharding a database involves breaking up a big database into many, much smaller and unique databases that can be stored across multiple servers with one or more replicas. Sharding implies the data is spread across multiple computer nodes while partitioning does not. For spatial data, there are many different partitioning and sharding algorithms that have been developed, such as GeoHash (Wikipedia, 2022a), z-order curve (Wikipedia, 2022b), and the Hilbert Curve (Wikipedia, 2022c). The common thread among these algorithms is that they all transform the spatial 2-D coordinate (point) into a 1-D integer number, which can then be used to decide the partition and/or sharding location of the given feature. The algorithms can be applied to geographic feature types, such as Point, Polygon (via centroid), and Polylines (via center point, though less common than Point/Polygon). Most of the open source datastores do not have good strategies for spatial sharding and partitioning. Currently, there is no clear or agreed upon best partitioning and sharding strategy one should use for any given spatial temporal big dataset—it is still an open research question.

It is quite a challenge to select the best data store for a SaaS product from at least half dozen candidates, one must evaluate each of them with real geospatial data and decide which one meets the specific requirements for the product. For example, ESRI's ArcGIS Velocity and GeoEvent Server chose Elasticsearch as its Big Data Store due to its capability to partition the data into smaller indexes based on time dimension and at the same time it could shard the data across multiple data nodes to provide the kind of distributed and parallel processing power to get the performance required for the products. It could greatly benefit both academic researchers and the GIS industry if various metrics could be developed to evaluate the performance of those open-source data stores with well-defined open-source geospatial datasets, similar to the ImageNet (Deng et al., 2009) which plays a critical role in pushing deep learning from early academic research to today's widely adopted status.

When data volumes go beyond what these Big Data Stores can handle, one may need to look for cloud data warehouses such as Amazon Redshift, Google BigQuery and Snowflake to host billions of geospatial data points. Integrating this kind of data stores into one's SaaS system poses different engineering challenges than big data stores that are tightly integrated with the SaaS system.

With the rapid accumulation of real time data, the data stores discussed will have a hard time making all the data available all the time. For example, in some IoT (Internet of Things) use cases, the daily data collection would exceed hundreds of millions of data points. In these use cases, the geospatial data must be archived into some cloud storage like Amazon S3 or Microsoft Azure Block Storage. Data stored in these cloud storage systems cannot be processed in real time (hence sometimes they are called "Cold Store" while the regular data stores discussed above are called "Hot Store").

15.2.7 Big Data Processing and GPU Database

Real-time geospatial data processing with big data is another great challenge that GIS SaaS developers are facing. Here we consider two categories of geospatial data capabilities: aggregating data on-the-fly and dynamic geofencing to illustrate the challenges surrounding big data processing.

When data size exceeds hundreds of millions or even billions of records, the raw data must be aggregated first before any realistic analysis is done. In these cases, analyzing individual data points is not only unrealistic but also unnecessary. Therefore, aggregating data into certain geographies becomes the critical step in the geospatial analytics workflow.

For example, to analyze the billions of data points they collect every day, Uber developed a Hexagonal Hierarchical Spatial Index called H3 to index the geospatial data they collect in real time (Brodsky, 2022). They use the H3 system to optimize their ride pricing and dispatch, and to visualize and explore the spatial data in real time. They even use the aggregated hexagon data points to feed into their own deep learning model to predict riding traffic within each hexagon. H3 is available as an open-source project, though integrating it into any data store mentioned above is still a challenge for GIS SaaS developers.

As another example, ESRI realized the importance of real-time data analysis very early on and developed its own proprietary aggregation technology used in its GeoEvent server and ArcGIS Velocity products. This fulfilled the requests of some of its institutional customers, who needed to process geospatial data in real- and semi-real time. This aggregation strategy (using hexagon indexing or others) should be part of the evaluation criteria when one chooses a spatial temporal big data store.

When it comes to real-time geospatial data analysis, there are certain use cases where even on-the-fly aggregation technology would not be enough. In these instances, a GPU based data store would be a great addition to the system.

In recent years, GPU technology has been popularized by the AI/Deep Learning community. GPU data store technology takes advantage of GPU processing powers and can handle billions of points in sub-seconds. For example, it can aggregate billions of point data into a hexagon grid at any detail level and render them as a heatmap on top of a basemap within a second. It can also return the aggregated hexagon grid as a set of features to feed other analytics for further analysis. Currently, there are multiple vendors offering GPU-based geospatial capabilities: OmniSci, Kinetica and SQream. These GPU data stores are sometimes called "Super-Hot Stories." The geospatial algorithms using GPU data stores are still in their early stages and there are a lot of opportunities for both academic and industry researchers and developers.

For GIS SaaS product developers, the question of how to integrate this kind of GPU based technology into the system and workflow will be critical to remain competitive. In this type of SaaS system, the real-time geospatial data would flow through three types of data stores in real time: from incoming sources to super-hot for instant analysis, to hot-store for real time analysis, and to cold-store for archiving and batch-analytics.

Apache Spark has become the industry standard for big data processing, which provides capabilities for both real-time and batch processing of big data. The unique challenge for GIS developers comes from the fact that the geospatial functions are usually much more computationally intensive than non-spatial data. For instance, in a real-time geospatial data processing use case like dynamic geofencing, the real-time data points most likely need to be enriched with some existing datasets which need to be available all the time. In computational terms, it is stateful processing versus the more common stateless one, which is much easier to handle in terms of scaling the application to meet sudden change conditions like spiking of incoming data events, or failure of certain nodes in a cluster. For a stateful use case, both the big data processing engine itself and the stateful states would require scaling up. There are many such operations in geospatial data processing such as a spatial join of incoming events with existing data layers. Developing a unified framework to handle these stateful use cases will be very important to differentiate one SaaS from others offering similar services.

15.2.8 Production System Monitoring

Before the SaaS becomes live to serve customers, the production system monitoring subsystem must be ready to monitor and collect all the metrics regarding the production system performance and stability. This will not only quickly and efficiently solve any issues arising in the production system, but will also help the development team resolve bugs and improve performance of the production system in the next iteration. Production monitoring should never be an afterthought and developed after the SaaS goes live. It should be part of the SaaS system development from the start. Collection of a series of performance and system load metrics should be part of the core functionality for a SaaS product. Once the system becomes live, all these metrics need to be analyzed to support further system improvements. One way to approach this analysis would be to use machine learning models to predict possible system outages and to decide when to take preventative action.

15.2.9 Integration of GeoAI and Machine Learning

Most GIS vendors already provide some traditional machine learning models such as Random Forest, Support Vector Machine (SVM), various clustering models (such as KMeans), and many spatial statistics tools.

The integration here refers to the recent deep learning phenomena. By using transfer learning, the GIS industry has already made tremendous progress. For example, ESRI has released dozens of pre-trained deep learning models for various use cases and geography. ESRI customers can already use those models—via ArcGIS Pro or Python scripts to do many things they could not do before or much more efficiently than before.

However, integrating these models into a Big Data workflow still posts a great challenge. When integrating a set of machine models into a Big Data processing pipeline, it will add many additional pre-processing and data transformation functions before the data can be fed into the model. And outputs from the models also have to go through another set of transformations (post-processing) before they can be fed or saved into an existing geospatial data store.

For instance, in a real-time GIS system running as a SaaS, those pre-post processes can easily become the bottleneck of the system, which can degrade the performance of the services. Another likely challenge comes from the fact that these models are designed for desktop and Notebook users, and most likely will not be able to be used directly in a SaaS system. One solution to this challenge could be for machine learning developers to create an inferencing sub-system that would provide a set of predefined services based on those pre-trained models. However, even current state-of-the-art deep learning models can only answer the question of "what" but not "why," which is still an open research question. This will pose tremendous challenges when customers ask for explanations about the outputs from these GeoAI models.

Most current GIS systems are *descriptive* (such as, possible causes for higher crime rate in certain areas) and *prescriptive* (such as, measures to reduce crime rate after analyzing the historic data). Future GIS systems will certainly have *predictive* capability, of which machine learning models will be a part. The specific kinds of predictive functionality that will be integrated into a SaaS system is another challenge facing the current generation of developers and architects.

15.2.10 Open-Source Strategy

In today's world, no software product can be developed without using any open-source products, especially with SaaS systems. As we have discussed above, the cloud service orchestration platform, Kubernetes, and data stores like Elasticsearch, OpenSearch, PostgreSQL-Citus, are all open-source products.

However, not all open-source projects are created equal. Some of the open-source projects are truly open with Apache 2.0 license, so that any vendor can use them freely without any license cost. On the other hand, ones like Elasticsearch and PostgreSQL-Citus can be used in an enterprise setting but cannot be used in a SaaS offering without licensing costs. The license fees for a SaaS system may be prohibitive enough that a technically less capable alternative is adopted.

Some open-source licenses, like GPL, even require derivative works to be released under the same license. Programs linked with a library released under the GPL must also be released under the GPL. This restriction alone could prevent it from being used as part of a full offering. It is a delicate balance, as one mistake in the early-stage development phase can have impactful ramifications down the road.

Open-source projects with one or more vendors behind them are usually more stable than ones without. A problem could still arise if the main vendor behind the open-source project changes its license. A recent license change for Elasticsearch by Elastic.com illustrates this issue: Elastic, the company behind the Elasticsearch project, changed the Elasticsearch license from Apache 2.0 to less open Elastic License after version 7.10. Any company using the Elasticsearch version after 7.10 in a SaaS system has to pay a license fee.

Therefore, a SaaS product developer must contemplate an open-source strategy. Product managers must consider the ramifications of various open-source projects when they are deciding whether to utilize them as part of the SaaS system. Are these open-source projects truly open? If not, what kind of license costs could affect the specific SaaS system trying to be developed? Is the cost prohibitive in terms of the overall SaaS system expenses? And developers must aim to design the system and services such that open-source products can be replaced with alternatives without causing a huge re-engineering of the system.

15.2.11 Geospatial Functionality Development

As we have discussed in the previous section, big data processing requires GIS vendors to develop new functionality like aggregating data into square, triangle, and hexagon grids to make it possible to visualize and analyze hundreds of millions and even billions of features.

Furthermore, existing geospatial capabilities like buffering, spatial overlay and spatial statistics must still be made available in the new SaaS system. Migrating these functionalities into the new microservice-based architecture poses another significant challenge.

First, traditional GIS systems usually process in a sequential fashion and mostly in a single machine but in a SaaS system, both data and processing are distributed. A batch job could be accomplished via technology like Apache Spark framework, but the same technology may not be enough to handle a real-time job as discussed above.

Second, some well-defined workflows may not fit the big data processing requirement. For example, it is proven that a hexagon grid for aggregating data is the best gridding system for most spatial analysis tasks (ESRI, 2022). However, determining the best grid size for a given dataset and specific problem is a very computationally intensive process. It may work for an offline, batch process workflow but may not be feasible for a given SaaS offering for real-time analysis without expensive GPU-based technology. An alternative is to use the aggregation strategy like H3 and similar ones developed by ESRI, which can compute stats for a hexagon grid at a given level of detail quickly and efficiently enough to make real-time spatial analysis possible.

But there are limitations with these spatial indexing schemes. For one, the grid systems and hexagon grid are fixed at a given detail level (Uber, 2022). A proper level of detail should be selected for the specific dataset and problem set, which may lead you to face the old MAUP (modifiable areal unit problem) issue. It would be a great contribution to the GIS community if someone could develop a strategy and/or algorithm to help users select a most appropriate level of detail grid for a given dataset and the question(s) to be answered. This is similar to choosing hyper-parameters for deep learning. In the early days, extensive training and background were required to be able to select the right set of hyper-parameters but nowadays, most deep learning models have very good default values for all the hyper-parameters.

Breaking down traditional monolithic GIS functionality into much smaller services requires collaboration between system architects and GIS specialists. A gradual migration is probably the best approach, starting with well-defined and most commonly used functions like buffering and spatial join. Then more sophisticated capabilities like networking and routing can be layered on top.

Another important trend in geospatial analysis is real-time or near real-time analytics. With increased deployment of IoT devices, there is a strong demand for real-time algorithms for spatio-temporal analysis of IoT streaming data. A related challenge is to apply existing geospatial algorithms to real-time data, where one may have to improve or even reinvent the existing algorithms so that they can be used in real-time workflow.

15.2.12 Development Team Building

All great software products are developed by great teams. No software product development can succeed without an effective team where everyone contributes from the product visionary, product managers, engineering leads to product engineers and developers. The team visionary must be an effective communicator to convey the vision to both upper-level stakeholders (usually non-technical) and all team members (most of which are technical). Both groups must buy into the vision. The product manager(s) and engineering leader(s) must excel at daily management of detailed workflows. Architects must either be masters of microservice architecture or geospatial capability design. Developers and testing engineers must have talents spreading multiple domains: both at a systems level and in the realm of geospatial functionality.

15.3 Concluding Remarks

In this chapter, we briefly discussed the challenges in developing a SaaS GIS product and what kinds of actions, design methodologies, and best practices should be adopted. Some of these challenges require collaborations between academic researchers and industry professionals. As a GIS developer who can adopt the methodology and best practices outlined here, will have a better chance to develop a successful SaaS product. The whole information technology industry is moving to the cloud platform, and the GIS industry must do the same. It will not be easy to keep current customers satisfied while convincing them to move to the new platform despite its benefits. The key is to have an open mind and to not be afraid of improving the current system based on the changing IT landscape.

References

Beck, K., Beedle, M., van Bennekum, A., Cockburn, A., Cunningham, W., Fowler, M., Martin, R. C., Mellor, S., Thomas, D., Grenning, J., Highsmith, J., Hunt, A., Jeffries, R., Kern, J., Marick, B., Schwaber, K., & Sutherland, J. (2022, February 14). The Agile Manifesto. https://www.agilealliance.org/agile101/the-agile-manifesto/

Brodsky, I. (2022, February 14). H3: Uber's hexagonal hierarchical spatial index. https://eng.uber.com/h3/

Deng, J., Dong, W., Socher, R., Li, L.-J., Li, K., & Fei-Fei, L. (2009). ImageNet: A large-scale hierarchical image database. In *2009 IEEE Conference on Computer Vision and Pattern Recognition* (pp. 248–255). https://image-net.org/

ESRI. (2022, February 14). Why hexagons? https://pro.arcgis.com/en/pro-app/latest/tool-reference/spatial-statistics/h-whyhexagons.htm

Fielding, R. T. (2000). *Architectural styles and the design of network-based software architectures*. University of California.

Newman, L. W. (2022, February 14). 'The internet is on fire'. https://www.wired.com/story/log4j-flaw-hacking-internet/

Nicholas, W. (2022, February 14). What's the Deal with the Log4Shell Security Nightmare? https://www.lawfareblog.com/whats-deal-log4shell-security-nightmare

RisingStack. (2022, February 14). The history of Kubernetes. https://blog.risingstack.com/the-history-of-kubernetes/

Uber. (2022, February 14). Table of cell areas for H3 resolutions. https://h3geo.org/docs/core-library/restable

Wikipedia. (2022a, February 14). Geohash. https://en.wikipedia.org/wiki/Geohash.

Wikipedia. (2022b, February 14). Z-order curve. https://en.wikipedia.org/wiki/Z-order_curve

Wikipedia. (2022c, February 14). Hilbert curve. https://en.wikipedia.org/wiki/Hilbert_curve

Chapter 16
Spatial Thinking of Computational Intensity in the Era of CyberGIS

Shaowen Wang

Abstract The transformation of spatial data into knowledge and understanding through spatial analysis has become an important and ubiquitous element of research and education in numerous fields, especially with support provided by geographic information science and systems (GIS). However, as the complexity and size of spatial data and sophistication of associated analysis approaches have significantly increased, spatial analysis has become increasingly computationally intensive. The focus of this chapter is to address the fundamental challenge of representing and evaluating computational requirements for optimal use of cyberGIS to enable computationally intensive spatial analysis. The chapter describes a computational intensity map (CIM) approach to representing computational requirements of spatial analysis and guiding cyberGIS-enabled spatial analysis. Computational intensity maps (CIMs) are conceptualized to apply the analytical capabilities of cartographic maps and critical spatial thinking to the representation of computational requirements. This map-based formalization allows for the exploitation of critical spatial thinking to evaluate computational requirements for cyberGIS-enabled spatial analysis.

Keywords Computational intensity map · cyberGIS · Geographic information science and systems (GIS) · Spatial analysis

16.1 Introduction

The transformation of spatial data into knowledge and understanding based on spatial analysis has become an important and ubiquitous element in numerous fields (Wang, 2016). Spatial analysis, broadly defined to encompass spatial analytical and modeling approaches that include but are not limited to artificial intelligence and deep learning (Openshaw & Openshaw, 1997; Xu et al., 2018), heuristics and optimization (Lin et al., 2015; Murray, 2021), simulation modelling (Benenson & Torrens, 2004; Davis & Wang, 2018), spatial statistics (Anselin, 1995; Gao et al., 2018), and

S. Wang (✉)
Department of Geography and Geographic Information Science, CyberGIS Center for Advanced Digital and Spatial Studies,, University of Illinois at Urbana-Champaign, Urbana, IL 61801, USA
e-mail: shaowen@illinois.edu

B. Li et al. (eds.), *New Thinking in GIScience*,
https://doi.org/10.1007/978-981-19-3816-0_16

visual analytics (Andrienko et al., 2007; Zhao et al., 2013) for geospatial problem-solving and decision-making (Jankowski & Nyerges, 2001; Yin et al., 2019). As geographic information science and systems (GIS) have been developed to empower spatial analysis in digital environments (Fotheringham & Rogerson, 1993), these methods of spatial analysis have been extensively applied in many domains (Goodchild & Janelle, 2004) but are faced with substantial computational challenges (Wang, 2013). CyberGIS is defined as cyber-based GIS (Wang, 2010) and integrates high-performance and distributed computing resources in a service-oriented and scalable fashion, necessary to resolve these challenges (Wang & Zhu, 2008).

CyberGIS departs from conventional GIS approaches as it represents a paradigm shift for harnessing big data, integrating advanced cyberinfrastructure resources and digital services, and solving complex scientific problems. CyberGIS has emerged as a new generation of GIS for seamlessly integrating cyberinfrastructure, GIS, and spatial analysis capabilities to enable widespread research advances and broad societal impacts (Wang & Goodchild, 2019). Theoretical foundations of cyberGIS have been developed to harness geospatial big data and enable computationally intensive spatial analysis (Wang, 2017). For example, a theoretical construct of spatial computational domain has been developed to guide development of generic methods and efficient algorithms for multi-scale spatial analysis (Shook et al., 2013; Wang et al., 2013). Formalizing the spatial computational domain enables us to address the following research questions. *How to harness massive, shared cyberinfrastructure resources for solving computationally intensive geospatial problems? What is the nature of computational intensity of spatial analysis? Why is spatial special for evaluating computational intensity of spatial analysis?*

These questions in the context of cyberGIS revolving around the computational challenges of spatial analysis are important because representations of computational requirements must capture varying characteristics (e.g., spatial relationships inherent in spatial data) of spatial analysis problems. Furthermore, evaluation of computational requirements must be sufficiently accurate to enable efficient use of cyberGIS capabilities in a scalable way. This chapter focuses on transforming spatial characteristics of data and analysis into computational requirement information that is crucial for cyberGIS to make optimal decisions on conducting spatial analysis tasks with varying computational requirements.

Consider the following scenario of spatial analysis that needs to engage evaluation of computational requirements beyond personal computers. A team of scientists needs to perform a suite of spatial interpolation analyses on a large collection of geospatial datasets for comparative assessment of spatial patterns of environmental impacts. This type of spatial analysis is often exploratory as it requires the tuning of parameters (e.g., interpolation resolution, and the number of nearest neighbors), and is computationally intensive (Wang & Armstrong, 2003). Personal computers may not be able to meet the computational requirements of such analyses to achieve results of adequate quality within reasonable time. Consequently, these scientists need to use high-performance and distributed computing resources made available as integrated cloud computing and supercomputing environments through advanced cyberinfrastructure. Most cyberinfrastructure environments are shared by users for simultaneously solving numerous problems (Wang & Liu, 2009). Therefore, it is imperative to allocate computational resources efficiently for satisfying the computational requirements of each problem and user. The purpose of this chapter is to address the fundamental challenge of representing and evaluating computational requirements of cyberGIS-enabled spatial analysis to assure optimal use of computational resources and achieve desirable analysis performance.

The process of conducting spatial analysis tasks enabled by cyberGIS can be abstracted as the problem of optimally mapping a set of tasks onto high-performance and distributed computing resources. This problem has been shown, in general, to be *NP*-complete, requiring the development of heuristic techniques (Wang, 2008). These heuristics have an objective to minimize the total execution time of the tasks based on the assumption that the estimated *Expected Time to Compute* (*ETC*) for each individual task on each available computational resource must be given as a priori knowledge. *ETC* estimation is also vital for achieving desirable spatial analysis performance, and often done based on task profiling and analytical benchmarking approaches (Smith et al., 2004). These approaches have the following two major drawbacks: (1) task profiling requires extra computational resources, and (2) benchmarking often cannot cover all possible scenarios. This chapter describes an approach to exploiting spatial characteristics for the representation of computational requirements such as *ETC* and argues that the advancement of cyberGIS-enabled spatial analysis needs rigorous evaluation of spatial characteristics in order to facilitate its wide adoption and foster broad applicability of cyberGIS to diverse spatial analysis problems.

16.2 Computational Intensity Map

The formalization of computational intensity maps (CIMs) serves the purpose of representing and evaluating computational requirements of spatial analysis. Computational intensity, collectively referring to computing, data, and communication intensity, is defined as the magnitude of computational requirements of a spatial analysis based on the evaluation of characteristics of the analysis, its input and output, and computational complexity (Wang, 2008). This definition has a different and broader scope than the concept of computational intensity in the literature of computer science that was used to characterize the ratio of the number of computing operations executed to the number of memory accesses (Dongarra & Dunigan, 1997). The importance and necessity of this spatially explicit approach are reflected in the focus of cyberGIS on efficiently solving scientific problems in a scalable way, the limitations of existing theoretical and experimental approaches (Worboys & Duckham, 2004), and computational challenges of data-intensive sciences (Wang et al., 2014).

CIM provides a means to fill the knowledge gap between computational complexity notations and experimental approaches. For example, for the purpose of deriving theoretical bounds in computational complexity assessment, spatial characteristics such as spatial distributions of point patterns may be generalized to satisfy a particular assumption of a balanced tree data structure (Samet, 2006). Although, this type of generalization is necessary for computational complexity assessment, it is not designed to capture spatial characteristics for the derivation of computational requirement information that is needed for optimal spatial analysis enabled by cyberGIS. Wang and Armstrong (2009) formulated a spatial computational domain that is composed of a set of two-dimensional computational intensity surfaces and defined computational transformation to derive these surfaces. CIM exploits cartographic modeling to integrate the capabilities of spatial computational domain and computational transformation. Based on this integrated approach, I argue for fundamental new research based on critical spatial thinking to resolve the computational intensity challenge of spatial analysis.

CIMs are constructed as a cartographic map-based abstraction that synthesizes the representation and evaluation of computational intensity. CIMs, rooted in cartographic modeling, are virtual maps that characterize computational intensity (Tomlin, 1990). Mapping functions are essential to transform spatial characteristics of computational intensity into CIM. This map-based formalization allows for the exploitation of well-established spatial representation and related analytical capabilities to evaluate computational intensity (Egenhofer & Mark, 1995; Miller & Wentz, 2003; Peuquet, 2002). Methodologically, a CIM instance can be modeled as a graph of objects, an array of cells on a field, or both using integrated object- and field-based representations (Goodchild et al., 2007). Each cell or object is associated with a vector of computational intensity values. Such values represent computational requirements in the aspects of computing time, memory/storage, input/output, and communication that are derived from transforming spatial characteristics of data and analysis operations.

16.3 Summary

While many researchers have long attempted to tackle the challenge of understanding computational requirements of spatial analysis to enable efficient and scalable use of high-performance and distributed computing resources, existing approaches lacking a systematic evaluation of spatial characteristics are often not applicable to resolving the practical computational intensity challenge as framed in this chapter. Therefore, fundamental research is called upon to address spatial characteristics in the representation and evaluation of computational intensity, as the exploitation of spatial knowledge is also crucial to engage the development of new methods and necessary tools for computationally intensive scientific problem solving. Spatial thinking, fundamental to numerous types of scientific problem solving, is ideal to be leveraged in this regard. The primary focus of this research direction examines the formalization of CIM through the exploitation of spatial thinking in the representation and evaluation of computational intensity. Specifically, cartographic modeling facilitates spatial thinking of computational intensity based on the development and evaluation of CIM.

CIM fills the knowledge gap between the computational complexity and experimental benchmarking approaches by systematically evaluating the spatial characteristics of computational intensity. The CIM formalization suggests the following cyberGIS law: "*computational intensity map units are related to each other, but near units are more related than distant units*" because CIM functions that transform spatial characteristics of data and analysis operations into computational intensity information reflect spatial autocorrelation (Wang & Armstrong, 2009). This law reflects the first law of geography: "*Everything is related to everything else, but near things are more related than distant things*" (Sui, 2004; Tobler, 1970). This cyberGIS law aims to pave a new path to synergistically apply spatial and computational thinking to cyberGIS-enabled spatial analysis while achieving optimal use of high-performance and distributed computing resources, which lays a foundation for theory-guided cyberGIS and, thus, empowers widespread scientific advances. Furthermore, CIM may be used to visually communicate the knowledge of computational intensity by leveraging the universal nature of spatial thinking, which promises to facilitate the education and workforce development for cyberGIS across disciplinary boundaries.

References

Andrienko, G., Andrienko, N., Jankowski, P., Keim, D., Kraak, M. J., MacEachren, A. M., & Wrobel, S. (2007). Geovisual analytics for spatial decision support: Setting the research agenda. *International Journal of Geographical Information Science, 21*(8), 839–857.

Anselin, L. (1995). Local indicators of spatial association—LISA. *Geographical Analysis, 27*, 93–115.

Benenson, I., & Torrens, P. M. (2004). *Geosimulation: Automata-based modeling of urban phenomena.* Wiley.

Davis, A., & Wang, S. (2018). Geoexpression: A petri network framework for representing geographic process concurrency. *Transactions in GIS, 22*, 1390–1405.

Dongarra, J., & Dunigan, T. (1997). Message-passing performance of various computers. *Concurrency: Practice Experience, 9*(10), 915–926.

Egenhofer, M. J., & Mark, D. M. (1995). Naive geography. In *Spatial information theory: A theoretical basis for GIS*. Lecture notes in computer sciences (Vol. 988, pp. 1–15). Springer.

Fotheringham, A. S., & Rogerson, P. A. (1993). GIS and spatial analytical problems. *International Journal of Geographical Information Systems, 7*(1), 3–19.

Gao, Y., Li, T., Wang, S., Jeong, M.-H., & Soltani, K. (2018). A multidimensional spatial scan statistics approach to movement pattern comparison. *International Journal of Geographical Information Science, 32*(7), 1304–1325.

Goodchild, M. F., & Janelle, D. G. (Eds.). (2004). *Spatially integrated social science.* Oxford University Press.

Goodchild, M. F., Yuan, M., & Cova, T. J. (2007). Towards a general theory of geographic representation in GIS. *International Journal of Geographical Information Science, 21*(3), 239–260.

Jankowski, P., & Nyerges, T. (2001). *Geographic information systems for group decision making: Towards a participatory geographic information science.* Taylor & Francis.

Lin, T., Wang, S., Rodríguez, L. F., Hu, H., & Liu, Y. Y. (2015). CyberGIS-enabled decision support platform for biomass supply chain optimization. *Environmental Modelling and Software, 70*, 138–148.
Miller, H. J., & Wentz, E. A. (2003). Representation and spatial analysis in geographic information systems. *Annals of the Association of American Geographers, 93*(3), 574–594.
Murray, A. T. (2021). Spatial analysis and modeling: Advances and evolution. *Geographical Analysis, 53*(4), 647–664.
Openshaw, S., & Openshaw, C. (1997). *Artificial intelligence in geography*. Wiley.
Peuquet, D. J. (2002). *Representation of space and time*. The Guildford Press.
Samet, H. (2006). *Foundations of multidimensional and metric data structures*. Morgan-Kaufmann.
Shook, E., Wang, S., & Tang, W. (2013). A communication-aware framework for parallel spatially explicit agent-based models. *International Journal of Geographical Information Science, 27*(11), 2160–2181.
Smith, W., Foster, I., & Taylor, V. (2004). Predicting application run times with historical information. *Journal of Parallel and Distributed Computing, 64*(9), 1007–1016.
Sui, D. Z. (2004). Tobler's first law of geography: A big idea for a small world? *Annals of the Association of American Geographers, 94*(2), 269–277.
Tobler, W. R. (1970). A computer movie simulating urban growth in the Detroit region. *Economic Geography, 46*, 234–240.
Tomlin, C. D. (1990). *Geographic information systems and cartographic modeling*. Prentice Hall.
Wang, S. (2008). Formalizing computational intensity of spatial analysis. In *Proceedings of the 5th International Conference on Geographic Information Science*, Park City, Utah, USA (pp. 184–187).
Wang, S. (2010). A cyberGIS framework for the synthesis of cyberinfrastructure, GIS, and spatial analysis. *Annals of the Association of American Geographers, 100*(3), 1–23.
Wang, S. (2013). CyberGIS: Blueprint for integrated and scalable geospatial software ecosystems. *International Journal of Geographical Information Science, 27*(11), 2119–2121.
Wang, S. (2016). CyberGIS and spatial data science. *GeoJournal, 81*(6), 965–968.
Wang, S. (2017). CyberGIS. In D. Richardson, N. Castree, M. F. Goodchild, A. L. Kobayashi, W. Liu, & R. Marston (Eds.), *The international encyclopedia of geography: People, the earth, environment, and technology*. Wiley-Blackwell and the American Association of Geographers.
Wang, S., Anselin, L., Bhaduri, B., Crosby, C., Goodchild, M. F., Liu, Y., & Nyerges, T. L. (2013). CyberGIS software: A synthetic review and integration roadmap. *International Journal of Geographical Information Science, 27*(11), 2122–2145.
Wang, S., & Armstrong, M. P. (2003). A quadtree approach to domain decomposition for spatial interpolation in grid computing environments. *Parallel Computing, 29*(10), 1481–1504.
Wang, S., & Armstrong, M. P. (2009). A theoretical approach to the use of cyberinfrastructure in geographical analysis. *International Journal of Geographical Information Science, 23*(2), 169–193.
Wang, S., & Goodchild, M. F. (2019). *CyberGIS for geospatial innovation and discovery*. Springer.
Wang, S., Hu, H., Lin, T., Liu, Y., Padmanabhan, A., & Soltani, K. (2014). CyberGIS for data-intensive knowledge discovery. *ACM SIGSPATIAL Newsletter, 6*(2), 26–33.
Wang, S., & Liu, Y. (2009). TeraGrid GIScience Gateway: Bridging cyberinfrastructure and GIScience. *International Journal of Geographical Information Science, 23*(5), 631–656.
Wang, S., & Zhu, X.-G. (2008). Coupling cyberinfrastructure and geographic information systems to empower ecological and environmental research. *BioScience, 58*(2), 94–95.
Worboys, M., & Duckham, M. (2004). *GIS: A computing perspective* (2nd ed.). CRC Press.
Xu, Z., Guan, K., Casler, N., Peng, B., & Wang, S. (2018). A 3D convolutional neural network method for land cover classification using LiDAR and multi-temporal Landsat imagery. *ISPRS Journal of Photogrammetry and Remote Sensing, 144*, 423–434.

Yin, D., Liu, Y., Hu, H., Terstriep, J., Hong, X., Padmanabhan, A., & Wang, S. (2019). CyberGIS-Jupyter for reproducible and scalable geospatial analytics. *Concurrency and Computation: Practice and Experience, 31*(11), e5040.

Zhao, Y., Padmanabhan, A., & Wang, S. (2013). A parallel computing approach to viewshed analysis of large terrain data using graphics processing units. *International Journal of Geographical Information Science, 27*(2), 363–384.

Chapter 17
GeoAI and the Future of Spatial Analytics

Wenwen Li and Samantha T. Arundel

Abstract This chapter discusses the challenges of traditional spatial analytical methods in their limited capacity to handle big and messy data, as well as mining unknown or latent patterns. It then introduces a new form of spatial analytics—geospatial artificial intelligence (GeoAI)—and describes the advantages of this new strategy in big data analytics and data-driven discovery. Finally, a convergent spatial analytical framework is suggested as a potential future pathway for spatial analysis.

Keywords Spatial analysis · GeoAI · Artificial intelligence · Deep learning · Data-driven discovery

17.1 Challenges in Spatial Analytics

As a set of quantitative and computational approaches for analyzing geospatial data, spatial analytics is the core of Geographic Information Science (GIScience) for exploration, knowledge discovery, and decision making in the spatial realm. Identified by Golledge (2009) as the unique contribution by geographers to the scientific community, spatial analysis is defined as the methods developed exclusively for analyzing location-based information. Location-based data need specialized analytics to handle spatial dependence, scale dependence, and ecological fallacy, which are not sufficiently accounted for using conventional statistical methods. In the past decades, as spatial theory and computing technology advanced, spatial analysis expanded considerably to cover spatial statistics (for example, exploratory spatial data analysis and spatial regression), spatial simulation (such as agent-based modeling and

W. Li (✉)
School of Geographical Sciences and Urban Planning, Arizona State University, Tempe, AZ 85287-5302, USA
e-mail: wenwen@asu.edu

S. T. Arundel
U.S. Geological Survey, Center of Excellence for Geospatial Information Science (CEGIS), Rolla, MO 65401, USA
e-mail: sarundel@usgs.gov

B. Li et al. (eds.), *New Thinking in GIScience*,
https://doi.org/10.1007/978-981-19-3816-0_17

microsimulation), spatial optimization (Murray, 2021), and data-driven techniques, such as data mining and artificial intelligence (Li, 2020).

Despite covering remarkable breadth, spatial analytics still faces substantial challenges. Goodchild (2009) identified notable issues that spatial analysis is facing. From the perspective of technology, the trend towards the migration of spatial analytical functions to the Web necessitates new business models. New models would ideally handle server-client communication and interoperability and manage data innovatively for online parallel processing services that require use of server-client communication. They also would ideally promote transparency in spatial analysis modules available online. From the science perspective, a (re)formulation of GIScience based on how spatial analytics are being used in scientific and practical problem solving would be beneficial. Over a decade later, we ask "how has the research landscape of spatial analysis changed, how well were Goodchild's challenges addressed, and what new challenges are emerging?".

The last 10 years have witnessed revolutionary advances in technology. Although the term 'cloud computing' was new a decade ago, it has become prevalent today to support the storage, computing, and analysis of geospatial data and its applications (Li et al., 2016). Instead of maintaining a dedicated server, geographic information system (GIS) users and developers have increasingly used cloud infrastructure based on highly reliable virtualized cloud machines capable of elastic computing to meet the different needs of end users. For example, Google Earth Engine, Google's cloud platform that hosts multi-decades of remote sensing images, offers the public rapid access to massive geospatial data and planetary-scale spatial analytics (Gorelick et al., 2017; Yang et al., 2018). The emergence of cyberinfrastructure and CyberGIS has also revolutionized the landscape of spatial analysis to allow collaborative data sharing, analytics, and decision-making (Anselin & Rey, 2012; Li et al., 2016, 2019a, 2019b; Wang, 2010; Yang et al., 2017).

Despite these advances, spatial analytics still have existing and new challenges. Here we present a few examples of these challenges from the computational and data science perspectives.

17.1.1 The Size Challenge of Big Data

Big data have changed nearly every aspect of our lives and the way we conduct science. Datasets, such as earth observation and remote sensing images, images from unmanned aerial vehicles (UAVs), and georeferenced data from social media platforms and sensors for the Internet of Things (IoT) have yielded the production and availability of geospatial data at unprecedented spatial and temporal coverage, resolution, and collection frequency (Li et al., 2020). Handling these data at high throughput and in real-time has presented considerable challenges for traditional analytical methods designed for processing small, clean datasets (Li et al., 2022). Spatial statistical methods, for instance, often require an abstraction of raw data to point data in tabular forms to identify clustering patterns or the associations between certain numerical attributes through linear regression. These methods have reached limitations when it comes to analyzing big data, which are, by definition, large, noisy, diverse, and complex. Although redesigning existing statistical methods to handle big data has been attempted (Laura et al., 2015; Li et al., 2019a, 2019b), many widely used spatial statistical software, such as PySAL (Python Spatial Analysis Library) (Rey et al., 2015) and Geographically Weighted Regression (GWR) (Oshan et al., 2019), continue deployment in desktop computing environments and lack the utilization of advanced computing devices, such as Graphics Processing Units (GPUs). This is likely because the focus of innovation remains on methodology rather than computational performance. In addition, to handle big data, sampling approaches are often introduced. However, in a large dataset with an unknown distribution, it is difficult to guarantee that conventional sampling does not introduce bias into the data, for example in sub-setting training and test sets.

17.1.2 Navigating Through the Messiness of Big Data

Conventionally, big data equals messy data. At the rates data are generated today, the diversity in data collection methods makes (timely) quality control difficult. For example, very fast sampling of some phenomena, such as an event of interest that occurs sporadically, can lead to many empty records. Data reduction can introduce problems, such as when stacking large numbers of raster images over time and then computing a mean or median response in co-located pixels, one can end up with a median image that is too dark in areas of dense cloud cover. Resampling issues result in less accurate results when images are not registered uniformly, and their pixels are aligned. Such issues are easier to detect in small datasets than in large ones. Hence, the ability to navigate through big, complex data becomes a new challenge that calls for innovative techniques designed for big data analytics. Census data for the 2020 Census alone cost the U.S. Census Bureau over $14 billion for compilation and delivery (GAO, 2021). This is one example of high quality, official data managed by governments. However, many other big datasets are created from social media and crowd sourcing platforms, such as Twitter, which have been increasingly used for research because of their broad spatial coverage, richness of content, and low collection cost. However, data from these platforms inevitably contain a substantial amount of noise due partially to their openness, which allows anyone to say anything at any time. In Bayesian statistics, where random variables are introduced, determining the proper prior distribution is often needed to make the estimated posterior distribution match with reality. In such cases, data noise will impede the accurate estimation of a prior distribution. The resulting errors will propagate to later stages of the inference process and lead to imprecise results.

17.1.3 Hypothesis Test Versus Knowledge Mining

Besides relying on well processed data, the traditional spatial analytical approach also requires an accurate understanding and prior knowledge of the underlying process. For instance, in agent-based modeling, heuristic rules need to be defined to guide how an agent moves in space and interacts with the environment and other agents (Li et al., 2020). When applying regression analysis, one needs to specifically define both the independent (X) and dependent variables (y) when building the model, which means we should have knowledge about how X are affecting y. The goal of the analytics is to explain whether and how these independent variables (for example, income or climate) affect the dependent variable (such as housing price) in a geographical region. To incorporate geographically varying effects resulting from spatial heterogeneity, local modeling, such as GWR, is introduced to determine the variation of effects across space. These analyses belong in general to the testing of a hypothesis or identifying the degree of effect between X and y in a predefined model. Whereas such methods are known to be effective in identifying patterns that are expected, their ability to discover or learn unknown relations is weak.

Confronting these challenges requires new spatial analytical methods capable of mining new knowledge from large datasets containing unanticipated or previously unknown patterns, as well as being tolerant to noise. The methods also would ideally be able to learn to model the process itself rather than relying on definitions drawn from prior knowledge. GeoAI has emerged as a new arena for attacking these challenges.

17.2 GeoAI: A New Form of Spatial Analytics

GeoAI, or geospatial artificial intelligence, is a transdisciplinary research area integrating cutting edge AI to solve geospatial problems (Li, 2020). In the past decade, amazing progress has been made in the field of AI, particularly in machine learning and deep learning. The convolutional neural network (CNN) framework is a milestone development (Reichstein et al., 2019). The CNN framework adopts the novel concept of artificial neural network (ANN) in building a computer model mimicking the biological neural network of the human brain even as it brings transformative changes through the introduction of the convolution modules (Fukushima, 2007; Li, 2021; Li et al., 2012; Zhang, 1988). Such modules can conduct information extraction (also known as feature extraction, with each feature treated being the independent variable X in a regression process) from the raw data. CNN-based techniques, therefore, can directly act on the raw data and uncover hidden patterns through deep mining and iterative learning. This kind of data-driven analysis relaxes the constraint in traditional spatial analytics for assuming any predefined rules or relationships between the data (input) and the objective (output), thus supporting discovery and pattern recognition directly from data. This is also known as data-driven discovery (Miller & Goodchild, 2015; Yuan et al., 2004).

Another breakthrough in the design of CNNs is that each convolution layer (Albawi et al., 2017) performs local operations on the data, making parallel computation possible. This design lifts the computation constraint in traditional ANN that has high dependency among the artificial neurons across the fully connected layers. The recent development of high-speed GPUs that contain a few hundred to several thousand micro-processing units allows the high-performance training of CNNs, even with complex structures, on its computing units running in parallel. This also empowers a deep learning model to process big data, furthering its ability to detect new patterns, extract useful information, and create high-quality foundational datasets to aid the elucidation of important scientific questions (Arundel et al. 2020).

Moreover, deep learning models are arguably better at handling noise in training labels than traditional statistical methods (Rolnick et al., 2017). Because many such models are designed to learn complex relations, they tend to overfit the training data. Overfitting occurs when a model fits the training data exactly. When this happens, the model's performance on unseen data will be inferior. One solution is to add noise to the training data such that the model will fit less perfectly, reducing the likelihood of overfitting, and increasing predictive accuracy. In addition, strategies, such as increasing the batch size and thus exposing the model to more samples for updating its parameters during the iterative learning process, lowering learning rates, allowing a more thorough search for the optimal solutions, and providing enough correctly labeled samples, will enable a deep learning model to tackle even extremely noisy data (Rolnick et al., 2017). Although noise in big data is inevitable, the way deep learning is designed and how it handles the data makes deep learning more robust towards dealing with noise than traditional spatial analytical approaches. On the other hand, deep learning requires thousands to billions of training examples to develop abstractions that the human brain can easily intuit through explicit, verbal definition (Marcus, 2018). Interpretability of the results and extension beyond the scope of the training data are also limitations to deep learning systems (Reichstein et al., 2019) that must be overcome.

17.3 Concluding Remarks

As a new form of spatial analytics, GeoAI is exciting because of its outstanding performance in big data analytics, especially in classification, prediction, and pattern recognition. However, the GeoAI domain is still in its infancy and more research is needed for it to become a well-established scientific field. The role of GeoAI in (re)formulating GIScience also needs to be more clearly defined. This need echoes insights shared by Goodchild (2009) in terms of the challenges of spatial analytics in general. We know that the complexity and black-box nature of GeoAI models render the model's reasoning process more difficult to explain than that of traditional spatial analytical approaches (Goodchild & Li, 2021). But this also offers an opportunity to create an even more powerful analytical framework by combining GeoAI and traditional methods. GeoAI can serve as a data pre-processing module that directly interacts with raw big data to achieve high-yield analysis and data filtering (Li et al., 2022).

For instance, a GeoAI-based analytical framework can achieve near real-time processing of satellite remote sensing imagery to create a national to global scale database characterizing natural and human-made features on Earth (Li & Hsu, 2020). This dataset, for which scientists and researchers have waited decades, can be integrated into subsequently processed statistical models to understand crucial environmental and climate change problems (Reichstein et al., 2019). The data and models may jointly contribute to a convergent research agenda for spatial analytics.

Clearly, the development and refinement of existing and future spatial analytics (GeoAI and beyond) should consider fundamental geospatial principles, such as location, scale, spatial autocorrelation, spatial heterogeneity, and geographic similarity. As data and systems become more open, they are less likely to follow fundamental principles and best practices. This concern is like that expressed by scholars during the early years of the development of GIS. Concerns included whether users would utilize the correct projection for the variable studied, correct their statistical analyses for bias in location, or analyze error by combining the variables of the spatial themes.

Whereas some elements of these potential problems are now controlled inherently by software systems, other problems persist or may not be envisioned in the present. Like GIS, GeoAI and subsequent technologies would ideally balance the accessibility of the approach with its applicability, the enforcement of the principles with the flexibility of application. This is the grand challenge of the spatial science community: to not only create and disseminate new tools towards the goal of empowering more vast and ethical utilization, but more importantly to leverage these tools to improve analysis of spatial information to address critical global, regional, and local problems.

Acknowledgements This work is supported in part by National Science Foundation (NSF) under grant BCS-1853864. Li acknowledges additional funding support from NSF (BCS-1455349, GCR-2021147, PLR-2120943, and OIA-2033521). Any use of trade, firm, or product names is for descriptive purposes only and does not imply endorsement by the U. S. Government.

References

Albawi, S., Mohammed, T. A., & Al-Zawi, S. 2017. Understanding of a convolutional neural network. In *2017 International Conference on Engineering and Technology (ICET)* (pp. 1–6).

Anselin, L., & Rey, S. (2012). Spatial econometrics in an age of CyberGIScience. *International Journal of Geographical Information Science, 26*(12), 2211–2226. https://doi.org/10.1080/13658816.2012.664276.S2CID942116

Arundel, S., T., Li, W., & Wang, S. (2020). GeoNat v1.0: A dataset for natural feature mapping with artificial intelligence and supervised learning. *Transactions in GIS, 24*(3), 556–572.

Fukushima, K. (2007). Neocognitron. *Scholarpedia, 2*(1), 1717.

GAO (2021). *2020 Census: Innovations helped with implementation, but Bureau can do more to realize future benefits*. United States Government Accountability Office (GAO). https://www.gao.gov/assets/gao-21-478.pdf

Goodchild, M. F. (2009). Challenges in spatial analysis. In A. S. Fotheringham, & P. A. Rogerson (Eds.), *The SAGE handbook of spatial analysis* (pp. 465–480). SAGE Publishing.

Goodchild, M. F., & Li, W. (2021). Replication across space and time must be weak in the social and environmental sciences. *Proceedings of the National Academy of Sciences, 118*(35).

Golledge, R. G. (2009). The future for spatial analysis. In A. S. Fotheringham, & P. A. Rogerson (Eds.), *The SAGE handbook of spatial analysis* (pp. 465–480). SAGE.

Gorelick, N., Hancher, M., Dixon, M., Ilyushchenko, S., Thau, D., & Moore, R. (2017). Google Earth Engine: Planetary-scale geospatial analysis for everyone. *Remote Sensing of Environment, 202*, 18–27.

Laura, J., Li, W., Rey, S. J., & Anselin, L. (2015). Parallelization of a regionalization heuristic in distributed computing platforms—A case study of parallel-p-compact-regions problem. *International Journal of Geographical Information Science, 29*(4), 536–555.

Li, W. (2020). GeoAI: Where machine learning and big data converge in GIScience. *Journal of Spatial Information Science, 20*, 71–77.

Li, W. (2021). GeoAI and deep learning. *International Encyclopedia of Geography: People, the Earth, Environment and Technology*, 1–6.https://doi.org/10.1002/9781118786352.wbieg2083

Li, W., Batty, M., & Goodchild, M. F. (2020). Real-time GIS for smart cities. *International Journal of Geographical Information Science, 34*(2), 311–324.

Li, Z., Fotheringham, A. S., Li, W., & Oshan, T. (2019a). Fast Geographically Weighted Regression (FastGWR): A scalable algorithm to investigate spatial process heterogeneity in millions of observations. *International Journal of Geographical Information Science, 33*(1), 155–175.

Li, W., Goodchild, M. F., Anselin, L., & Weber, K. T. (2019b). A smart service-oriented CyberGIS framework for solving data-intensive geospatial problems. In *CyberGIS for geospatial discovery and innovation* (pp. 189–211). Springer.

Li, W., & Hsu, C. Y. (2020). Automated terrain feature identification from remote sensing imagery: A deep learning approach. *International Journal of Geographical Information Science, 34*(4), 637–660.

Li, W., Liu, Y., & Wang, S. (2022). Real-time GIS and geocomputation. In J. P. Wilson (Ed.), *The geographic information science & technology body of knowledge* (3rd Quarter 2021 Edition) (in press).

Li, W., Raskin, R., & Goodchild, M. F. (2012). Semantic similarity measurement based on knowledge mining: An artificial neural net approach. *International Journal of Geographical Information Science, 26*(8), 1415–1435.

Li, W., Shao, H., Wang, S., Zhou, X., & Wu, S. (2016). A2CI: A cloud-based, service-oriented geospatial cyberinfrastructure to support atmospheric research. In *Cloud Computing in Ocean and Atmospheric Sciences* (pp. 137–161). Academic Press.

Marcus, G. (2018). Deep learning: A critical appraisal (pp. 1–27). arXiv preprint arXiv:1801.00631.

Miller, H. J., & Goodchild, M. F. (2015). Data-driven geography. *GeoJournal, 80*, 449–461. https://doi.org/10.1007/s10708-014-9602-6

Murray, A. T. (2021). Significance assessment in the application of spatial analytics. *Annals of the American Association of Geographers, 111*(6), 1740–1755.

Oshan, T. M., Li, Z., Kang, W., Wolf, L. J., & Fotheringham, A. S. (2019). mgwr: A Python implementation of multiscale geographically weighted regression for investigating process spatial heterogeneity and scale. *ISPRS International Journal of Geo-Information, 8*(6), 269.

Reichstein, M., Camps-Valls, G., Stevens, B., Jung, M., Denzler, J., & Carvalhais, N. (2019). Deep learning and process understanding for data-driven Earth system science. *Nature, 566*(7743), 195–204.

Rey, S. J., Anselin, L., Li, X., Pahle, R., Laura, J., Li, W., & Koschinsky, J. (2015). Open geospatial analytics with PySAL. *ISPRS International Journal of Geo-Information, 4*(2), 815–836.

Rolnick, D., Veit, A., Belongie, S., & Shavit, N. (2017). Deep learning is robust to massive label noise. arXiv preprint arXiv:1705.10694.

Wang, S. (2010). A cyberGIS framework for the synthesis of cyberinfrastructure, GIS, and spatial analysis. *Annals of the Association of American Geographers, 100*(3), 535–557.

Yang, C., Huang, Q., Li, Z., Liu, K., & Hu, F. (2017). Big Data and cloud computing: Innovation opportunities and challenges. *International Journal of Digital Earth, 10*, 13–53. https://doi.org/10.1080/17538947.2016.1239771

Yang, Z., Li, W., Chen, Q., Wu, S., Liu, S., & Gong, J. (2018). A scalable cyberinfrastructure and cloud computing platform for forest aboveground biomass estimation based on the Google Earth Engine. *International Journal of Digital Earth, 12*(9), 995–1012.

Yuan, M., Buttenfield, B. P., Gahegan, M. N., & Miller, H. (2004). Geospatial data mining and knowledge discovery. In *A research agenda for geographic information science* (p. 24). CRC Press.

Zhang, W. (1988). Shift-invariant pattern recognition neural network and its optical architecture. In *Proceedings of Annual Conference of the Japan Society of Applied Physics.*

Chapter 18
Deep Learning of Big Geospatial Data: Challenges and Opportunities

Guofeng Cao

Abstract With rapid advances of geospatial data acquisition technologies, spatiotemporal data have become increasingly available. As the geography and spatial science community is shifting rapidly to embrace the data-rich era, the long-standing challenges facing the spatiotemporal analysis remain not only unsolved but of increasing prominence in producing geographic knowledge out of the rich data. This chapter reviews these challenges posed by the big spatiotemporal data and discusses the recent progresses in addressing them with a particular focus on the promises of deep learning and GeoAI methods. The chapter is then concluded with a discussion on possible future directions.

Keywords Machine learning · GeoAI · Big data · GIScience · Spatial statistics

18.1 Introduction

In the past two decades, the landscape of geospatial science and technology has shifted dramatically driven by the recent advances in computing and information technology. What is most remarkable in the shift is the advent of the 'Big Data' revolution and the rise of machine intelligence, defined by the increasing availability of data sources with fine scales and the advance of smart algorithms and computing resources. These technology advancements enable enormous opportunities for GISciences and spatiotemporal analysis and at the same time pose significant challenges technically and methodologically.

Geospatial data have become important contributing sources to the Big Data (Manyika et al., 2011) and play increasingly important roles in scientific discovery and practical decision-making. Several drivers are behind the data explosion. The first one is the rapid advances in communication and locational technologies, which make it possible to integrate highly accurate spatial and temporal information with virtually any observations. Example data sources include location-based social media, mobile

G. Cao (✉)
Department of Geography, University of Colorado, Boulder, CO 80309, USA
e-mail: guofeng.cao@colorado.edu

B. Li et al. (eds.), *New Thinking in GIScience*,
https://doi.org/10.1007/978-981-19-3816-0_18

phone call logs and human movement trajectories. The second one is the growing varieties of remote sensors, particularly the ones that are satellite-borne sensors, making it possible to observe the geographic environments with global coverage and with high resolutions. The remote sensors can be further complemented with networks of ground sensors for a more reliable and comprehensive picture of the processes at study. The above two advancements also enable citizens to perform observations of their own about the surrounding environments and to share the observations via different discourses with communities. The so-called *citizen science* provides a unique approach to address the challenges in scientific research and practical applications. The crowd-sourcing geospatial data or volunteered geographic information (Goodchild, 2007) filled an important gap in the spectrum of big spatial data. Thirdly, the ideas of open data and open source were more adopted by governments and data vendors in the past few years, marked by the Landsat imagery collection made freely available in 2008 (Zhu et al., 2019a; Zhu et al., 2019b) and the pass of the open-data law in New York City in 2012 (The New York City Council, 2012). These new sources of geospatial data are complementary to each other and together provide an unprecedented comprehensive view of the physical and socioeconomic environments. In addition to the novel sources of data, new modes of geospatial analysis, such as cyberGIS (Wang, 2010) and GeoAI (Janowicz et al., 2020) buoyed by the advances of high-performance computing and machine learning, are emerging to enable the effective analysis of big geospatial data. The advances of big data and spatially explicit machine intelligence make it possible to examine the geographic and socioeconomic environments at fine scales that were deemed impossible before.

Despite the promises, to fully take advantages of the technology advancements poses significant challenges to GIScientists. In particular, the fundamental methodological and theoretical challenges posed by the highly complex nature and unique characteristics of geospatial data, such as *geospatial pattern complexity*, *geospatial uncertainty*, and *geospatial data heterogeneity*, remain largely unsolved, and become more prominent in the Big Data era as we look to fully take advantage of the geographic information buried in the sea of heterogeneous geospatial data. The remainder of the chapter will revisit these challenges and discuss the opportunities and recent progresses in addressing them with the advent of deep learning.

18.2 Challenges in Geospatial Analysis of Big Geospatial Data

18.2.1 Complex Geospatial Patterns

The analysis of geospatial patterns lies in the very heart of GIScience. One can argue that most if not all the theories, methods, and tools in GIScience are developed to represent, recognize, characterize, and model geospatial patterns. As is well-recognized in the literature, a key property of geospatial data is that spatial information, such as geographic locations and boundaries, scales and proximity or configurations, plays a prominent role. Geospatial data hence exhibits distinctive spatial characteristics compared to aspatial data. These characteristics have been well documented by geographers and GIScientists (Goodchild, 2004). For example, the most well-known Tobler's First Law of geography describes the ubiquity of the similarity or autocorrelation of spatial observations (Tobler, 1970). It indicates the statistical methods developed under the assumption of independence might not work best for spatial patterns (Anselin, 1990). The less well-known second law of geography describes the spatial heterogeneity and the non-stationary nature of spatial observations (Goodchild, 2004), which highlights the importance of incorporating local conditions in modeling and understanding spatial patterns. Furthermore, a recent version of the third law of geography was proposed (Zhu et al., 2018) to extend the spatial patterns into more general geographic contexts including spatial configurations (e.g., spatial scales) and environmental conditions. These characteristics of spatial observations render methods and analytic originally developed for aspatial data might not be best suitable for spatial data. How to effectively represent, characterize and consider such characteristics of geospatial data is one of the most long-standing problems in GIScience and geospatial analysis.

A geospatial pattern can become complex as it involves increasing number of locations or objects and interactions. In GIScience, many theoretical frameworks have been proposed to represent the complex spatiotemporal patterns (e.g., Goodchild et al., 2007; Takeyama, 1997; Yuan, 2001) and a plethora of methods have been devised from different perspectives for modeling of spatial patterns in maps and imagery. Most of these methods typically assume geospatial observations regularly distributed over a Euclidean space and tend to model the complex spatiotemporal patterns as functions of pairwise interactions such as Euclidean distances and time lags among local neighbor- hoods (Besag et al., 1974; Cressie & Wikle, 2011). For example, in the well-known kriging family of methods in geostatistics (Chiles & Delfiner, 1999), multivariate statistics of spatial patterns (multiple-point or higher order statistics) are simplified in terms of the so-called covariograms or covariance functions of distances (two-point or second order statistics) that essentially measure the pairwise dissimilarity (or similarity) between two locations. The two-point statistics are then combined to estimate the needed higher order statistics for modeling the spatiotemporal patterns. Similarly in Markov-based statistical methods (e.g., Markov random fields) (Besag et al., 1974; Clifford & Hammersley, 1971; Tso & Mather, 2009) and geographically weighted methods (Brunsdon et al., 1998; Fotheringham et al., 2015; Huang et al., 2010), the pattern statistics are often simplified as a log-linear combination of pairwise potential functions.

This simplification-based statistical paradigm has gained widespread popularity in the research and practices of diverse disciplines over the years. Due to the simplification, these methods tend to perform best at *homogeneous* geospatial patterns but are limited to complex geospatial patterns. Figure 18.1 gives several examples of these geographic patterns that are common in practice but well beyond the capabilities of the current statistical methods to effectively model. The problem can become even

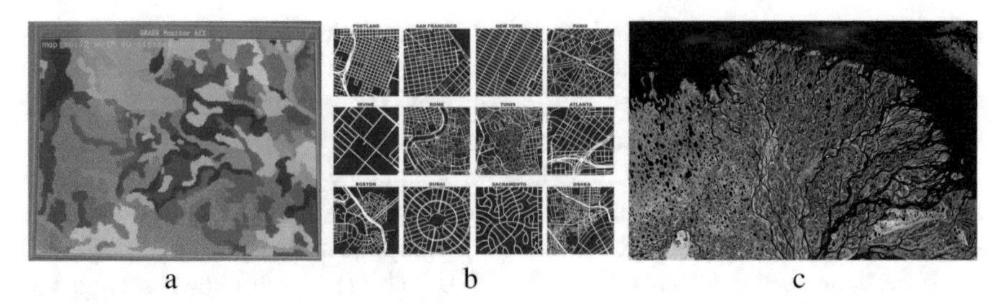

Fig. 18.1 Examples of common geographic patterns that are difficult for existing spatial statistical methods to model: **a** is a soil map, where soil types demonstrate complex geometric and juxtaposition (adjacency) patterns; gray areas in **b** represent the urban neighborhoods, and the blue areas in **c** represent the water in the delta of a river

more daunting when multiple spatial variables with complex patterns are involved and when it is extended into spatiotemporal settings for changing dynamics.

18.2.2 Heterogeneous Data Sources

The concept of Big Data is often characterized by the *volume*, *velocity, and variety*. As a specific type of Big Data, the large volume of geospatial data often demonstrates a large amount of *variety*, which tends to fragment the data-rich environment. Most recent advances in GIScience and spatial analysis (e.g., Wang, 2010; Wright & Wang, 2011; Yang et al., 2010, 2017) have focused on the issues of volume and velocity, buoyed by advances in computing resources. Much less attention has been given to the heterogeneity of geospatial data sources, despite the wide recognition of its importance (Goodchild, 2016).

The two main aspects comprising geospatial data are well known: the non-spatial attribute measurements, and the associated spatial units or *spatial support* indicating where the attribute is measured and how the attribute measurements are aggregated geographically. Both the attribute variables and spatial support vary in terms of data types resulting in a wide range of heterogeneity in geospatial data. The attribute variables can be continuous, categorical (or binary) and count variables, while the spatial support varies in terms of size, shape (point, line, area, and surface in the format of lattice or uniform grid), orientation and map scale. These different types of spatial data have distinct statistical properties and often require specific methods for statistical modeling and analysis. For different types of attribute variables (linear or non-linear), different flavors of kriging methods (Chiles & Delfiner, 1999) and Bayesian hierarchical models (Banerjee et al., 2004; Diggle et al., 1998) have been developed for point-referenced data. Failure to consider the distinction of the data characteristics can result in severe bias. Similarly for different types of spatial support, methods developed for point-referenced data cannot directly be used for areal data. Existing statistical developments for each type of spatiotemporal data have been well documented in the existing literature (Chiles & Delfiner, 1999; Cressie, 1993; Cressie & Wikle, 2011; Schabenberger & Gotway, 2017).

The heterogeneity of geospatial data and the associated methods makes the joint use of available data exceedingly difficult. Given a research problem for a specific geographic region, the geospatial data available are typically collected by diverse sources of providers with differing emphases and purposes, and with a large variety of incompatible data characteristics (e.g., incompatible spatial support, scales, misaligned spatial and temporal units, differing intrinsic attribute types). As highlighted by van der Putten et al. (2002), in spite of the exponential growth of the available data, the number and diversity of data sources, over which the information is fragmented, grow at an even faster rate, thus making more difficult the joint use of the data. Despite the heterogeneity, each source of data often offers a partial yet complementary view of the research problem at hand. It therefore highlights the need for statistical methods that can effectively reconcile the heterogeneity of spatial data for a comprehensive view in geographic analysis.

Many statistical data fusion or transformation methods have been developed to address the differences in attribute and spatial support separately. To account for the attribute difference, the statistical venue of Bayesian maximum entropy (Christakos, 2000, 2002) has been trimmed as a data fusion framework to combine hard and soft data for spatial prediction (Bogaert & Fasbender, 2007; Fasbender et al., 2007), while the spatial mixed model (Cao et al., 2011) has been developed as a data fusion framework for the integration of Gaussian and non-Gaussian spatial measurements (Cao et al., 2014; Yoo et al., 2013). The problem dealing with the differences in spatial support is often referred to as *change of support problem* (COSP) (Gelfand et al., 2001; Gotway & Young, 2002). The problem has been encountered in several fields of study, with numerous terms introduced to describe one or more facets of the problem and associated solutions (Gotway & Young, 2002). For example, the well-known the ecological inference problem (Robinson, 1950), the modifiable areal unit problem (MAUP) (Openshaw & Taylor, 1979) and the scaling problem (e.g., downscaling) can be taken as special cases of this incompatible problem (Gotway & Young, 2002).

18.2.3 Geospatial Uncertainty

Geospatial uncertainty describes the disagreements between the geospatial data and the corresponding true phenomena or processes they represent. Since it is impossible to create a perfect representation of the infinitely complex real world, all geospatial data are subject to uncertainty (Goodchild, 2008). New sources of uncertainty are introduced in every step of the map derivation processes, and the exact characteristics of this uncertainty usually are not known. If multiple sources of data are available for a geographic process, in a stochastic sense, each geospatial dataset can be regarded as a sample of the 'true' process it represents. The core question is then how to model the spatiotemporal uncertainty and evaluate the impact of data uncertainty in practical applications and scientific modeling.

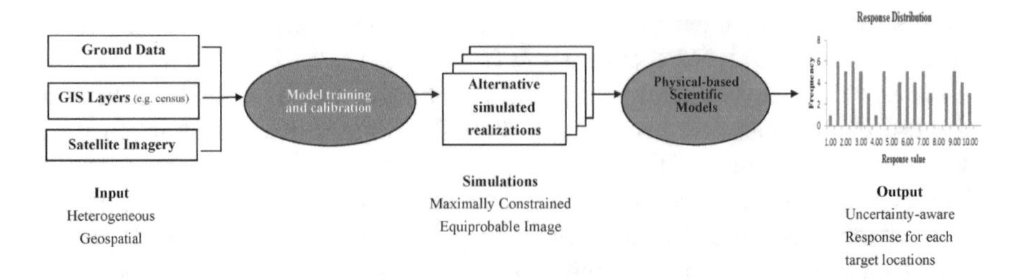

Fig. 18.2 A flowchart of Monte Carlo-based methods in evaluating the propagation of geospatial uncertainty; modified based on Kyriakidis and Dungan (2001)

The proliferation of geospatial data has spurred considerable interests in the problem of spatiotemporal uncertainty and the needs are shared by many fields, including climate science, atmospheric science and GIScience, for effective statistical methods in the quantification of spatiotemporal uncertainty. Geospatial uncertainty inherits many of the characteristics of the measurements themselves, including the issues inherent to spatiotemporal effects and spatial support. Therefore, the previous discussion on spatiotemporal pattern and heterogeneity can also be applied to spatiotemporal uncertainty. See Zhang et al. (2002) and Shi et al. (2018) for the numerous methods developed for spatial uncertainty modeling and characterization. The commonly used geostatistical methods can be used for quantifying spatiotemporal uncertainty by adding a simulated noises on measurements (Chiles & Delfiner, 1999). The uncertainty and the impact on decision marking can then be characterized through the Monte Carlo-based geostatistical simulations (see Fig. 18.2 for an example workflow of the Monte Carlo-based approaches). Bayesian hierarchical model (Banerjee et al., 2004; Cressie & Wikle, 2011) is another venue for characterizing and modeling spatiotemporal uncertainty by assuming a prior distribution on the geospatial measurements and specification parameters. However, existing methods for spatiotemporal uncertainty characterization and modeling share the above-mentioned pitfalls for data analysis and modeling; that is, they tend to oversimplify the complex spatiotemporal patterns and cannot effectively deal with the wide range of geospatial heterogeneity.

18.3 The Promises of Deep Learning

Deep neural network-based methods, namely deep learning, have dramatically improved the state-of-the-art in pattern recognition (LeCun et al., 2015). With a deep neural network with multiple levels of processing layers, deep learning-based methods have been shown to excel at discovering intricate complex patterns from high-dimensional data. Deep learning-based methods have been developed to model complex spatiotemporal dynamic systems, such as weather (e.g., Ravuri et al., 2021) and climate systems (e.g., Stengel et al., 2020). GIScientists are on the frontiers of the adoption and development of the new deep learning paradigm. A plethora of spatially explicit deep learning methods have been developed for geospatial pattern analysis and modeling, and the collective efforts led to the emergence of the new interdisciplinary field of GeoAI (Janowicz et al., 2020).

Great progresses were made in exploiting the power of deep learning in addressing the above-mentioned challenge of complex geospatial patterns. Most of the progresses focus on remote sensing imagery analysis and understanding [see Ma et al. (2019) and Zhang et al. (2016) for recent reviews of deep learning in remote sensing]. It makes sense considering the similarity of the challenges in remote sensing and computer vision (e.g., segmentation, classification, and change detection) where most of the deep learning innovations were started. As mentioned previously, geospatial analysis and modeling rely on the effective characterization and modeling of complex spatial patterns and for different types of geospatial data, geospatial patterns exhibit differently. Many works were reported to go beyond image analysis and to apply recent deep learning innovations in general tasks of geospatial analysis. For the traditional task of spatial interpolation, for example, the idea of conditional GAN (generative adversarial neural network) has been successfully applied for filling gaps of elevation data (Zhu et al., 2019a; Zhu et al., 2019b). The graph neural networks that integrate the graph theory and deep learning have been used for spatiotemporal interpolation of observations made at irregular locations (Amato et al., 2020; Wu et al., 2020). The recurrent neural networks that excel at learning long-term dependencies, such as LSTM (long short-term memory) and GRU (gated recurrent unit), were used for modeling complex spatiotemporal dynamics like human movement trajectories (Rao et al., 2020) and weather systems (Shi et al., 2015; Sonderby et al., 2020).

More efforts are needed to exploit the power of the deep learning paradigm in the other challenges discussed previously (modeling of geospatial uncertainty and the integration of heterogeneous geospatial data). To quantify the uncertainty in geospatial data and the deep learning models, a promising and active direction is the Bayesian deep learning that aims to integrate the concepts of Bayesian approach (e.g., *priors* and *Bayesian inferences*) with deep neural networks. One conceptually straightforward approach is to apply *Monte Carlo dropout* (MC-dropout) (Gal, 2016; Gal & Ghahramani, 2016) to the deep neural networks. MC-dropout is based on the simple idea of randomly dropping out the links in a deep neural network (Mele & Altarelli, 1993), and can be taken as an approximation of Bayesian inferences. To integrate the heterogeneous geospatial data, most of deep learning related progresses are made in analyzing image or gridded geospatial data. Particularly, the GAN model has shown superior performance to traditional statistical methods in downscaling (or super-resolution) and fusing satellite images (e.g., Tsagkatakis et al., 2019) or climate predictions (e.g., Stengel et al., 2020). Both the above mentioned spatial interpolation and downscaling work can be taken as special cases of heterogeneous data integration. Given its effectiveness in capturing complex patterns, the potential of deep learning in addressing the general problem of geospatial heterogeneity needs more exploration.

18.4 Discussions

This chapter revisited the traditional challenges in geospatial analysis and modeling (i.e., complex geospatial patterns, geospatial heterogeneity, and geospatial uncertainty) in the context of Big Data and briefly reviewed the recent progresses of deep learning in addressing such challenges. These challenges become more prominent as geospatial data with fine spatiotemporal scales become increasingly available. The power of deep learning in expressing complex structures of high dimensional data makes it a promising paradigm for geospatial analysis and modeling. In the past few years, great progresses have been made in exploring this power in addressing the traditional challenges in geospatial analysis and modeling. The potentials of deep neural network structures are far from being fully exploited particularly in modeling geospatial uncertainty and integrating heterogeneous geospatial data.

As in other technical advances, deep learning is not free of issues. For geospatial analysis and modeling, one of its major issues is the lack of model interpretability. While the models can perform well for the tasks of spatial estimations and predictions, due to the high model complexity, one often finds it difficult to interpret the model and to link the model parameters with the domain knowledge. Also related to the model complexity, the training process usually requires a large amount of training data, a requirement often hard to satisfy in geographic research. The lack of interpretability also raises interesting issues from the perspectives of education, curriculum design, and public outreaching. Tackling these issues may require a geography-informed deep learning or a deep integration of the geographic domain knowledge and the deep learning structures. It becomes critically important as the GeoAI algorithms are permeating in every aspect of GIScience as well as people's everyday life.

References

Amato, F., Guignard, F., Robert, S., & Kanevski, M. (2020). A novel framework for spatio-temporal prediction of environmental data using deep learning. *Scientific Reports, 10*(1), 1–11.

Anselin, L. (1990), What is special about spatial data? Alternative perspectives on spatial data analysis. In *Spatial Statistics, Past, Present and Future*. Institute of Mathematical Geography.

Banerjee, S., Carlin, B. P., & Gelfand, A. E. (2004). *Hierarchical modeling and analysis for spatial data*. CRC Press.

Besag, J., Society, S., & Methodological, S. B. (1974). Spatial interaction and the statistical analysis of lattice systems. *Journal of the Royal Statistical Society. Series B (Methodological), 36*(2), 192–236.

Bogaert, P., & Fasbender, D. (2007). Bayesian data fusion in a spatial prediction context: A general formulation. *Stochastic Environmental Research and Risk, 21*, 695–709.

Brunsdon, C., Fotheringham, S., & Charlton, M. (1998). Geographically weighted regression. *Journal of the Royal Statistical Society: Series D (The Statistician)*.

Cao, G., Kyriakidis, P. C., & Goodchild, M. F. (2011). A multinomial logistic mixed model for the prediction of categorical spatial data. *International Journal of Geographical Information Science, 25*(12), 2071–2086.

Cao, G., Yoo, E.-H., & Wang, S. (2014). A statistical framework of data fusion for spatial prediction of categorical variables. *Stochastic Environmental Research and Risk Assessment, 28*(7), 1785–1799.

Chiles, J. P., & Delfiner, P. (1999). *Geostatistics: Modeling spatial uncertainty* (Vol. 136). Wiley.

Christakos, G. (2000). *Modern spatiotemporal geostatistics.* Oxford University Press.

Christakos, G. (2002). On the assimilation of uncertain physical knowledge bases: Bayesian and non-Bayesian techniques. *Advances in Water Resources, 25*(8–12), 1257–1274.

Clifford, P., & Hammersley, J. M. (1971). *Markov fields on finite graphs and lattices.* University of Oxford.

Cressie, N. A. C. (1993). *Statistics for Spatial Data (revised edition).* Wiley.

Cressie, N., & Wikle, C. K. (2011). *Statistics for spatio-temporal data.* Wiley.

Diggle, P. J., Tawn, J. A., & Moyeed, R. A. (1998). Model-based geostatistics. *Applied Statistics, 47*(3), 299–350.

Fasbender, D., Obsomer, V., Radoux, J., Bogaert, P., & Defourny, P. (2007). Bayesian data fusion: spatial and temporal applications. In *2007 International Workshop on the Analysis of Multi-temporal Remote Sensing Images.*

Fotheringham, A. S., Crespo, R., & Yao, J. (2015). Geographical and temporal weighted regression (GTWR). *Geographical Analysis, 47*(4), 431–452.

Gal, Y. (2016). *Uncertainty in deep learning* (PhD Thesis).

Gal, Y., & Ghahramani, Z. (2016). Dropout as a Bayesian approximation: Representing model uncertainty in deep learning. In *International Conference on Machine Learning* (pp. 1050–1059).

Gelfand, A. E., Zhu, L., & Carlin, B. P. (2001). On the change of support problem for spatio-temporal data. *Biostatistics (Oxford, England), 2*(1), 31–45.

Goodchild, M. F. (2004). The validity and usefulness of laws in geographic information science and geography. *Annals of the Association of American Geographers, 94*(2), 300–303.

Goodchild, M. F. (2007). Citizens as voluntary sensors: Spatial data infrastructure in the world of Web 2.0. *International Journal of Spatial Data Infrastructures Research, 2*, 24–32.

Goodchild, M. F. (2008). Statistical perspectives on geographic information science. *Geographical Analysis, 40*(3), 310–325.

Goodchild, M. F. (2016). GIS in the era of big data. *Cybergeo: European Journal of Geography*, 1–25.

Goodchild, M. F., Yuan, M., & Cova, T. J. (2007). Towards a general theory of geographic representation in GIS. *International Journal of Geographical Information Science, 21*(3), 239–260.

Gotway, C. A., & Young, L. J. (2002). Combining incompatible spatial data. *Journal of the American Statistical Association, 97*(458), 632–648.

Huang, B., Wu, B., & Barry, M. (2010). Geographically and temporally weighted regression for modeling spatio-temporal variation in house prices. *International Journal of Geographical Information Science.*

Janowicz, K., Gao, S., McKenzie, G., Hu, Y., & Bhaduri, B. (2020). GeoAI: Spatially explicit artificial intelligence techniques for geographic knowledge discovery and beyond. *International Journal of Geographical Information Science, 34*(4), 625–636.

Kyriakidis, P. C., & Dungan, J. L. (2001). A geostatistical approach for mapping thematic classification accuracy and evaluating the impact of inaccurate spatial data on ecological model predictions. *Environmental and Ecological Statistics, 8*(4), 311–330.

LeCun, Y., Bengio, Y., & Hinton, G. (2015). Deep learning. *Nature, 521*(7553), 436–444.

Ma, L., Liu, Y., Zhang, X., Ye, Y., Yin, G., & Johnson, B. A. (2019). Deep learning in remote sensing applications: A meta-analysis and review. *ISPRS journal of photogrammetry and remote sensing, 152*, 166–177.

Manyika, J., Chui, M., Brown, B., Bughin, J., Dobbs, R., Roxburgh, C., & Byers, A. H. (2011). Big data: The next frontier for innovation, competition, and productivity. *McKinsey Global Institute*, 1–137.

Mele, B., & Altarelli, G. (1993). Lepton spectra as a measure of b quark polarization at LEP. *Physics Letters B, 299*(3–4), 345–350.

Openshaw, S., & Taylor, P. J. (1979). A million or so correlation coefficients: Three experiments on the modifiable areal unit problem. *Statistical applications in the spatial sciences* (pp. 127–144).

Rao, J., Gao, S., Kang, Y., & Huang, Q. (2020), LSTM-TrajGAN: A deep learning approach to trajectory privacy protection. In Leibniz International Proceedings in Informatics, LIPIcs' (Vol. 177). Schloss Dagstuhl- Leibniz-Zentrum fur Informatik GmbH, Dagstuhl Publishing.

Ravuri, S., Lenc, K., Willson, M., Kangin, D., Lam, R., Mirowski, P., Fitzsimons, M., Athanassiadou, M., Kashem, S., Madge, S., Prudden, R., Mandhane, A., Clark, A., Brock, A., Simonyan, K., Hadsell, R., Robinson, N., Clancy, E., Arribas, A., & Mohamed, S. (2021, February). Skillful precipitation nowcasting using deep generative models of radar. *Nature, 597*(7878), 672–677.

Robinson, A. H. (1950). Ecological correlation and the behaviour of individuals. *American Sociological Review, 15*, 351–357.

Schabenberger, O., & Gotway, C. A. (2017). *Statistical methods for spatial data analysis.* CRC Press.

Shi, W., Zhang, A., Zhou, X., & Zhang, M. (2018). Challenges and prospects of uncertainties in spatial big data analytics. *Annals of the American Association of Geographers, 108*(6), 1513–1520.

Shi, X., Chen, Z., Wang, H., Yeung, D. Y., Wong, W. K., & Woo, W. C. (2015). Convolutional LSTM network: A machine learning approach for precipitation nowcasting. In *Advances in Neural Information Processing Systems* (Vol. 2015, pp. 802–810).

Sonderby, C. K., Espeholt, L., Heek, J., Dehghani, M., Oliver, A., Salimans, T., Agrawal, S., Hickey, J., & Kalchbrenner, N. (2020). *MetNet: A neural weather model for precipitation forecasting* (pp. 1–17).

Stengel, K., Glaws, A., Hettinger, D., & King, R. N. (2020). Adversarial super-resolution of climatological wind and solar data. *Proceedings of the National Academy of Sciences of the United States of America, 117*(29), 16805–16815.

Takeyama, M. (1997). *Geo-Algebra: A mathematical approach to integrating spatial modeling and GIS*. Doctoral dissertation, University of California, Santa Barbara.

The New York City Council. (2012). A Local Law to amend the administrative code of the city of New York, in relation to publishing open data.

Tobler, W. R. (1970). A computer movie simulating urban growth in the Detroit region. *Economic Geography, 46*, 234–240.

Tsagkatakis, G., Aidini, A., Fotiadou, K., Giannopoulos, M., Pentari, A., & Tsakalides, P. (2019). Survey of deep-learning approaches for remote sensing observation enhancement. *Sensors, 19*(18), 3929.

Tso, B., & Mather, P. (2009). *Classification methods for remotely sensed data.* CRC Press.

van der Putten, P., Kok, J. N., & Gupta, A. (2002). Data fusion through statistical matching. *Available at SSRN 297501.*

Wang, S. (2010). A CyberGIS framework for the synthesis of cyberinfrastructure, GIS, and spatial analysis. *Annals of the Association of American Geographers, 100*(3), 535–557.

Wright, D. J., & Wang, S. (2011). The emergence of spatial cyberinfrastructure. *Proceedings of the National Academy of Sciences, 108*(14), 5488–5491.

Wu, Y., Zhuang, D., Labbe, A., & Sun, L. (2020). Inductive graph neural networks for spatiotemporal kriging. ArXiv Preprint ArXiv:2006.07527

Yang, C., Raskin, R., Goodchild, M., & Gahegan, M. (2010). Geospatial cyberinfrastructure: Past, present and future. *Computers, Environment and Urban Systems, 34*(4), 264–277.

Yang, C., Yu, M., Hu, F., Jiang, Y., & Li, Y. (2017). Utilizing Cloud Computing to address big geospatial data challenges. *Computers, Environment and Urban Systems, 61*, 120–128.

Yoo, E. H., Hoagland, B. W., Cao, G., & Fagin, T. (2013). Spatial distribution of trees and landscapes of the past: a mixed spatially correlated multinomial logit model approach for the analysis of the public land survey data. *Geographical Analysis, 45*(4), 419–440.

Yuan, M. (2001). Representing complex geographic phenomena in GIS. *Cartography and Geographic Information Science, 28*(2), 83–96.
Zhang, J., Goodchild, M. F., & Shaw, S.-L. (2002). *Uncertainty in geographical information* (Vol. 93). CRC Press.
Zhang, L., Xia, G. S., Wu, T., Lin, L., & Tai, X. C. (2016, June). Deep learning for remote sensing image understanding. *Journal of Sensors, 2016*. https://doi.org/10.1155/2016/7954154
Zhu, A., Lu, G., Liu, J., Qin, C., & Zhou, C. (2018). Spatial prediction based on Third Law of Geography. *Annals of GIS, 24*(4), 225–240.
Zhu, D., Cheng, X., Zhang, F., Yao, X., Gao, Y., & Liu, Y. (2019a). Spatial interpolation using conditional generative adversarial neural networks. *International Journal of Geographical Information Science, 34*(4), 735–758.
Zhu, Z., Wulder, M. A., Roy, D. P., Woodcock, C. E., Hansen, M. C., Radeloff, V. C., Healey, S. P., Schaaf, C., Hostert, P., Strobl, P., Pekel, J. F., Lymburner, L., Pahlevan, N., & Scambos, T. A. (2019b). Benefits of the free and open Landsat data policy. *Remote Sensing of Environment, 224*, 382–385.

Chapter 19
Towards Domain-Knowledge-Based Intelligent Geographical Modeling

Cheng-Zhi Qin and A-Xing Zhu

Abstract Geographical modeling has been recognized as a powerful way to solve complex geographic problems. However, its wide applicability is increasingly hindered by its complexity in domain knowledge required and the procedures involved. In this chapter, we argue that domain knowledge plays a key role in making geographical modeling intelligent. Domain-knowledge-based intelligent geographical modeling would not only solve wide geographical problems in an easy-to-use manner on the premise of the effectiveness of the built model specific to the application context, but also contribute to research in artificial intelligence.

Keywords Intelligent geographical modeling · Artificial intelligence (AI) · Domain knowledge · Knowledge-based modeling

19.1 Complexity in Geographical Modeling

Currently, geographical modeling is the fundamental means of conducting geospatial analysis and simulation in a quantitative and computer-aided manner. It has been recognized as a powerful way to solve diverse complex problems related to geography, from scientific research to decision-making problems in application domains with multiple stakeholders (including watershed management and urban planning) (Chen et al., 2020). Users of geographical modeling extend from those modeling experts to end-users in application domains (such as decision-makers) who are often

C.-Z. Qin (✉) · A.-X. Zhu
State Key Laboratory of Resources and Environmental Information System, Institute of Geographic Sciences and Natural Resources Research, Chinese Academy of Sciences, Beijing 100101, China
e-mail: qincz@lreis.ac.cn

A.-X. Zhu
e-mail: azhu@wisc.edu

A.-X. Zhu
Department of Geography, University of Wisconsin-Madison, Madison, WI 53706, USA

School of Geography, Nanjing Normal University, Nanjing 210023, Jiangsu, China

B. Li et al. (eds.), *New Thinking in GIScience*,
https://doi.org/10.1007/978-981-19-3816-0_19

non-experts in modeling. New progress in model research, data collection abilities, and advanced computing technologies has made it possible to model and solve complex problems more effectively. However, for many users (especially non-expert modelers), geographical modeling is becoming increasingly complicated due to the following issues.

- Problem definition. Note that end-users and modeling experts are often concerned with different aspects of modeling. The end-users, especially those non-experts in modeling, are concerned with the model outcomes to their respective applications. For example, a resource manager would be more interested in knowing how to manage the landuse in a watershed to maximize the economic and environmental benefits. In contrast, the modelers concern themselves with modeling conditions, steps, and details. They need to define the modeling problems in a technical way such as which processes to consider and what optimization techniques to use. The successful transformation between the end-users' application problems and modelers' modeling problems is highly dependent on the domain (both application domain and modeling domain) knowledge. The gap in such a transformation in problem definition remains a non-trivial issue, despite increasing attention being given to participatory modeling (Hedelin et al., 2021).
- Model structure determination and algorithm selection. For a specific modeling problem, diverse model structures can be adopted to define which geographic factors and/or (sub-)processes should be considered, and how to organize them logically. Model structures can be classified into different types, including probability-based or process-based, lumped or spatially distributed, deterministic or stochastic, loosely coupled or tightly coupled. For each component within the model, different algorithms can be adopted. Each of the model structures and algorithms for the same model component have different application conditions, mixing diverse dimensions in a spatially heterogeneous manner. The determination of a proper model structure and corresponding algorithms often requires implicit and/or empirical knowledge that requires a steep study curve, which is complicated for many end users.
- Input data preparation and parameter settings. Different model structures and algorithms for a specific modeling problem may largely diverge on their input data, together with the metaphor of "Garbage-in-garbage-out." Therefore, the availability and quality of input data largely determine whether a model structure can be not only executable but also effective for modeling problems. In many situations, most modeling efforts are on input data preparation. A typical example is the input data preparation for a watershed model, such as SWAT model. Such preparation involves the collection of the original dataset available for the study area, a series of transformation processes from the original dataset to the input data ready for the model, and the parameter settings for each algorithm in the model. Not only is the collection of the original dataset often tedious, but the transformation process is complicated, multidisciplinary, and often leads to errors during user operations. This process can also be regarded as an iterative secondary modeling problem.

- Efficient execution of the model. Even when a model with all algorithms, input data, and parameter settings is ready for execution, the model execution also faces the diversity of computation resources with different efficiencies for applications. Currently, advanced cyberinfrastructures (e.g., parallel computing and cloud services; Wang, 2010) are increasingly available and necessary for executing geographic models with increasingly high data throughput and large computational complexity for large-area, long-term simulations under high spatio-temporal resolution. Consideration of the computational resources for a model application often tangles with the modeling stage. For example, the data services from the data portal on the web bring not only wider data availability, but also the requirement for the corresponding algorithm implementation for the model. This is particularly challenging for users with less experience and knowledge of programming or cyberinfrastructure.

The above-mentioned complexities in geographical modeling create a bottleneck that limits the wider applicability of geographical modeling among different application domains by diverse users.

19.2 Intelligent Geographical Modeling

Intelligent geographical modeling has been proposed to handle the complexities of geographical modeling by automatic means instead of manual user operations. It can minimize the dependence on users' modeling knowledge and skills on the premise of the reasonableness and effectiveness of the models constructed. Thus, users, especially non-expert users, can apply geographical modeling in an easy-to-use manner to build application-problem-specific models to solve their geographical problems in applications (Zhu et al., 2021). This means that it may break the bottleneck and foster wide applicability among different application domains by diverse users.

Intelligent geographical modelling is not an entirely new approach. In a broad sense, since the first geographical information system (GIS) was proposed, the automation of geographical models and algorithms for easy-to-use geographical modeling has received the attention of GIS researchers and developers. These efforts include, but are not limited to:

- Implementing and integrating algorithms/models into algorithm/model libraries (e.g., ArcGIS Toolbox), which can be called and executed in a consistent manner.
- Providing secondary programming tools for calling algorithm/model libraries to customize application-specific models. This requires users to possess sufficient programming skills.
- Combining a model with its necessary data-processing workflow to be a GIS-coupled tool, for example ArcSWAT for the watershed model SWAT. This method is often limited to specific models and difficult for end-users to extend to wide applications.

- Visualizing the modeling process of building a geographical model (often as a workflow), for example ArcGIS Modelbuilder. This user-friendly interactive method ensures that the built models are executable; however, this does not mean that the models are reasonable or effective for specific application problems.

Note that the key difference between intelligent geographical modeling and automated geographical modeling is that the former highlights the built models as reasonable and effective for specific application problems when the latter highlights fewer manual user operations during the modeling process. Current GIS software is still far from intelligent geographical modeling, which has become a research frontier in GIS.

Similar to the growth of GIS, which has closely followed the development of computer science, the growth of intelligent geographical modeling follows development within the domain of artificial intelligence (AI). Currently, the prevailing trend in AI is data-driven machine learning, especially its representation of deep learning, which has dramatically succeeded in processing diverse problems across wide application domains (Bergen et al., 2019), particularly classification problems related to text, voice, and image. Advanced AI methods have also been widely explored for solving geographical problems (such as classification or clustering related to remote sensing images, point cloud, map layout, and geospatial citizen data, Reichstein et al., 2019). In recent efforts, geospatial characteristics in geographical patterns and processes (including spatial autocorrelation, spatial heterogeneity, and spatial configuration, Goodchild, 2004) have been considered within existing data-driven AI frameworks, which is now encapsulated as GeoAI (Janowicz et al., 2020; Li, 2020).

The data-driven and black-box characteristics of prevailing machine learning methods make them comparatively incompatible with knowledge-driven tradition in geographical modeling (Reichstein et al., 2019). The prevailing machine learning methods and GeoAI still cannot effectively solve geographical modeling problems. How can geographical modeling be made more intelligent?

19.3 Domain Knowledge and Operation of Intelligent Geographical Modeling

Think about how domain experts conduct geographical modeling to solve application problems. Domain experts exhaustively apply domain knowledge to every step of geographical modeling to address modeling complexity and then achieve satisfactory solutions of geographic problems. Domain knowledge includes different types, including the following:

- Knowledge of the concepts and relations between concepts of the application domains and related disciplines. Such discipline knowledge is used to transform application problems into modeling problems, consider the model structure at the conceptual modeling level, and transform the final model results to their non-expert end-users or stakeholders.

- Knowledge of model and algorithm implementation (such as the semantics of input/output and parameters, valid data type(s), and value ranges). At either the logical or numerical modeling level, such model-algorithm knowledge is used to ensure that the model, algorithms, and data can be coupled as an executable application model (but unnecessary to be a proper model for a specific modeling problem).
- Knowledge of application context suitable for each model and algorithm (including parameter settings). Knowledge of application context is key to ensuring that an application model built is not only executable but also appropriate for specific modeling problems (Qin et al., 2016).

Domain knowledge should play a key role in intelligent geographical modeling. The more thoroughly the domain knowledge can be used by computer-understandable means, the more intelligent geographical modeling. An imaginary example of intelligent geographical modeling when thoroughly using domain knowledge could behave as described in the following stages:

- Problem-defining stage. The user's description of the application problem is parsed as the modeling problem (for the application goal, i.e., the solution for the application problem) based on discipline knowledge related to the application domain. An example, as mentioned in Sec. 19.1, is "which kind of a watershed model with which coupled processes should be built for a watershed to conduct scenario analysis and optimization under a given set of conditions?" The application context of the application problem (such as study area characteristics, data availability, and other user-assigned restrictive conditions) can also be formalized as part of the modeling problem for the following stages of intelligent geographical modeling. Note that when the study area is located, information on the application context may be automatically derived using existing spatial analysis methods (such as digital terrain analysis, remote sensing image processing, and map algebra) with increasingly open-access spatial datasets (including DEM, remote sensing images, and land use maps).
- Goal-driven iterative modeling stage. The start of this stage is to answer the question: Which model fits the above-defined modeling problem (i.e., achieving the application goal) in its application context? Model–algorithm knowledge is used to filter out candidate models with outputs that semantically match the application goal (Jiang et al., 2019). Then, a semantic match between the necessary input of each candidate model and the available dataset list of the local database or/and open data portal is conducted to identify those missing inputs for the candidate model (Hou et al., 2019). Application context knowledge is used to ensure that the model, as well as the corresponding algorithm and parameter settings, are appropriate for the application context (Qin et al., 2016). The question waiting for the answer is then updated: Which model/algorithm may output data for the unsettled goal, that is, the missing input for the candidate model under its application context? Under the guidance of both model-algorithm knowledge

and application context knowledge, this modeling process iterates automatically in a goal-driven manner. In each iteration, the candidate model(s) is updated with each of its necessary inputs being satisfied with existing data or output from other models/algorithms coupled to the former version of the candidate model. Candidate models whose necessary inputs cannot be satisfied are discarded. Such a goal-driven iterative modeling stage continues until each candidate model has all necessary inputs ready and has set its algorithms and parameter settings properly as the application context fits.

- Model-submission and result-interaction stages. The candidate model chosen can be seen as an application model that is not only executable but also suitable for specific modeling problems. It is ready for submission and execution. The execution results and intermediate results are fed back to users, typically through visualization for user-friendly interaction. Users may adjust the application problem (such as restrictive conditions) to conduct further geographical modeling.

19.4 How to Realize Intelligent Geographical Modeling?

The key to realizing intelligent geographical modeling is to extract and formalize diverse types of geographical modeling knowledge to be computer-understandable (such as external knowledge base directly accessed by a program, and solidification within an algorithm). This is similar to the situation in knowledge-based modeling in other disciplines. What makes it particularly difficult in intelligent geographical modeling is not only that the geographical modeling knowledge is multidisciplinary, but that the application-context knowledge is highly spatially varied, application-dependent, often non-systematic, and tacit.

Both discipline knowledge of concepts and relations between concepts and the model-algorithm knowledge have been explicitly described in textbooks, model manuals, and scientific papers. Advanced machine learning methods are suitable for text processing. Semantic web technologies (including ontology, knowledge graphs, and resource description frameworks) have shown promising results in formalizing and using these types of knowledge. However, application-context knowledge, which is crucial for intelligent geographical modeling, is difficult to extract and formalize. This originates from the characteristics of this knowledge, including:

- highly spatially varied (having weak replication across space; Goodchild & Li, 2021);
- application-dependent (thus, inconsistent with those factors considered for different applications);
- often non-systematic (thus hard to formalize in a top-down manner);
- often empirical and tacit (thus hard to be formalized as rules).

Application-context knowledge is often implicitly (rather than explicitly as for other knowledge types) contained in application cases that have been recorded in many scientific papers and application reports. Case and case-based reasoning, as an intuitive way to apply modeling, are good at formalizing and using such knowledge. Some preliminary studies have shown that case-based methods are promising for intelligent geographical modeling (Liang et al., 2020; Qin et al., 2016). However, a main challenge to its success is how to efficiently build large-scale case bases for diverse application domains when those papers, representing the knowledge source, are often weak structures and many necessary case contents are only implicitly contained and unspecified in the text.

19.5 Potential Contributions to AI

Since the beginning of AI in the 1950s, symbolic AI (or rule-based AI) has been extensively and successfully applied, for example, to expert systems. However, symbolic AI requires experts to encode knowledge in advance to formulate computer-readable rules (crisp or fuzzy), which is a non-trivial task. Case-based methodology is also not new. The difficulty in building a comprehensive case base impedes its applicability. Along with the large leap of advanced computing power and big data, data-driven AI currently prevails and symbolic AI is comparatively reduced. Recent calls for explainable AI (Reichstein et al., 2019) may indicate the renovation of symbolic AI and case-based methodologies, probably in some form of mixing of symbolic and data-driven AI.

Along with such trends of AI research, intelligent geographical modeling will not only solve wide geographical problems effectively and efficiently but also contribute to the broader domain of AI research. What makes "geographical" special also makes "geographical modeling" special. The unique characteristics of geographical modeling knowledge (particularly application-context knowledge) provide probably the most challenging testbed for explainable AI. This is similar to the fact that in GIS history, both the unique characteristics of geospatial data and the requirement of efficient geospatial data processing fostered geospatial database and also largely contributed to database research.

19.6 Concluding Remarks

Intelligent geographical modeling can solve a wide range of geographical problems in an easy-to-use manner on the premise of reasonableness and effectiveness of the built model specific to the application context. This should be an indispensable functionality of next-generation GIS (Zhu et al., 2021). Geographical modeling has been model-directed for a long period in GIS and has become increasingly data-driven in recent years. Although facing many methodological and technical challenges, the domain-knowledge-based method might move towards intelligent geographical modeling, which could further innovate geographical computing based on the triplet of model-data-knowledge.

Acknowledgements This work was supported by the National Key R&D Project of China (Grant No. 2021YFB3900904), the National Natural Science Foundation of China (No. 41871362), and the 111 Program of China (No. D19002). Supports to A-Xing Zhu through the Vilas Associate Award, the Hammel Faculty Fellow Award, and the Manasse Chair Professorship from the University of Wisconsin-Madison are greatly appreciated.

References

Bergen, K. J., Johnson, P. A., de Hoop, M. V., & Beroza, G. C. (2019). Machine learning for data-driven discovery in solid Earth geoscience. *Science, 363*, eaau0323.

Chen, M., Voinov, A., Ames, D. P., Kettner, A. J., Goodall, J. L., Jakeman, A. J., Barton, M. C., Harpham, Q., Cuddy, S. M., DeLuca, C., Yue, S., Wang, J., Zhang, F., Wen, Y., & Lu, G. (2020). Position paper: Open web-distributed integrated geographic modelling and simulation to enable broader participation and applications. *Earth-Science Reviews, 207*, 103223.

Goodchild, M. F. (2004). The validity and usefulness of laws in geographic information science and geography. *Annals of the Association of American Geographers, 94*, 300–303.

Goodchild, M. F., & Li, W. (2021). Replication across space and time must be weak in the social and environmental sciences. *Proceedings of the National Academy of Sciences, 118*(35), e2015759118.

Hedelin, B., Gray, S., Woehlke, S., BenDor, T. K., Singer, A., Jordan, R., Zellner, M., Giabbanelli, P., Glynn, P., Jenni, K., Jetter, A., Kolagani, N., Laursen, B., Leong, K. M., Olabisi, L. C., & Sterling, E. (2021). What's left before participatory modeling can fully support real-world environmental planning processes: A case study review. *Environmental Modelling & Software, 143*, 105073.

Hou, Z.-W., Qin, C.-Z., Zhu, A.-X., Liang, P., Wang, Y.-J., & Zhu, Y.-Q. (2019). From manual to intelligent: A review of input data preparation methods for geographic modeling. *ISPRS International Journal of Geo-Information, 8*(9), 376.

Janowicz, K., Gao, S., McKenzie, G., Hu, Y., & Bhaduri, B. (2020). GeoAI: Spatially explicit artificial intelligence techniques for geographic knowledge discovery and beyond. *International Journal of Geographical Information Science, 34*(4), 625–636.

Jiang, J., Zhu, A.-X., Qin, C.-Z., & Liu, J. (2019). A knowledge-based method for the automatic determination of hydrological model structures. *Journal of Hydroinformatics, 21*(6), 1163–1178.

Li, W. (2020). GeoAI: Where machine learning and big data converge in GIScience. *Journal of Spatial Information Science, 20*, 71–77.

Liang, P., Qin, C.-Z., Zhu, A.-X., Hou, Z.-W., Fan, N.-Q., & Wang, Y.-J. (2020). A case-based method of selecting covariates for digital soil mapping. *Journal of Integrative Agriculture, 19*(8), 2127–2136.

Qin, C.-Z., Wu, X.-W., Jiang, J.-C., & Zhu, A.-X. (2016). Case-based knowledge formalization and reasoning method for digital terrain analysis—Application to extracting drainage networks. *Hydrology and Earth System Sciences, 20*, 3379–3392.

Reichstein, M., Camps-Valls, G., Stevens, B., Jung, M., Denzler, J., Carvalhais, N., & Prabhat. (2019). Deep learning and process understanding for data-driven Earth system science. *Nature, 566*, 195–204.

Wang, S. W. (2010). A CyberGIS framework for the synthesis of cyberinfrastructure, GIS, and spatial analysis. *Annals of the Association of American Geographers, 100*(3), 535–557.

Zhu, A.-X., Zhao, F.-H., Liang, P., & Qin, C.-Z. (2021). Next generation of GIS: Must be easy. *Annals of GIS, 27*(1), 71–86.

Chapter 20
Mitigating Spatial Bias in Volunteered Geographic Information for Spatial Modeling and Prediction

Guiming Zhang

Abstract VGI (volunteered geographic information) observations are often spatially biased, which degrades the quality of inferences drawn from field sample sets consisting of VGI observations. This chapter presents a novel representativeness-directed approach to mitigating spatial bias in VGI for spatial modeling and prediction. The approach, based on the Third Law of Geography (the similarity principle), defines the representativeness of a field sample set as the degree to which the field sample locations capture the spatial variability of environmental covariates in the study area. Sample representativeness is then quantified as the overlap between the probability distribution of covariate values over sample locations and the distribution over the whole study area. Adjusting the weights for individual sample locations towards increasing the overlap thus mitigates spatial bias in the sample locations and improves sample representativeness. Applications of the approach to species habitat suitability mapping and digital soil mapping demonstrate its effectiveness in mitigating spatial bias to improve the accuracy of spatial modeling and prediction.

Keywords Volunteered geographic information (VGI) · Sample representativeness · Spatial bias mitigation · Modeling and prediction · Predictive mapping

20.1 Introduction

Volunteered geographic information (VGI) refers to geographic information created by citizen volunteers (Goodchild, 2007). It has proliferated in recent years as advancements in geospatial and communication technologies enable the general public to contribute geographic data. With ubiquitous access to the Internet, ordinary citizens can now easily create and share geographic observations of the world, for example, by sharing geo-referenced photos of species observations in citizen science communities or on social media through location-aware smartphones. VGI broadly

G. Zhang (✉)
Department of Geography & the Environment, University of Denver, Denver 80208, USA
e-mail: guiming.zhang@du.edu

B. Li et al. (eds.), *New Thinking in GIScience*,
https://doi.org/10.1007/978-981-19-3816-0_20

encompasses geographic data generated by volunteer participants in citizen science, crowdsourcing, social media, etc. as they all involve voluntary and non-expert geographic data creation (Zhang, 2021). VGI is useful in many domains such as emergency response, environmental monitoring, land cover map validation, and biodiversity modeling (Yan et al., 2020). Exemplary VGI projects include OpenStreetMap (Haklay & Weber, 2008) that compiles an open and free geographic databases for the world, and iNaturalist (Unger et al., 2020) and eBird (Wood et al., 2011) which document species observations across the globe on a daily basis. VGI represents a paradigm shift in how geographic data is created and shared and in its content and characteristics (Elwood, 2008). In a broader context, VGI is an important source of geospatial big data (Yang, 2017) which is propelling geographic research towards emerging paradigms such as "data-driven geography" (Miller & Goodchild, 2014) and "data-intensive science" (Kelling et al., 2009).

VGI has become a supplementary or even alternative mechanism for acquiring geographic data due to its unique advantages. First, VGI contains rich local information because citizens as local experts and sensors (Goodchild, 2007) have long been accumulating knowledge of their local environments (Zhang et al., 2018; Zhu et al., 2015b). Second, VGI makes it feasible to collect geographic data over large areas given that potential VGI contributors are all over the world. Third, VGI can provide timely updated data that are difficult to obtain through other means. Lastly, VGI is much less expensive than traditional spatial data collection protocols (e.g., survey). As such, VGI has a great potential to reveal the spatiotemporal dynamics of geographic phenomena at high spatiotemporal resolutions over large areas.

Such potential can be realized through spatial modeling and prediction based on VGI observations. Nonetheless, VGI observations still represent only a set of sample observations regarding the phenomenon under concern, despite its seemingly extensive coverages (Zhang & Zhu, 2018). For instance, occurrence locations of a bird species reported by volunteers is a sample set from the population consisting of all possible species occurrence locations. In this respect, VGI observations are similar to field sample data collected through traditional geographic sampling. One of the significant differences, though, is that locations for designed geographic sampling are carefully chosen (e.g., following statistical sampling design) so that the sampled locations are representative of the spatial variabilities in the study area (Jensen & Shumway, 2010). In contrast, VGI contributors decide where (and when) to conduct observations at their own discretion without following a coordinated sampling scheme. This characteristic of voluntary data creation often results in spatial bias in VGI data, which has profound implications on drawing inferences about the target phenomenon (i.e., population) from VGI observations (i.e., sample). This chapter focuses on this issue and presents a novel representativeness-directed approach to mitigating spatial bias in VGI for spatial modeling and prediction.

20.2 Spatial Bias in VGI

Data quality of VGI is under constant scrutiny (Goodchild & Li, 2012), and spatial bias is a prominent concern when using VGI for mapping, modeling, and prediction (Zhang & Zhu, 2018). VGI observations in spatial distribution tend to concentrate in certain geographic areas as observations made by volunteers are opportunistic in nature, which results in spatial bias in sampling. Spatial distribution of the observation effort can be considered neither random nor regular in the sense of geographic sampling design, but 'ad hoc' (Zhu et al., 2015b). As a result, VGI observations are often of higher density in specific areas, for example, populous and accessible areas (Kadmon et al., 2004; Zhang, 2020).

Due to spatial bias, a field sample set consisting of VGI observations may not be representative of the spatial variabilities of the phenomena in the study area. Spatial bias, if not appropriately accounted for, would adversely affect the quality of inferences drawn from VGI observations (Leitão et al., 2011). Spatial bias is one form of sample selection bias (Zhang & Zhu, 2018). Many methods rely on information of the underlying observation process (e.g., selection probabilities) to correct for sample selection bias, but such information is often unavailable in VGI data.

Here a novel representativeness-directed approach was developed to mitigate spatial bias in VGI to improve the quality of spatial modeling and prediction (Zhang, 2018; Zhang & Zhu, 2019a, 2019b). Specifically, it is for mitigating spatial bias in field sample sets to improve the accuracy of predictive mapping, a framework for predicting the spatial variation of a target variable based on environmental covariate data and a model capturing the covariation relationship (f) between the target variable (T) and the covariates (E) (Fig. 20.1).

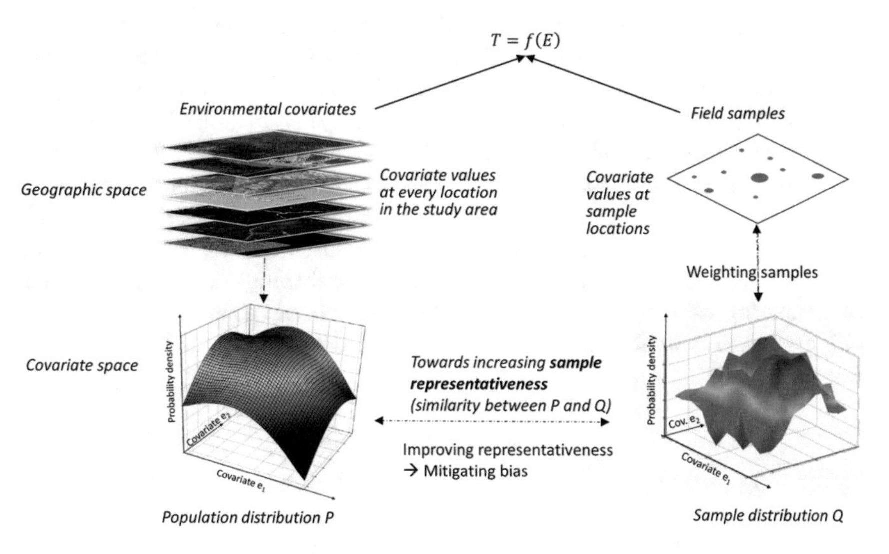

Fig. 20.1 Basic idea of representativeness directed spatial bias mitigation. Reused from Zhang and Zhu (2019b) with permission

20.3 A Representativeness-Directed Approach to Bias Mitigation

Spatial bias has adverse effects on spatial modeling and prediction as it impedes the representativeness of VGI-based field sample sets. In the context of predictive mapping, sample representativeness essentially is the degree to which the observations made at sample locations capture the spatial variability of the relationship between the target variable (e.g., species habitat suitability) and the environmental covariates (e.g., elevation, land cover, precipitation) in the area. This is achieved by capturing the variability in the target variable and that in covariates. With covariate data (raster layers), it is feasible to assess sample representativeness. Sample representativeness with respect to the target variable is hard to assess as its spatial variation is unknown (to be predicted). Nonetheless, according to the Third Law of Geography (Zhu et al., 2018; Zhu & Turner, 2022), which states that similar values of the target variable can be expected at locations with similar geographic configurations (e.g., environmental conditions), it can be reasonably expected that the representativeness measured on the covariates would approximate the representativeness on the target variable because the target variable and the covariates should correlate (Zhu et al., 2015a). Based on this idea, sample representativeness can be defined and measured to guide spatial bias mitigation.

20.3.1 Measuring Sample Representativeness

Sample representativeness is defined as the "goodness-of-coverage" of the sample locations in the covariate space, which in turn is measured as the similarity between the probability density distribution of the sample locations in the covariates space (i.e., sample distribution Q) and the probability density distribution of all spatial units (e.g., raster cells) in the area (i.e., population distribution P) (Fig. 20.1). Stronger spatial bias in the sample locations would lead to poorer sample representativeness.

Sample representativeness is computed as the similarity between the sample and population distributions over the covariate space (Zhang & Zhu, 2019b). Kernel density estimation was used to estimate probability density distributions for computing sample representativeness. First, sample and population distributions with respect to individual covariate were estimated as per Eqs. (20.1) and (20.2), respectively.

$$Q_l(v_l) = \sum_{i=1}^{n} w_i \frac{1}{h_{lQ}} K\left(\frac{v_l - V_{li}}{h_{lQ}}\right) \tag{20.1}$$

$$P_l(v_l) = \sum_{j=1}^{m} \frac{1}{h_{lP}} K\left(\frac{v_l - V_{lj}}{h_{lP}}\right) \tag{20.2}$$

In the above equations, $K(\cdot)$ is the Gaussian kernel function, n is the number of sample locations and m is the number of locations (cells) in the study area. Q_l and P_l are the estimated sample and population distributions on covariate l (denoted as v_l), respectively. V_{li} is the value of v_l at sample location i and w_i is a normalized sample weight ($\sum_{i=1}^{n} w_i = 1$) associated with location i. V_{lj} is the value of v_l at cell j in the study area. h_{lQ} and h_{lP} are kernel bandwidths. Second, the similarity between Q_l and P_l (S_l) was computed as the overlapping area between the two distributions (Eq. 20.3) (Zhu, 1999):

$$S_l = \frac{2 \times A_{Q_l} \cap A_{P_l}}{A_{Q_l} + A_{P_l}} \tag{20.3}$$

where A_{Q_l} and A_{P_l} are the areas under the sample and population distribution curves, respectively and $A_{Q_l} \cap A_{P_l}$ is the overlapping area (Fig. 20.2). S_l reflects the goodness-of-coverage of the sample regarding this covariate. Finally, sample representativeness was computed as the overall similarity between the sample and population distributions with respect to all covariates. It is a weighted average of the similarities with respect to individual covariates (Eq. 20.4):

$$R = \sum_{i=1}^{L} \frac{\lambda_i}{\sum_{j=1}^{L} \lambda_j} S_i \tag{20.4}$$

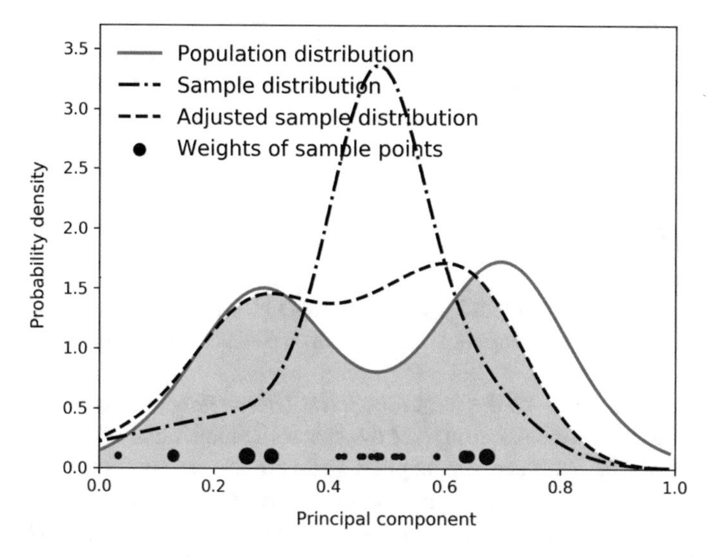

Fig. 20.2 An illustration of the effects of representativeness-directed spatial bias mitigation. Reused from Zhang and Zhu (2019b) with permission

where R is sample representativeness with a larger value indicating higher sample representativeness, and λ_i the weight associated with covariate i indicating covariate importance in measuring sample representativeness.

20.3.2 Representativeness-Directed Bias Mitigation

Spatial bias in field sample sets can then be mitigated by improving sample representativeness. Weights of the sample locations (Eq. 20.2) affect the estimated sample distributions and hence sample representativeness. Therefore, improving sample representativeness is achieved by adjusting the sample distribution towards increasing its similarity to the population distribution through weighting sample locations (Zhang & Zhu, 2019b). That is, sample locations in under-represented areas would receive larger weights and be treated as more important in training models. Weighting the sample locations in this way is expected to mitigate spatial bias and improve sample representativeness. The key is to determine the optimal weights. This can be conceived as an optimization problem, where the goal is to find a set of optimal weights associated with the sample locations that maximizes sample representativeness. A Genetic Algorithm was adopted to search for the optimal weights using sample representativeness as the objective function.

The weighted sample locations were used to train models to establish the relationships between the target variable and the covariates. Weights can be incorporated in the model training process by weighting the error term associated with each sample location (e.g., training a regression model using weighted ordinary least square) (Zhang & Zhu, 2019a, 2019b). The trained models can be applied to the covariate data layers (cell-by-cell) to predict spatial variation of the target variable.

20.4 Applications

The representativeness-directed approach to spatial bias mitigation was evaluated through case studies in two application domains: species habitat suitability mapping, and digital soil mapping.

Occurrence locations of the red-tailed hawk (*Buteo jamaicensis*) obtained from eBird were used to model and predict the species habitat suitability in Wisconsin, United States. The approach was applied to determine weights for species occurrence locations (Fig. 20.3) to train a habitat suitability model with logistic regression. Validation shows that the accuracy of predicted suitability map (Fig. 20.4) improved with weighted occurrence locations. Additionally, a positive relationship between sample representativeness and prediction accuracy was observed (Fig. 20.5), suggesting that sample representativeness is a valid indicator of suitability prediction accuracy (Zhang & Zhu, 2019b).

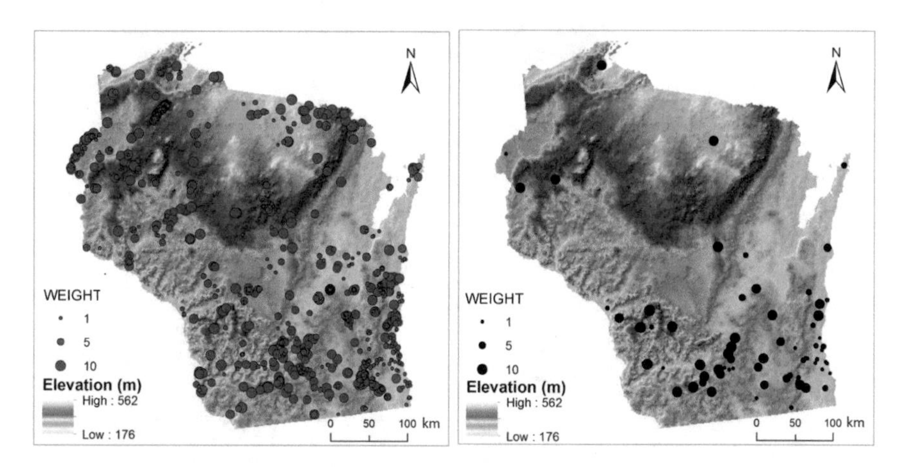

Fig. 20.3 Optimal weights determined through the representativeness-directed approach for all eBird observation locations (left) and for the red-tailed hawk occurrence locations (right). Reused from Zhang and Zhu (2019b) with permission

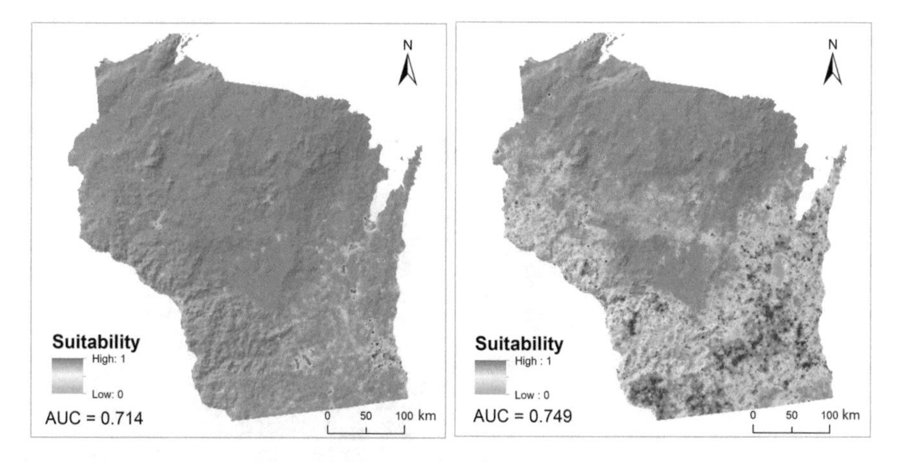

Fig. 20.4 Predicted habitat suitability maps based on unweighted species occurrence locations (left) and weighted occurrence locations (right). Higher AUC (area under the receiver operating characteristic curve) indicates higher prediction accuracy. Reused from Zhang and Zhu (2019b) with permission

The representativeness-directed approach was also applied to mitigate spatial bias in existing soil samples for digital soil mapping in Heshan study area, northeastern China. Existing soil samples in the study area were pooled from various sources and subject to spatial bias. Quantitative evaluations show that weighting soil samples using the weights determined from the approach (Fig. 20.6) improved A-horizon soil organic matter content prediction accuracy with either the iPSM method (Zhu et al., 2015a) or multiple linear regression (Fig. 20.7). A positive relationship between

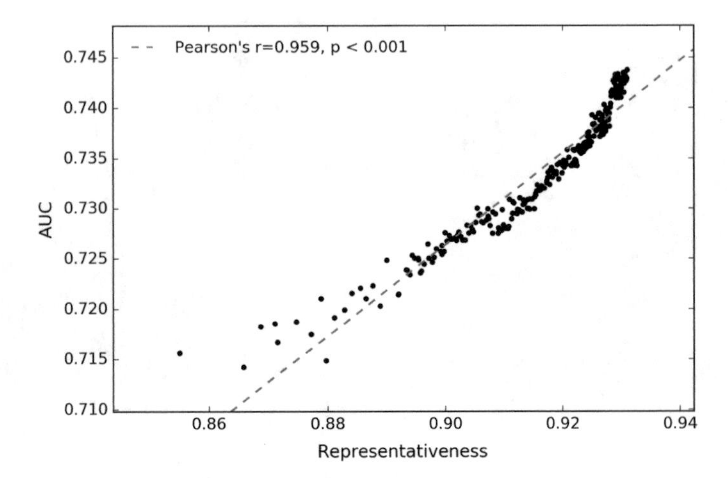

Fig. 20.5 The relationship between sample representativeness and prediction accuracy over the generations of the genetic algorithm. Reused from Zhang and Zhu (2019b) with permission

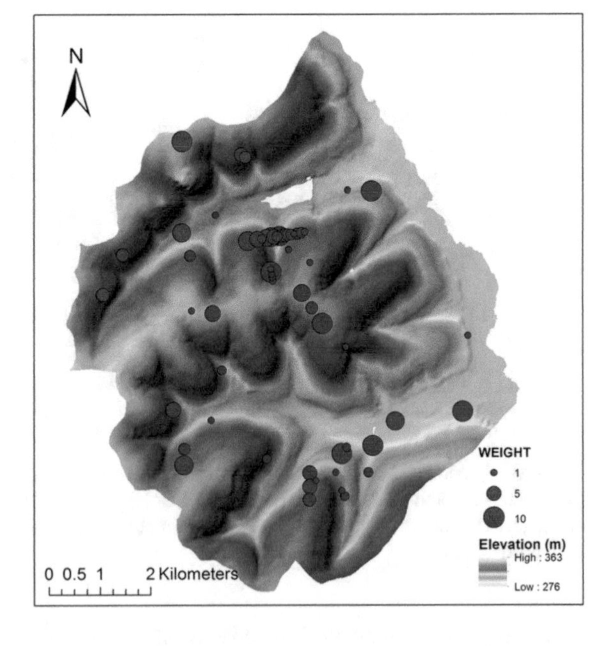

Fig. 20.6 Weights of the soil samples determined through the representativeness-directed approach. Reused from Zhang and Zhu (2019a) with permission

sample representativeness and prediction accuracy was again observed (Fig. 20.8). Moreover, the weights were informative of individual sample importance and thus can be used as guidance to filter soil samples to improve soil prediction accuracy (Zhang & Zhu, 2019a).

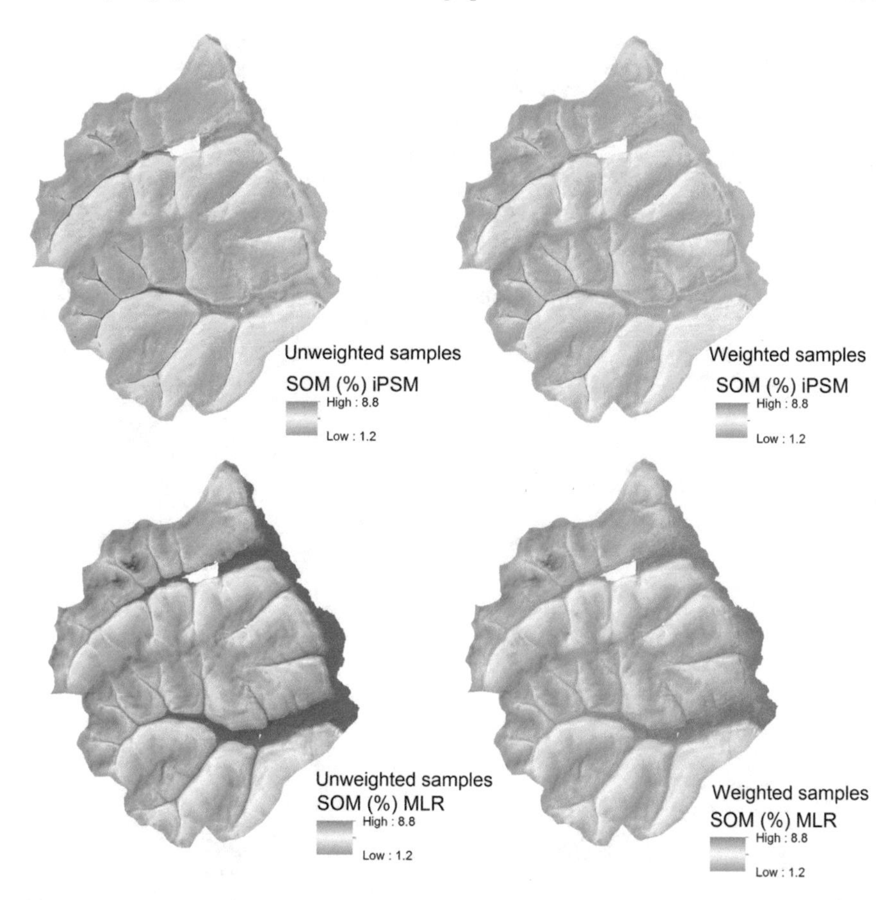

Fig. 20.7 A-horizon soil organic matter content predicted with iPSM (top row) and multiple linear regression (bottom row) based on unweighted soil samples (left column) and weighted soil samples (right column). Lower RMSE (root mean squared error) indicates higher prediction accuracy. Reused from Zhang and Zhu (2019a) with permission

20.5 Outlook on Future Research

Beyond the two application case studies, the idea of the representativeness-directed approach should apply to sample selection bias mitigation in general for spatial modeling and prediction. Specifically, beyond its applicability to global modeling methods, the approach can be extended to train localized models (e.g., modeling based on sample locations within a neighborhood of the prediction location) that account for spatial non-stationarity. It would also be interesting to examine the applicability of the approach for classification problems (e.g., soil class prediction) in addition to regression tasks explored. Lastly, spatial bias in a field sample set

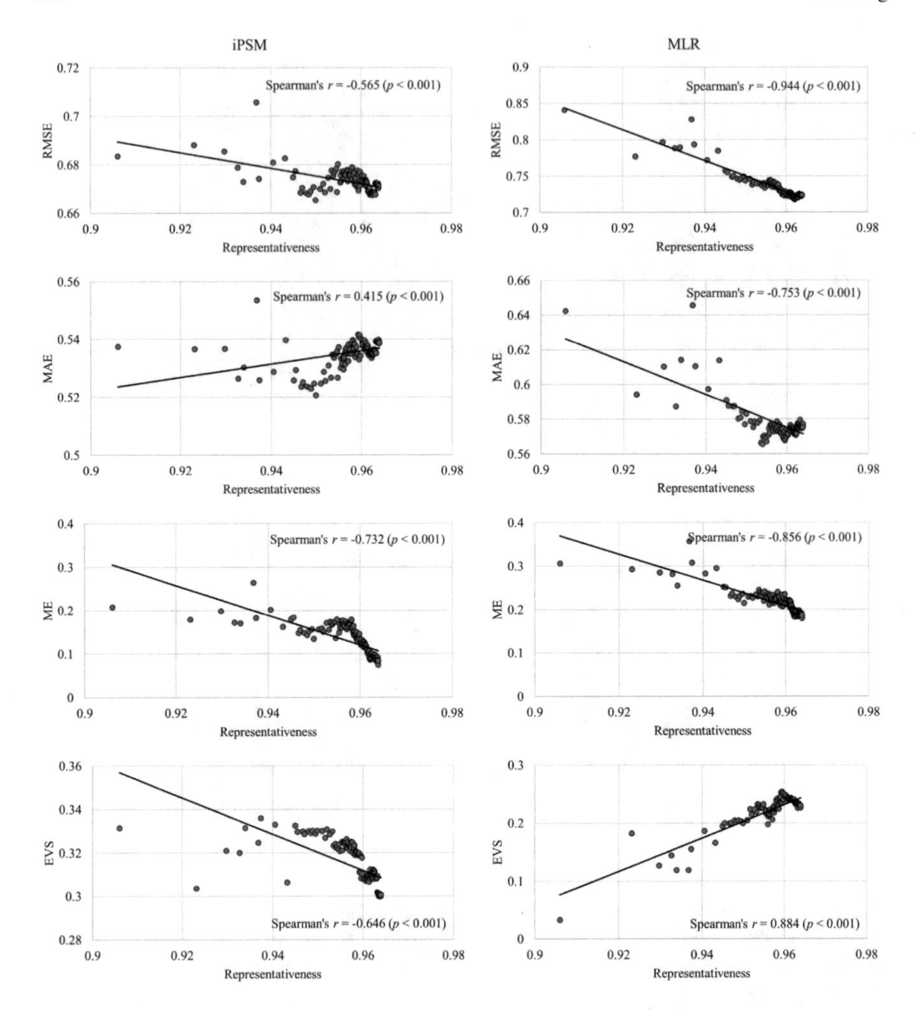

Fig. 20.8 Relationship between sample representativeness and prediction accuracy (measured by root mean squared error—RMSE, mean absolute error—MAE, mean error—ME, and explained variance score—EVS) over the generations of the genetic algorithm. Reused from Zhang and Zhu (2019a) with permission

may not be mitigated completely. It is thus worth exploring how to utilize information on sample representativeness to quantify modeling and prediction uncertainties, preferably in a spatially explicit manner.

At the core of the approach is defining and measuring sample representativeness in the covariate space. The idea can be translated to the social space. For example, it may be used to quantify demographic and socio-economic biases embedded within social media users to inform to what extent inferences drawn from social media data truly reflect the status of the population at large. In a broader sense, big data often suffers from biases. The concept of defining and quantifying representativeness offers a novel perspective on how to appropriately deal with biases in big data so that more accurate insights can be gained from them.

Acknowledgements The author sincerely thanks the editors team Drs. Bin Li, A-Xing Zhu, Xun Shi, Cuizhen Wang and Hui Lin for their invitation to contribute a chapter and their tireless effort to organize and edit this book series.

References

Elwood, S. (2008). Volunteered geographic information: Key questions, concepts and methods to guide emerging research and practice. *GeoJournal, 72*(3), 133–135.

Goodchild, M. F. (2007). Citizens as sensors: The world of volunteered geography. *GeoJournal, 69*(4), 211–221.

Goodchild, M. F., & Li, L. (2012). Assuring the quality of volunteered geographic information. *Spatial Statistics, 1*, 110–120.

Haklay, M., & Weber, P. (2008). OpenStreetMap: User-generated street maps. *Pervasive Computing, IEEE, 7*(4), 12–18.

Jensen, R. R., & Shumway, J. M. (2010). Sampling our world. In B. Gomez, & J. P. Jones III (Eds.), *Research methods in geography: A critical introduction* (pp. 77–90).

Kadmon, R., Farber, O., & Danin, A. (2004). Effect of roadside bias on the accuracy of predictive maps produced by bioclimatic models. *Ecological Applications, 14*(2), 401–413.

Kelling, S., Hochachka, W. M., Fink, D., Riedewald, M., Caruana, R., Ballard, G., & Hooker, G. (2009). Data-intensive science: A new paradigm for biodiversity studies. *BioScience, 59*(7), 613–620.

Leitão, P. J., Moreira, F., & Osborne, P. E. (2011). Effects of geographical data sampling bias on habitat models of species distributions: A case study with steppe birds in southern Portugal. *International Journal of Geographical Information Science, 25*(3), 439–454.

Miller, H. J., & Goodchild, M. F. (2014). Data-driven geography. *GeoJournal, 80*(4), 449–461.

Unger, S., Rollins, M., Tietz, A., & Dumais, H. (2020). iNaturalist as an engaging tool for identifying organisms in outdoor activities. *Journal of Biological Education*, 1–11.

Wood, C., Sullivan, B., Iliff, M., Fink, D., & Kelling, S. (2011). eBird: Engaging birders in science and conservation. *PLoS Biology, 9*(12), e1001220.

Yan, Y., Feng, C., Huang, W., Fan, H., & Wang, Y. (2020). Volunteered geographic information research in the first decade: A narrative review of selected journal articles in GIScience. *International Journal of Geographical Information Science, 34*(9), 1765–1791.

Yang, C. P. (2017). Geospatial cloud computing and big data. *Computers, Environment and Urban Systems, 61*, 119.

Zhang, G. (2018). *A representativeness directed approach to spatial bias mitigation in VGI for predictive mapping*. University of Wisconsin-Madison.

Zhang, G. (2020). Spatial and temporal patterns in volunteer data contribution activities: A case study of eBird. *ISPRS International Journal of Geo-Information, 9*(10), 597.

Zhang, G. (2021). Volunteered geographic information. In J. P. Wilson (Ed.), *The geographic information science & technology body of knowledge* (1st Quarter 2021 Edition).

Zhang, G., & Zhu, A. X. (2018). The representativeness and spatial bias of volunteered geographic information: A review. *Annals of GIS, 24*(3), 151–162.

Zhang, G., & Zhu, A. X. (2019a). A representativeness directed approach to spatial bias mitigation in VGI for predictive mapping. *International Journal of Geographical Information Science, 33*(9), 1873–1893.

Zhang, G., & Zhu, A. X. (2019b). A representativeness heuristic for mitigating spatial bias in existing soil samples for digital soil mapping. *Geoderma, 351*, 130–143.

Zhang, G., Zhu, A. X., Huang, Z. P., Ren, G., Qin, C. Z., & Xiao, W. (2018). Validity of historical volunteered geographic information: Evaluating citizen data for mapping historical geographic phenomena. *Transactions in GIS, 22*(1), 149–164.

Zhu, A. X. (1999). A personal construct-based knowledge acquisition process for natural resource mapping. *International Journal of Geographical Information Science, 13*(2), 119–141.

Zhu, A. X., Liu, J., Du, F., Zhang, S., Qin, C. Z., Burt, J., Behrens, T., & Scholten, T. (2015a). Predictive soil mapping with limited sample data. *European Journal of Soil Science, 66*(3), 535–547.

Zhu, A. X., Lu, G., Liu, J., Qin, C., & Zhou, C. (2018). Spatial prediction based on Third Law of Geography. *Annals of GIS, 24*(4), 225–240.

Zhu, A.-X., & Turner, M. (2022). How is the third law of geography different? *Annals of GIS, 28*(1), 57–67. https://doi.org/10.1080/19475683.2022.2026467.

Zhu, A. X., Zhang, G., Wang, W., Xiao, W., Huang, Z. P., Dunzhu, G. S., Ren, G., Qin, C. Z., Yang, L., Pei, T., & Yang, S. (2015b). A citizen data-based approach to predictive mapping of spatial variation of natural phenomena. *International Journal of Geographical Information Science, 29*(10), 1864–1886.

Chapter 21
Dealing with Unstructured Geospatial Data

Huayi Wu and Zhaohui Liu

Abstract Unstructured geospatial data become more and more important in the big data era. Compared with classic spatial data and other big data, the special characteristics of the unstructured geospatial data are investigated and summarized. The key technologies and challenges in data storage, management, analysis, mining, and high-performance computing are evaluated. Finally, future GIS are characterized as smart GIS with real-time sensing, ubiquitous interconnection, deep integration, and intelligent services integrated.

Keywords Unstructured geospatial data · Smart GIS · Data management · High-performance computing

21.1 Introduction

With the rapid development of earth observation, wireless sensor networks, Internet and communication technologies, and the popularity of social media platforms, the collection tools for geographic data have changed from traditional surveying instruments to ubiquitous sensors. The development of collection tools has promoted that the acquisition mode of geospatial data is changed from "purposely data acquisition for predefined questions" to "ubiquitously data acquisition for undefined questions" and the research paradigm is changed from "seeking data for predefined questions" to "seeking questions for ubiquitous data".

The active observation data with simple objectives have the predefined data units, attributes, and relationships among them, and they have a clear mapping relationship with the entities in the real world, thus they are structured data. The ubiquitous monitoring data are flexible but vague sometimes for the data units, attributes, and

H. Wu (✉) · Z. Liu
State Key Lab of Information Engineering in Surveying, Mapping and Remote Sensing, Wuhan University, Hongshan District, Wuhan 430079, Hubei, China
e-mail: wuhuayi@whu.edu.cn

Z. Liu
e-mail: liuzhaohui@whu.edu.cn

B. Li et al. (eds.), *New Thinking in GIScience*,
https://doi.org/10.1007/978-981-19-3816-0_21

relationships among them; therefore, the data structure cannot be predefined, which is also defined as semi-structured and unstructured.

The development of the theory and technology for data acquisition and processing facilitates the changing of the connotation and scope of unstructured data. From the traditional perspective, the data that can be stored in relational databases are the structured data, and geospatial data, especially vector data, are unstructured data. However, classical geospatial data, such as vector data, are also considered as structured data with the development of object-oriented data model and databases.

At the same time, many new unstructured geospatial data have appeared. Text, images, videos, and other data are generated by various business and entertainment applications on the Internet, and some of them contain geographical information. Internet of Things sensors such as radio frequency identification devices, infrared sensors, global positioning system, and laser scanners constantly generate the flow data for environmental detection, which also include dynamic geographical information. In the last decades, new surveying and mapping technologies such as oblique photogrammetry, laser scanning survey, satellite video remote sensing, and nighttime light remote sensing are developing vigorously, and original oblique photogrammetry photos, high-density point clouds, and street view photos produced by these techniques are also important unstructured geospatial data (Zhong et al., 2020).

21.2 Characteristics of the Unstructured Geospatial Data

Different from classical geospatial data and unstructured non-geospatial data, unstructured geospatial data has some unique characteristics:

1. **Implicit geographic information and fuzzy semantics**

Compared with classical spatial data, unstructured geospatial data appears in multimodal forms such as text, audio, pictures, videos. Some unstructured geospatial data are geo-tagged, such as Weibo check-in data, tracks of vehicles and ships, and surveillance videos, while other data have implicit spatial information, such as social media texts, news, and geospatial images, which contain location. In addition, these unstructured geospatial data also record the semantic descriptions of attributes, themes, and emotions about all kinds of geographical entities, geographical phenomena, and geographical events. The implied spatial and fuzzy semantic information could be extracted artificially; however, they cannot be input to spatial models directly given that their structures are flexible, irregular, and diverse.

2. **Dynamic space–time correlation**

Compared with other big data, unstructured geospatial data has unique spatio-temporal characteristics. The unstructured geospatial data records the dynamic evolution and correlation of geographical entities, geographical phenomena, and geographical events through spatio-temporal sampling with different granularity. There is abundant spatial and semantic information about the geographic entities, geographic phenomena, and geographic events in unstructured geospatial data, but these data are disorganized with ambiguous relationships. Therefore, linking and aggregating unstructured geospatial data to form a comprehensive description of spatial and semantic attribute information of geographical entities, geographical phenomena and geographical events are the key issues for unstructured geospatial data applications.

3. **Big data characteristics**

Unstructured geospatial data is a type of big data, inheriting the 5Vs of big data, e.g., volume, velocity, variety, veracity, and value (Zhong et al., 2020). Volume and velocity refer to a surge of data scale and generation speed. Variety is about the multiple forms of big data and various semantic information behind these data, which are difficult to describe with uniform data models and data structures. Veracity deals with the data certainty and trustworthiness, which is a key issue in big data processing. Value is an important feature of big data, turning big data to big value is the goal of big data applications.

21.3 Technologies and Challenges of Unstructured Geospatial Data

Opportunities and challenges coexist in the big data era. On the one hand, the massive unstructured geospatial data drastically increase the availability of data resources for research in geography and across academic disciplines. On the other hand, unstructured geospatial data require processing functionality for both conventional GIS data and big data. Traditional GIS aims at processing and analyzing fixed classical spatial data, which cannot deal with unstructured geospatial data with implicit geographic information and fuzzy semantics. Specifically, traditional GIS faces challenges in data storage and management, analysis and mining, as well as other technical domains.

1. **Storage and management**

Since the turn of the century, traditional spatial data collected by surveying, mapping and remote sensing are managed by relational databases. Designing a spatial data engine based on a mature relational database to achieve centralized storage and management of data was a hot topic for spatial data storage. Relational databases perform well in processing structured data. To adapt to the relational database model, traditional spatial data are always split and encapsulated into BLOB (Binary Large Object). However, the BLOB model may lead to efficient access and analysis; and the relational database model has problems with schema definition and scalability. With the emergence of unstructured geospatial data, data storage and management based on the relational database has reached a bottleneck, which limits the scalability of storage capacity and cannot support the low-latency and high-concurrency access for large-scale application (Guan, 2020).

Distributed non-relational databases, with a flexible, scalable, and portable data model, have been widely employed on the Internet, providing a new option for the storage and management of unstructured geospatial data. Height-balanced tree index based on dynamic partitioning and spatiotemporal index based on static rule provide support for efficient storage and management. Spatiotemporal coding technologies such as tandem coding and spatiotemporal cross coding reduce the dimension of multi-dimensional spatiotemporal coordinates, supporting the fast retrieval of spatiotemporal objects. These key technologies have promoted the transformation of GIS application from traditional data processing, mapping, and spatial analysis to online analysis and location service.

With the scope of GIS data expanding from 2 to 3D, outdoor to indoor, surface to underground, and gradually establishing the digital world corresponding to the physical world, future GIS will be all-encompassing. The digital world serves the physical world, making it potentially more efficient and orderly. Bidirectional mapping, real-time connection, and dynamic interaction between the physical world and the digital world will generate ubiquitous unstructured geospatial data, which will pose a great challenge to data storage and management.

2. **Analysis and mining**

Traditional GIS analysis develops algorithms for vector, raster, and other classical data for numerical calculation directly. However, unstructured data have flexible forms and embed implicit spatial and semantic information. An essential work with unstructured geospatial data is to mine the implied location and semantic information. Extracting the spatiotemporal semantic information with dynamic change of geographical entities, geographical phenomena, and geographical events from massive, dynamic, and fuzzy data poses a challenge to developing intelligent analysis and mining algorithms.

The rise of artificial intelligence technology (AI) provides a powerful tool for the analysis and mining of unstructured geospatial data, enabling the development of geospatial natural language processing, geospatial computer vision, geospatial data mining. AI technology represented by deep learning can be applied to the research of geospatial artificial intelligence (Janowicz, 2020), and many representative works have emerged. For example, mining spatial and semantic information about geographical entities, geographical phenomena, and geographical events, from geospatial texts such as social media texts, blogs, news, and historical archives, and knowledge inferences are hot topics (Hu, 2018). Some researches focus on automatic mapping and environmental perception based on oblique photographic images, high-density point clouds, street photos, which are obtained by mapping technology based on oblique photogrammetry, laser scanning measurement, satellite remote sensing (Zhong et al., 2020). In addition, some researchers attempted to understand the characteristics of social activities from taxi tracks, and mobile phone cellular signaling data (Wu et al., 2019).

With the development of the analysis and mining algorithms, the application scope of unstructured geospatial data will gradually expand from automatic information acquisition, environmental perception, and social activities understanding to spatial information service and decision-making supporting of digital twin, which provide the means to monitor, understand, and optimize the functions of physical entities (Francesco, 2020). Future GIS calls for a higher demand for analysis and mining algorithms of multimodal data including unstructured geospatial data.

3. **High performance computing**

Parallel computing based on multi-thread and multi-process technology, enabled by GPU and software platforms such as CUDA and OpenCL, makes it possible to build high performance GIS computing engines that achieve unprecedented computing capability (Guan, 2020). However, in the big data era, the volume, velocity, and scale of unstructured data pose new challenges to the computational performance, resulting a surge of demand for computational capacity and efficiency.

Cloud computing, characterized as large scale, virtualization, high reliability, versatility, high scalability, and on-demand, provides a new approach for massive unstructured data processing. However, network delay and bandwidth limitation may reduce service performance and reliability in centralized cloud applications. As a supplement to cloud computing, edge computing can improve the performance of the entire system using edge servers deployed near the terminal to share processing functions. At the same time, GIS integrated cloud-edge-terminal provides management and scheduling of computing resources based on virtualization and container technology, achieving the processing capabilities for big data through a dynamically scalable distributed infrastructure (Zhong et al., 2020). Dynamically allocating computing resources according to the tasks, employing dynamic expansion of computing nodes, and maximizing the use of computing resources in the entire distributed system are effective approaches to improving the performance of distributed processing of massive unstructured data.

With the continuous expansion of application scope, the service field of GIS will expand from public service, enterprises service, and government service to other industries in the society, and the future GIS will be ubiquitous. Distributing all kinds of information, knowledge, functions, and even resources to users in the form of services and providing users with comprehensive, convenient, efficient, and intelligent services, require a higher standard for high-performance service capability of GIS.

21.4 Conclusion

Unstructured geospatial data are a new data source and a new approach to data acquisition, with combined characteristics of traditional geospatial data and big data. Originating from a wide range of sources and taking diverse forms, unstructured geospatial data greatly expanded the data pool but introduced theoretical and technical challenges as they are in large volume, updated in high frequency and often have spatial and semantic information implicitly embedded, which requires new ways for management, processing, and analysis. The GIS academic and industrial communities have taken advantages of unstructured geospatial data and greatly expanded the scope of GIS. The emergence of digital twins, which aims to establishing bidirectional and real time mapping of the physical world and enabling dynamic interactions between the two worlds, and the introduction of smart GIS, which is equipped with real-time sensing, ubiquitous interconnections among objects, deep integration of hardware and software, and intelligent services functions, represent the future trend of GIS, with unstructured geospatial data being an integral component.

References

Francesco, G. (2020). A digital twin for distant visit of inaccessible contexts. In *IMEKO TC-4 International Conference on Metrology for Archaeology and Cultural Heritage, Trento.*

Guan, X. F. (2020). *High Performance Space-time Computing and Applications* (pp. 113–114). Science Press.

Hu, Y. J. (2018). Geo-text data and data-driven geospatial semantics. *Geography Compass, 12*(11), e12404.

Janowicz, K., Gao, S., Mckenzie, G., Hu, Y. J., & Bhaduri, B. (2020). GeoAI: Spatially explicit artificial intelligence techniques for geographic knowledge discovery and beyond. *International Journal of Geographical Information Science, 34*(6), 1–13.

Wu, H. Y., Huang, R., You, L., & Xiang, L. G. (2019). Recent progress in taxi trajectory data mining. *Acta Geodaetica Et Cartographica Sinica, 48*(11), 1341–1356.

Zhong, E. S., Song, G. F., & Tang, G. A. (2020). *Principles, technologies, and applications of big data geographic information system* (pp. 6–14). Tsinghua University Press.

Chapter 22
Green Cartography and Energy-Aware Maps: Possible Research Opportunities

Mingguang Wu, Guonian Lv, and Linwang Yuan

Abstract Cartography's roles in environmentally sustainable development are twofold: first, expressive maps can communicate and narrative environmentally sustainable development; second, digital maps themselves, as device-dependent digital tools, can be more energy efficient in minimizing our impact on the environment. In this chapter, we discuss the concept of green cartography that encourages aligned map design and use for *energy awareness*. First, we investigate how map design and use impact energy consumption. Then, we discuss the possible ways in which digital maps can be *energy-aware*, including how to make and how to use *energy-aware* maps, outlining a series of possible research opportunities.

Keywords Digital maps · Green cartography · Energy consumption

22.1 Introduction

Cartography intersects environmentally sustainable development with two trends. First, informative and expressive maps are needed to communicate information about environmentally sustainable development. Many wonderful atlases and thematic maps have been made, and more are coming with the topic of sustainable development. Second, as digital maps themselves consume energy, they can be more *energy-aware* to minimize greenhouse gas emissions, which is considered a major concern for climate change (Erickson, 2017). In this chapter, we focus on the later.

Discussion on digital maps' energy consumption is limited. The current discussion on digital maps' energy consumption is dedicated to an urgent problem with mobile maps: battery life. While mobile maps provide us with incredible mobility by allowing one to explore spaces and experience places, they also face a series of physical constraints resulting from the limitations of mobile devices, such as small screen size and limited bandwidth and processing power, and battery life is one of the most critical among these constraints (Roth et al., 2018), imaging a map user

M. Wu (✉) · G. Lv · L. Yuan
School of Geographic, Nanjing Normal University, Nanjing 210023, Jiangsu, China
e-mail: wmg@njnu.edu.cn

B. Li et al. (eds.), *New Thinking in GIScience*,
https://doi.org/10.1007/978-981-19-3816-0_22

who wants to use navigation in an unfamiliar place on a mobile device whose battery is dying (Han et al., 2021).

We argue here that the idea of *energy awareness* intersects with cartography not only on mobile maps with battery life as a physical constraint but also on *energy awareness*, which impacts almost all digital maps, including server-side data storage and client-side map display. With the rapid development of information and communications technology (ICT), an increasing number of helpful digital tools have been developed, such as the Internet and, recently, the Internet of Things. The ICT industry also accounts for an enormous amount of greenhouse gas emissions. Emissions from the global ICT industry accounted for approximately 2% of global greenhouse gas emissions in 2008 (Mingay & Pamlin, 2008) and could reach 23% in 2030 in the worst case (Andrae & Elder, 2015), suggesting a new trend: the efficient creation, use, and disposal of computing resources to limit environmental impacts, called green computing (Williams & Curtis, 2008).

Digital maps involve green computing for two major reasons. First, there are a large number of map users. Digital maps are now indispensable visual media on various displays: globally, using a map or directions service was one of the top three online activities in 2019 (Statista, 2020). An enormous amount of energy is used to digital maps on billions of client-side devices. Second, massive data volume. Behind front-end cartographic representation, a huge amount of data, such as aerial imageries, trajectories from GPS trackers, geo-tagged textual data from social media, and data streams from sensor networks, are collected (still contiguously) to make content-rich and up-to-data maps. Storing and processing those data and transforming the raw data into readable maps also consume large amounts of energy. Discussion on green cartography remain limited.

In this chapter, we use the term green cartography to describe the knowledge, methods, and attitudes regarding making and using digital maps with a specific concern for their carbon footprint. In the following, we investigate why we need *energy-aware* maps in Sect. 22.2. We then explore the possible ways in which digital maps can be *energy-aware*, outlining a series of possible research opportunities in Sect. 22.3. Finally, we summarize our discussion in Sect. 22.4.

22.2 Should Digital Maps Be *Energy-Aware*?

We address *energy-aware* maps from two aspects: map content and form. Here, map content refers to the underlying geospatial data of digital maps, including three major components: geographic, attribute, and temporal. Map form refers to the visual appearance of a map, including scale, projection, symbolization, typography, visual hierarchy, and interfaces and interactions. For energy consumption, through specific devices and applications, we distinguish three primary energy-intensive operations: 1) displaying, 2) computing, and 3) storage and transmission. In the following, we analyze the carbon footprint of digital maps' content and form in terms of those three operations.

22.2.1 *Map Content with Energy Consumption*

22.2.1.1 Geographic Components

One major factor that dominates map content is cartographic generalization. Cartographers often apply generalization, such as *selection, simplification, smoothing,* and *aggregation*, to modify the details of features and the overall information density. Its impacts on energy consumption are twofold: first, it impacts the visual look of maps and therefore influences the energy consumption on displaying. Generalization operators may change the shape and density of map objects that need to be displayed. As shown in Fig. 22.1, there are two test maps with the same area but hold different mapping themes. They cost different amount of energy on display. Second, generalization itself is computationally intensive and therefore energy hungry; the more mapping features are involved, the more energy is needed.

As generalization is data- and computing-intensive, tiling, and prerendering maps into tiles, tiling techniques are now widely used to make sloppy maps (Sack & Roth, 2017). Ramm (2012) conducted a test on the tiling OpenStreetMap (OSM) dataset in which the average render time per metatile was approximately 0.2 s at zoom 12. It would take more than 15 h to render every tile at these zoom levels; the rendering time will exponentially increase at higher zoom levels and therefore will consume a much larger amount of energy. Importantly, tiling will also exponentially increase the data volume, requiring substantial energy consumption to store the map tiles. For example, the volume of the raw OSM dataset is approximately 50G, and prerendering all tiles would use approximately 54 TB of storage (Wiki, 2021a).

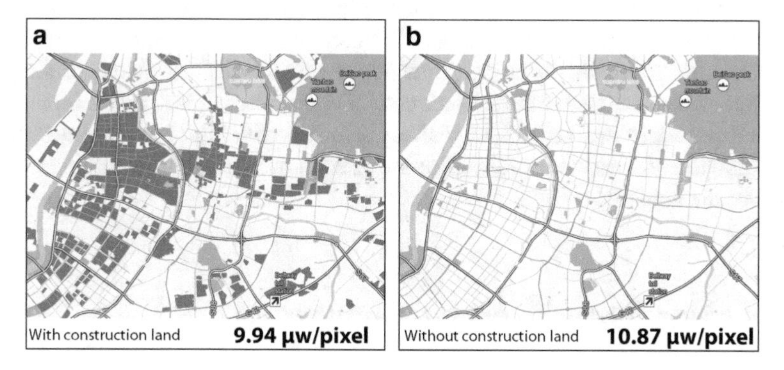

Fig. 22.1 A comparison of the energy consumption of two maps with the same area but hold different mapping themes. **a** With construction land; **b** without construction land

22.2.1.2 Attribute Components

When making maps, cartographers often use classification to reveal patterns on attributes. Classification methods' impact on energy consumption are twofold: first, classification impacts the visual look of maps. A series of classification methods are crafted for different data distributions, such as equal intervals, quantiles, standard deviations, and natural breaks. These classification methods lead to noticeable differences in visual looks, which require different energy consumption for display. Second, classification themselves could be computationally intensive. Much energy is needed to classify large datasets.

Interpolation is another attribute-related operation that impacts maps' energy consumption. Interpolation impacts energy consumption in a similar way as classification. First, it impacts the visual look of maps. A series of interpolation methods are suggested to estimate the attributes of uncovered sites by using samples, such as inverse distance weighting, kriging, natural neighbor, and spline. Certainly, they derive different estimations, resulting in different energy consumption when displaying the results. Second, interpolation methods themselves could also be computationally intensive.

22.2.1.3 Temporal Components

Temporal sampling with dynamic variables, such as display date, duration, order, rate-of-change, frequency, and synchronization, impacts the representation of spatiotemporal processes. With the development of sensors, many temporal processes, such as traffic flow, can be frequently sampled, which enables cartographers to directly capture detailed spatiotemporal changes. Different sampling techniques and dynamic variables may result in different visual looks of visualization of spatiotemporal processes, which require different amounts of energy to display. In addition, while temporal processes can now be intensively sampled with high temporal resolution, extra storage and bandwidth resources are needed to store and transmit streaming data over the network.

22.2.2 Map Form with Energy Consumption

22.2.2.1 Visual Look

A visual look of the map directly impacts the energy consumption on display. Dong and Zhong (2012) propose a model for estimating the energy consumption of a pixel according to the red, green, and blue components of the color of the pixel. Cartographic scale, projection, color, symbol, thematic map type and figure-ground separation are all involved in shaping the appearance of maps and therefore impact maps' energy consumption on display.

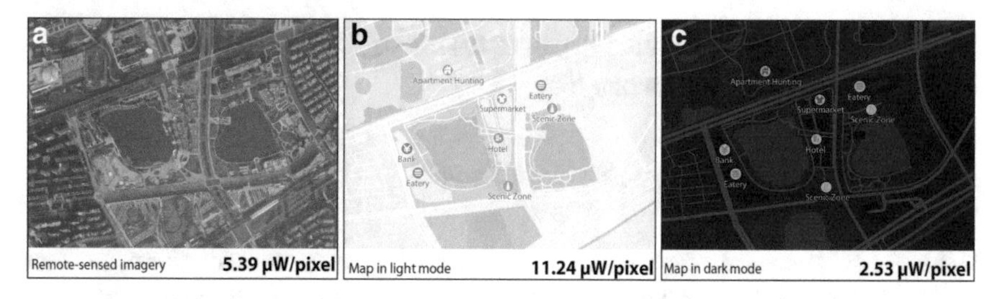

Fig. 22.2 A comparison of the energy consumption of map and remote-sensed imagery; **a** remote-sensed imagery; **b** a map in light mode; and **c** a map in dark mode with the same map data

To demonstrate the visual impact on energy consumption, we compare the two maps' energy consumption with paired remotely sensed imagery. As shown in Fig. 22.2, we use two testing maps with exactly the same underlying data but with different colors. When compared with the paired remotely sensed imagery, the map in light mode consumes more energy, the map in dark mode consumes far less, and the map in light mode consumes almost 4.5 times as much energy as the map in dark mode. Beyond this case, symbol, typography, and visual hierarchy also impact maps' energy consumption.

Although the energy consumption of a single map-use activity is rather small (see Fig. 22.2), it will accumulate over billions of devices and applications. For example, the total active user count of GaudMaps from China and Google Maps from the USA is approximately 926 million per month (BigData Research, 2021; Verto Analytic, 2021). The annual energy consumption of these two mobile map apps is approximately 906,964,648 kWh, equivalent to releasing 711,967 metric tons of greenhouse gas. For a national comparison, the energy consumption of these apps in 2019 was greater than 65 countries' energy consumption worldwide in 2020 (Wiki, 2021b).

22.2.2.2 User Interface

The impacts of the user interface and interaction on the energy consumption of displaying are also twofold. First, similar to the visual look of maps (as shown in Fig. 22.2), different user interface styles (e.g., light or dark) may consume different amounts of energy. Second, interaction through user interfaces, such as routing with road networks, may involve computing and performing data access and query operations, which consume extra energy for data retrieval and data transmission.

22.3 Possible Research Opportunities of Digital Maps Being *Energy-Aware*

22.3.1 Making Energy-Aware *Maps*

Making *energy-aware* maps requires compromising map content, form, and the carbon footprint with a map use context. Accordingly, there are three possible ways to make *energy-aware* maps: (1) reorganize map content; (2) adjust map form; and (3) refine map-use context to be *energy-aware.*

22.3.1.1 Opportunity #1: Reorganize Map Content to Be *Energy-Aware*

Map elements could be reorganized to be *energy-aware.* Involving both mapping data and map design decisions, upscaling, dimension reduction, and removing features does not necessarily reduce energy consumption, but replacing bright pixels does. In this sense, *energy awareness* encourages cartographers to derive energy-saving resolution, map scale, and map element dimensionality with a specific mapping context.

Map data can also be reorganized to save energy by saving storage space, removing redundancy, and minimizing data access. For example, as we mentioned before, prerendering all OSM tiles would use approximately 54 TB of storage, which would consume considerable energy to maintain them all. In practice, overall, 1.79% of tiles are viewed (Wiki, 2020), suggesting an opportunity to design an on-demand data organization schema that considers users' interests.

22.3.1.2 Opportunity #2: Adapt Map Form to Save Energy

One possible way of being *energy-aware* is to adjust map form to save energy for existing maps. For example, when crafting map color, designers tend to follow certain conventions, typically blue for water bodies, blue for low (cool) values and red for high (hot) *values* (e.g., in weather maps). Dong and Zhong's (2012) energy estimation model suggests that blue is the most energy-hungry primitive color, and white is the most energy-hungry mixed color. Furthermore, as shown in Fig. 22.2, figure-ground separation plays a critical role in energy consumption on map display. For existing maps, mapmakers can adjust visual looks, such as map color, symbols, and figure-ground separation, to save energy without sacrificing the underlying data. Thus, the aesthetic and communicative quality of the original maps should be considered for consistent map use.

Another possible way is to craft novel principles to make *energy-aware* maps. Generally, colorless (darker) and compact graphics (occupying fewer pixels) consume less energy, coinciding with Occam's Razor, which states that "entities should not be multiplied unnecessarily". While there has been much discussion of visual variables, gestalt organization, iconicity, and the cultural negotiation of map symbol design, with several new trends (such as pluralism and feminism) receiving much attention, the development of minimalism of maps to reduce the carbon footprint is limited.

22.3.1.3 Opportunity #3: Refine Map-Use Context to Be Greener

Specific map use context impacts the direction and level of trade-offs between content and form. Typically, individual differences, such as disability (e.g., visual impairment) and map-reading experience, impact the direction of these trade-offs. For example, for a navigation user who is familiar with the mapping area, more map content can be compromised; for a map user with visual impairment, such as color blindness, adapting map color should be carefully addressed to avoid misunderstanding. In practice, energy savings may not be the only concern when crafting maps. In this sense, energy savings should be compromised among other cartographic design concerns, such as consistency of map content and aesthetics of map form.

Ignoring all the differences among map use contexts, a research opportunity comes up with developing an 'energy saver' mode that maximizes energy reduction while minimizing the loss of map content and form. As map content and form are all involved, it would be extremely computationally intensive to find the best energy saving solution. Thus, the extra energy consumed to develop the 'energy saver' mode should also be compromised with the energy intended to save when using the resulting maps.

22.3.2 *Using* Energy-Aware *Maps*

Energy-aware maps can absolutely help to save energy, but they do not certainly contribute to improving the quality of maps.

22.3.2.1 Opportunity #4: Examine the Byproducts of *Energy-Aware* Maps

One possible byproduct of *energy-aware* maps is color shifting. Generally, *energy-aware* maps tend to be dark, as black consumes the least amount of energy. While people with impaired vision tend to perform better in dark mode (Sloan, 1977), general user performance is better in light mode (Aleman et al., 2018). Examination of the color shifting of making *energy-aware* maps is lacking: light mode and darker mode, which one is better for map reading, and for what kinds of map-use context? More broadly, color shifting not only introduces differences in physical stimuli with visual perception but also may result in more or less drift on affective responses, which is critical in specific mapping contexts, such as narratives or storytelling.

Another possible byproduct of *energy-aware* maps is inconsistency with map content. As we discussed in Sect. 22.3.1.2, a series of techniques, such as generalization, can potentially be used to reduce energy consumption. However, it also results in inconsistency with map content, such as changing map element shapes, which could be problematic. For example, when using maps in a cooperative context in which multiple users need to share a common map, if individual users use *energy-aware* maps with different shapes, then undesired misunderstandings could be introduced. Generally, byproducts of *energy-aware* maps should be systemically examined toward making and using energy effective yet communication effective maps.

22.3.2.2 Opportunity #5: Reshape the Attitude of Using *Energy-Aware* Maps

Energy-aware maps matter only when they gain a large population. As we discussed in Sect. 22.2, although the energy consumption of a single map-use activity is rather small, it will accumulate over billions of devices and applications. Even though cartographers can (hopefully) make *energy-aware* maps, using *energy-aware* maps to benefit our environment is not necessarily a well-known sense and therefore a default option. In other words, environmentally friendly organizational and social practices are needed, suggesting a gap between the techniques of making *energy-aware* maps and the end user's attitude toward using *energy-aware* maps. The question of how to encourage using *energy-aware* maps may be beyond current discussion in the field of cartography, but as digital maps are involved in the worldwide issue of sustainable development, we note that green cartography requires additional ideas from related disciplines, such as social psychology. For example, benefitting from education and legislation on environmental protection, plastic or paper are now a common sense and public option in our daily life; can *energy-aware* maps be a paper-or-plastic-like option in the future? If so, how.

22.4 Summary

As environmentally sustainable development is now an urgent concern worldwide, in this chapter, we discuss the concept of green cartography, which encourages aligned map making and use for *energy awareness*. We analyzed how map design decisions impact the energy consumption of digital maps; we then outlined the possible ways in which digital maps can be *energy aware*. Specifically, how to make and how to use *energy-aware* maps are discussed.

We also recognize that green cartography not only requires novel principles, techniques, and tools to be *energy-aware* but also requires environmentally friendly attitudes and individual, organizational, and social behavioral changes with regard to making and using *energy-aware* maps.

Acknowledgements This work was supported by the National Natural Science Foundation of China under Grant (No. 41971417).

References

Aleman, A., Wang, M., & Schaeffel, F. (2018). Reading and myopia: Contrast polarity matters. *Scientific Reports, 8*(1), 1–8. https://doi.org/10.1038/s41598-018-28904-x

Andrae, A. S., & Edler, T. (2015). On global electricity usage of communication technology: Trends to 2030. *Challenges, 6*(1), 117–157. https://doi.org/10.3390/challe6010117

BigData Research. (2021). China mobile map market research report for the 3rd quarter of 2019. http://www.bigdata-research.cn/yanjiubaogao/2.html. Accessed October 20, 2021.

Dong, M., & Zhong, L. (2012). Power modeling and optimization for OLED displays. *IEEE Transactions on Mobile Computing, 11*(9), 1587–1599. https://doi.org/10.1109/TMC.2011.167

Erickson, L. E. (2017). Reducing greenhouse gas emissions and improving air quality: Two global challenges. *Environmental Progress & Sustainable Energy, 36*(4), 982–988. https://doi.org/10.1002/ep.12665

Han, Y., Wu, M., & Roth, R. E. (2021). Toward green cartography & visualization: A semantically-enriched method of generating energy-aware color schemes for digital maps. *Cartography and Geographic Information Science, 48*(1), 43–62. https://doi.org/10.1080/15230406.2020.1827040

Mingay, S., & Pamlin, D. (2008). *Assessment of global low-carbon and environmental leadership in the ICT sector, by Gartner and WWF*. Gartner Inc. (2008). Tillgänglig Online: https://www.fujitsu.com/downloads/SVC/fs/whitepapers/assessmentenvironmental-leadership-gartner.pdf. Accessed October 14, 2021.

Ramm, F. (2012). *Optimising the Mapnik/osm2pgsql rendering toolchain Vortrag, SOTM*. https://www.geofabrik.de/media/2012-09-08-osm2pgsql-performance.pdf. Accessed October 15, 2021.

Roth, R. E., Young, S., Nestel, C., Sack, C., Davidson, B., Janicki, J., & Zhang, G. (2018). Global landscapes: Teaching globalization through responsive mobile map design. *The Professional Geographer, 70*(3), 395–411. https://doi.org/10.1080/00330124.2017.1416297

Sack, C. M., & Roth, R. E. (2017). Design and evaluation of an Open Web Platform cartography lab curriculum. *Journal of Geography in Higher Education, 41*(1), 1–23. https://doi.org/10.1080/03098265.2016.1241987

Sloan, L. L. (1977). *Reading aids for the partially sighted: A systematic classification and procedure for prescribing*. Williams & Wilkins.

Statista. (2020). *Share of global internet users who have performed selected online activities in the past month via mobile as of 1st quarter 2019*, https://www.statista.com/statistics/783357/leading-mobile-first-activities/. Accessed October 25, 2020.

Verto Analytic. (2021). *Leading US map and navigation smartphone apps, ranked by monthly unique users*, August 2019. https://www.emarketer.com/chart/234831/leading-us-map-navigation-smartphone-apps-ranked-by-monthly-unique-users-aug-2019. Accessed October 15, 2021.

Wiki. (2020). https://wiki.openstreetmap.org/wiki/Zoom_levels. Accessed October 15, 2020.

Wiki. (2021a). https://wiki.openstreetmap.org/wiki/Tile_disk_usage. Accessed October 20, 2021a.

Wiki. (2021b). https://en.wikipedia.org/wiki/List_of_countries_by_electricity_consumption. Accessed September 20, 2021b.

Williams, J., & Curtis, L. (2008). Green: The new computing coat of arms? *IT Professional, 10*(1), 12–16. https://doi.org/10.1109/MITP.2008.9

Chapter 23
Next Step in Vegetation Remote Sensing: Synergetic Retrievals of Canopy Structural and Leaf Biochemical Parameters

Jing M. Chen, Mingzhu Xu, Rong Wang, Dong Li, Ronggao Liu, Weimin Ju, and Tao Cheng

Abstract Shortwave remote sensing signals acquired from vegetation contain information not only for vegetation structure, such as leaf area index and clumping index, but also for leaf biochemical parameters, such as pigments, nitrogen content, water content, dry matter, etc. However, the retrievals of these two types of parameters are generally carried out separately without considering the influence of one type of parameters on the spectral signals used to retrieve the other type of parameters.

J. M. Chen (✉) · M. Xu · R. Wang
School of Geographical Science, Fujian Normal University, Fuzhou 350007, Fujian, China
e-mail: jing.chen@utoronto.ca

M. Xu
e-mail: xumz@fjnu.edu.cn

J. M. Chen
Department of Geography and Planning, University of Toronto, Toronto, ON M5G 3G3, Canada

D. Li
National Engineering and Technology Center for Information Agriculture, Nanjing Agricultural University, One Weigang, Nanjing 210095, Jiangsu, China

R. Liu
State Key Laboratory of Resources and Environmental Information System, Institute of Geographic Sciences and Natural Resources Research, Chinese Academy of Sciences, Beijing 100101, China
e-mail: liurg@igsnrr.ac.cn

W. Ju
International Institute of Earth System Science, Nanjing University, Nanjing 210046, Jiangsu, China
e-mail: juweimin@nju.edu.cn

T. Cheng
MOE Engineering Research Center of Smart Agriculture, MARA Key Laboratory of Crop System Analysis and Decision Making, Jiangsu Key Laboratory for Information Agriculture, National Engineering and Technology Center for Information Agriculture, Nanjing Agricultural University, One Weigang, Nanjing 210095, Jiangsu, China
e-mail: tcheng@njau.edu.cn

B. Li et al. (eds.), *New Thinking in GIScience*,
https://doi.org/10.1007/978-981-19-3816-0_23

Since green leaves would be very different from brown leaves in performing photosynthesis and transpiration, we suggest that a next step in vegetation remote sensing be directed towards synergetic retrievals of these two types of parameters for the purpose of improving regional and global carbon and water cycle estimation.

Keywords Vegetation structural parameters · Leaf biochemical parameters · Shortwave remote sensing · Photosynthesis · Transpiration

23.1 Introduction

Terrestrial ecosystems regulate the Earth's climate through their large energy and mass fluxes. They are also structurally and biologically diverse, and their fluxes are spatially and temporally variable and highly challenging to estimate. Satellite remote sensing data have played indispensable roles in retrieving vegetation information for regional and global carbon and water cycle estimation (Running et al., 1989; Xiao et al., 2019; Yan et al., 2012). The most widely used satellite-retrieved vegetation parameters have been leaf area index (LAI) (Knyazikhin et al., 1998; Liu et al., 2012; Verger et al., 2015) and Fraction of Photosynthetically Active Radiation (FPAR) absorbed by vegetation (Gitelson, 2019; Gobron et al., 2007). LAI, defined as one half the total all-sided leaf area per unit ground surface area (Chen & Black, 1992), is the most important vegetation structural parameter that determines the amount of radiation absorbed by the canopy under a given incident radiation that drives photosynthesis and transpiration, while FPAR can either be derived from LAI or from vegetation indices (Friedl et al., 1995; Gitelson, 2019). To characterize the diverse vegetation structure, the second structural parameter, clumping index (CI), has also been derived from multi-angle remote sensing (Chen et al., 2005; He et al., 2012; Jiao et al., 2016; Wei et al., 2019). CI quantifies the degree of foliage spatial distribution that deviates from random distribution. Using these structural parameters in conjunction with climate and soil data, global distributions of vegetation productivity as well as carbon and water fluxes have been produced in many studies (Chen et al., 2019; Prince & Goward, 1995; Yan et al., 2012; Zhao et al., 2005).

While vegetation structure forms the substrate for absorbing radiation that drives subsequent physical and biological activities within the structure, the physiological status of leaves in the canopy determines how the absorbed radiation is utilized for photosynthesis and its associated water vaporization. It has therefore been realized that carbon and water flux estimation based on vegetation structural parameters is only an approximation without knowing the leaf-level physiological conditions. This realization has prompted rapid development of remote sensing algorithms to retrieve leaf biochemical parameters related to leaf physiological conditions, including leaf chlorophyll content (LCC), leaf nitrogen content (LNC), leaf water content (LWC), and other leaf contents. These parameters are mutually related and determine in concert how plant leaves utilize radiative energy. LCC, for example, is responsible for harvesting photosynthetically active radiation (PAR) and creating excitation energy that is used in photosystems that assimilate air CO_2 for photosynthesis (Demmig-Adams & Adams, 1992). In leaves fully adapted to the environment, LCC is closely linked to the photosynthetic nitrogen pool in the leaf carboxylation enzyme controlling the leaf photosynthetic capacity (Croft et al., 2017; Lu et al., 2020; Xu et al., 2012). It is also related to LWC as the production and function of LCC need water. Recently, a global map series of LCC has been produced with multi-spectral satellite data (300 m, 7 days interval, 2003–2012) (Croft et al., 2020), and this map series has been converted to the maximum carboxylation rate (Vcmax) to improve global gross primary productivity (GPP) and evapotranspiration (ET) estimation (Luo et al., 2019).

Incident radiation on vegetation undergoes interception, absorption and scattering processes by leaves and woody materials in the canopy before exiting the canopy as reflected radiation, and therefore not only LAI and CI (structure) but also the optical properties (leaf internal structure and biophysical and biochemical contents) are responsible for the amount and spectral distribution of reflected radiation (Baret et al., 1992). Observed reflectance from vegetation, therefore, carries the signals of both canopy structure and leaf physiological status. Few efforts have been made to entangle the contributions of canopy structure and leaf biochemical parameters to spectral reflectance, and therefore it would not be accurate to retrieve one type of parameters without considering the influence of the other type. Houborg et al. (2015) pioneered a remote sensing algorithm to derive LAI and LCC simultaneously for two crop types, as a proof of concept for the possibility and usefulness of separating these two types of parameters using multi-spectral remote sensing data. As structural and biophysical parameters are both needed for accurate estimation of terrestrial carbon and water cycles, we suggest that synergetic retrievals of these two types of parameters be a next step in vegetation remote sensing methodological development. For the convenience of discussion, we focus on the canopy structural parameter LAI and the leaf biochemical parameter LCC in this chapter.

23.2 Synergetic Retrievals of Both Canopy Structural and Leaf Biochemical Parameters

The contributions of canopy structural and leaf biochemical parameters to optical reflectance vary greatly with wavelength (λ) because leaf reflectance and transmittance spectra vary greatly with λ due to variable spectral absorption by leaf constituents (pigments, water content, dry mass, etc.). The variations of structural parameter LAI and biochemical parameter LCC have similar but somewhat different effects on canopy spectral reflectance, as demonstrated in Fig. 23.1, which is simulated using the 5-Scale geometric optical model (Chen & Leblanc, 1997, 2001). As green leaves absorb strongly in blue (400–500 nm), red (650–680 nm) and shortwave infrared (SWIR, 1300–2400 nm) and weakly in near-infrared wavelengths (NIR, 800–1300 nm), increasing LAI leads to decrease in blue, red and SWIR reflectance

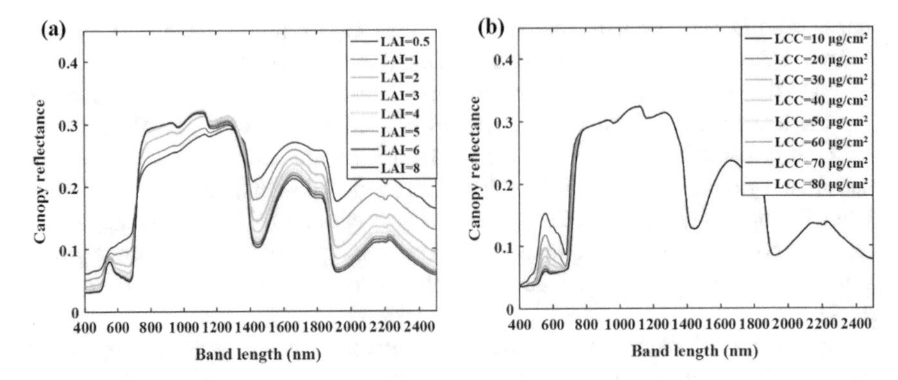

Fig. 23.1 Modelled canopy spectral reflectance using the 5-Scale model. **a** At different LAI values with fixed LCC of 40 $\mu g/cm^2$; **b** at different LCC values with fixed LAI of 3 m^2/m^2. The leaf reflectance and transmittance spectra used in the 5-Scale model were simulated using the PROSPECT-D leaf optical model, with fixed values of leaf structure parameter (1.5), total carotenoid content (10 $\mu g/cm^2$), total anthocyanin content (0), brown pigment content (0), LWC (0.021 g/cm^2) and dry matter content = 0.007 (g/cm^2). Other parameters of 5-Scale were fixed as: stand density = 1400 trees/ha, crown (spheroid) radius = 1.25 m, crown height = 8 m, stick height = 10 m, clumping index = 0.9, solar zenith angle = 30°, view zenith angle = 0°, and relative azimuth angle = 30°. A reflectance spectrum of soil was assumed as the background

and increase in NIR reflectance, as shown in Fig. 23.1a, which is simulated for an average LCC of 40 $\mu g/cm^2$ with other leaf parameters being fixed. The variation of LCC has similar effects on canopy reflectance in the visible part of the spectrum, e.g., as LCC increases, both blue and red reflectances decrease (Fig. 23.1b). NIR and SWIR reflectances are not directly affected by LCC but would also vary with LCC if LCC is correlated with other leaf parameters such as LNC, LWC, leaf dry matter, etc., but these correlations are not considered in Fig. 23.1.

It can be easily inferred from Fig. 23.1 that the retrieval of LAI cannot be independent of LCC because LCC would have strong effects on the red reflectance used in LAI retrieval. Conversely, LCC cannot be reliably retrieved without considering LAI as LAI would have the dominant contribution to canopy reflectance.

23.2.1 Major Issues in LAI Retrieval

Global LAI products have so far been exclusively derived from passive shortwave remote sensing data, although the usefulness of active sensors has also been explored (Manninen et al., 2013; Tang et al., 2014). The basic passive spectral information useful for LAI retrieval is in red and NIR bands, although SWIR bands are sometimes used to derive LAI (Brown et al., 2000; Chen et al., 2002). Empirical LAI algorithms are developed based on correlations of field-measured LAI with two-band (red, NIR) or three-band (red, NIR, SWIR) vegetation indices (VIs) (Nemani et al., 1993). Model-based LAI algorithms are generally developed with assumed average leaf spectral reflectance representing the average leaf physiological conditions (pigments, LWC, LNC, etc.) (Deng et al., 2006; Myneni et al., 2002). Both types of LAI algorithms do not consider the influence of LCC variation. Based on the discussion above and previous studies (Gobron et al., 1997; Myneni et al., 2002; Wang et al., 2017), we summarize the following major issues in the current LAI retrieval methodology:

1. Leaf biochemical conditions are not explicitly considered. Since these conditions vary spatially and seasonally, such as the large seasonal variation in LCC, we expect large distortions in the seasonal and spatial variations of retrieved LAI;
2. Vegetation background is not explicitly considered. Many algorithms assume soil as the background, but the background could consist of understory, litter, and moss in forests and sometimes snow. Since the optical properties of these background types differ greatly from soil, the retrieved LAI based on the contrast between vegetation and soil would be in large error (Pisek et al., 2010).
3. Random leaf spatial distributions within plant canopies are often assumed, although leaves in forests, shrubs and row crops are clumped, causing considerable underestimation of retrieved LAI (Garrigues et al., 2008).

While the first issue has not been addressed in any regional and global LAI algorithms and deserves our further attention, the second and the third issue have been considered in some LAI algorithms. For example, multiple angle data have been used to retrieve forest background regionally (Pisek et al., 2010) and globally (Jiao et al., 2014) and used for LAI retrieval. Clumping index has also been derived using multi-angle remote sensing data at the global scale (Chen et al., 2005; He et al., 2012; Jiao et al., 2016; Wei et al., 2019), and CI has been used to convert effective LAI obtained from mono-angle remote sensing data to the true LAI (Deng et al., 2006; Xiao et al., 2016). The MODIS LAI algorithm also considered the clumping index simulated using a radiative transfer model (Knyazikhin et al., 1998). Therefore, the major outstanding issue in LAI retrieval is how to overcome the influence of leaf biochemical parameters on spectral signals used for deriving LAI.

23.2.2 Major Issues in LCC Retrieval

During the growing season, spectral reflectance observed from healthy plant canopies is dominated by LAI while LCC modifies the reflectance in some spectral ranges. In order to retrieve LCC, the dominant influence of LAI needs to be fully considered. The major issues in LCC retrieval are:

1. LAI used in LCC algorithms is not accurate. Since LAI retrieval algorithms have not considered the influence of LCC, it is in considerable error especially when LCC is low at the beginning and end of the growing season;

2. The influence of other pigments in leaves on canopy spectral reflectance is difficult to consider in LCC algorithms with often insufficient spectral information;
3. Vegetation background (soil, litter, moss, grass, etc.) also interferes with the retrieval of LCC because the chlorophyll absorption over the visible spectrum could be mixed with spectral absorption by the background. The effect of background is particularly pronounced for sparse vegetation (e.g., low LAI).

Although the first issue has yet to be addressed in future LCC algorithm development for global applications, LAI has been the key parameter considered in existing LCC algorithms. Xu et al. (2019) developed an LCC inversion method based on a two-dimensional matrix constructed with two VIs, one sensitive to LAI and the other sensitive to LCC. In this way, LCC is estimated separately within different ranges of LAI, and hence the influence of LAI on LCC derivation is minimized. This algorithm is developed and validated for a cropland in China. Croft et al. (2020) developed a two-step model algorithm for global LCC retrieval. In the first step, a geometrical optical model 5-Scale (Chen & Leblanc, 2001) for forests and shrubs and the SAIL model (Verhoef, 1984) for crops and grasses are used to invert canopy-level reflectance to sunlit leaf reflectance based on the inputs of LAI and sun-target-sensor observation geometry. In the second step, a leaf radiative transfer model (PROSPECT) (Féret et al., 2017) is used to derive LCC from sunlit leaf reflectance. LAI therefore plays a critical role in this algorithm. Li et al. (2020) formulated an LAI-Insensitive Chlorophyll Index (LICI) for retrieving LCC for crops. It is demonstrated that LICI can greatly suppress the influence of LAI ranging from 1 to 6 on LCC retrieval. Such a method is simple, but its generality for other crops and other vegetation types is yet to be demonstrated. There are also empirical studies that relate VIs to either LCC (Croft et al., 2014; le Maire et al., 2008; Wu et al., 2008;) or canopy chlorophyll content (Gitelson et al., 2005; Inoue et al., 2016; Li et al., 2017). The empirical relationships developed without explicit consideration of the influence of LAI in these studies may be site and plant type specific and may not be reliable for regional and global applications.

There have been some studies on the second issue. Brown pigments in leaves have spectral absorption partly overlapping with that of LCC and were found to interfere with LCC retrieval (Houborg et al., 2009). The new versions of the PROSPECT model (Féret et al., 2017) are capable of simulating the effect of brown pigments on leaf spectral reflectance and would be a useful tool for developing LCC algorithms with consideration of the influence of brown and other pigments (Jiang et al., 2018).

The third issue is not yet systematically addressed in existing studies, although the global forest background dataset retrieved from Multi-angle Imaging SpectroRadiometer (MISR) (Jiao et al., 2014) would be useful for addressing this issue.

23.2.3 Synergetic Retrievals of LAI and LCC

From the discussion above, it becomes apparent that the current LAI and LCC algorithms developed separately cannot avoid the issue of the mutual influence of LAI and LCC on the canopy spectral reflectance used for their retrievals. This issue can cause large errors in retrieved LAI and LCC, in particular when LAI is low or when the seasonal variation of LCC is large. To avoid this issue, we therefore propose to retrieve LAI and LCC synergistically and simultaneously, as an important next step in vegetation remote sensing aiming at improving terrestrial carbon and water cycle estimation.

There are several ways to achieve synergetic retrievals of LAI and LCC. Some of them are:

1. To develop algorithms in the form of lookup tables that simultaneously consider the influences of LAI and LCC on canopy reflectance in various spectral bands (Houborg et al., 2015). Typically blue, red and NIR bands are influenced by both LAI and LCC, while red edge bands are more influenced by LCC than LAI (Gitelson et al., 2005; Jacquemoud, 1993). Green bands are influenced more by other pigments than by LCC and LAI, and SWIR bands are mostly sensitive to LAI and LWC (Cheng et al., 2014; Hunt et al., 2011). Because responses to LAI and LCC have large and subtle differences across the shortwave spectral range, it is possible to separate these two parameters with a set of spectral bands. Ideally, we need all blue, green, red, red edge, NIR and SWIR bands to retrieve both parameters simultaneously. However, the interactions among the signals from these spectral bands are complex, so we need an advanced canopy radiative transfer model to simulate the influence of LAI and LCC on the reflectance in these spectral bands. A model for this purpose would need to have the capability of considering the complex canopy structure and multiple scattering of radiation in plant canopies. DART (Gastellu-Etchegorry et al., 2004), 5-Scale (Chen & Leblanc, 1997, 2001) and GOST (Fan et al., 2014) are some of the models suitable for this purpose. These models can be used to construct large look-up tables with the mechanism to iteratively calculate LAI and LCC or to construct their cost functions in different bands because both have a strong influence on red and NIR reflectances in particular.
2. To develop machine learning methods to handle the complex interactions among different spectral bands over wide ranges of LAI and LCC. Advanced radiative transfer models are also needed to produce outputs as the basis for machine learning (Ali et al., 2021). Such machine learning for rather complex radiative transfer processes involving multiple canopy structural and leaf biochemical attributes would be a daunting task but would be possible if it is guided by physical principles in steps and the outcomes in each step have clear physical meanings.
3. To develop semi-empirical algorithms to estimate LAI and LCC either separately or simultaneously. The two-dimensional matrix approach (Xu et al., 2019) developed to retrieve LCC for crops has the potential for retrieving both LCC

and LAI simultaneously and could be expanded to consider other vegetation types. As concurrent empirical data for LAI and LCC are rather limited, the relationships of LAI and LCC with different VIs can be established with the aid of advanced canopy radiative transfer models. It is also possible to find some mathematical combinations of several spectral bands that are insensitive to LAI in LCC retrieval, such as LICI (Li et al., 2020), or insensitive to LCC in LAI retrieval (yet to be developed).

In all three types of methods mentioned above, the use of an advanced radiative transfer model is essential. However, models, although complex and sophisticated, are only abstract representations of reality and often produce results with unknown errors. It is therefore rather important to compile concurrent LAI and LCC field measurements for model validation purposes. Through the effort of the remote sensing community, a fairly large global LAI dataset is now available (Camacho et al., 2013), but the concurrent LAI and LCC dataset is rather small (Croft et al., 2020). We therefore need to make a concerted global effort in collecting and compiling concurrent LAI and LCC data, since we have demonstrated the feasibility in retrieving LCC globally using remote sensing data and its usefulness in carbon and water cycle studies.

23.3 Tradeoff of Canopy Structural and Leaf Biochemical Parameters in Terrestrial Ecosystem Models

The contributions of canopy structural and leaf biophysical parameters to canopy-level photosynthesis, quantified with gross primary productivity (GPP), can be simulated using a process-based ecosystem model. Figure 23.2 shows the simulated GPP and ET results at different LAI and LCC values using the Boreal Ecosystem Productivity Simulator (BEPS) (Chen et al., 1999, 2012; Ju et al., 2006). GPP increases with LAI due to increased PAR absorption by the canopy, while GPP also increases with LCC because LCC is positively related to the maximum leaf carboxylation rate. We understand that the errors in LAI retrieval would be positive without considering LCC if LCC is larger than the average, so the ultimate question is: would the positively biased LAI compensate for the ignorance of higher LCC in GPP estimation? Conversely, LCC retrievals would be overestimated if LAI used in the retrieval algorithm is underestimated, so the ultimate question would also be: would the increased LCC compensate for underestimated LAI in GPP estimation? It is necessary to ask these questions because if LAI and LCC mutually compensate each other in GPP estimation, we can treat the traditional LAI retrieval as a surrogate of LAI and LCC without the need to separate them (this is implicitly our current practice). In other words, are we troubling ourselves in separating LAI and LCC for something insignificant?

It has been demonstrated that in GPP estimation for plant canopies, sunlit and shaded leaves need to be separated using two-leaf models (Guan et al., 2021; Luo

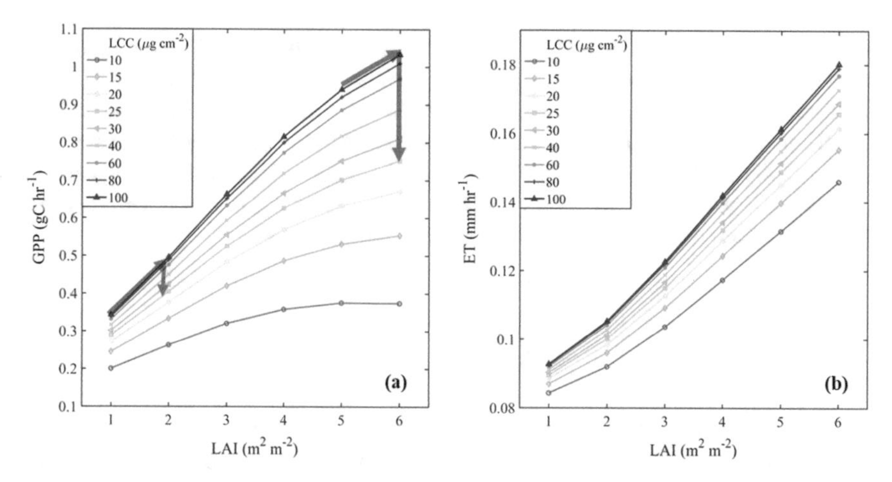

Fig. 23.2 Gross primary productivity (GPP) and evapotranspiration (ET) of a conifer forest modelled by BEPS at different LAI and LCC values, with CI = 0.66, Ta = 20 °C, shortwave radiation of 1000 W m^{-2} at a solar zenith angle of 30°, soil moisture at 60% of field capacity and Vcmax = 1.3 + 3.72LCC (in μmol m^{-2} s^{-1}) (Luo et al., 2019). **a** GPP variation with LAI at different LCC values. The arrows indicate the relative impacts of LAI and LCC on GPP at low (1–2) and high (5–6) LAI values; **b** ET variation with LAI at different LCC values

et al., 2018; Sprintsin et al., 2012), because these two leaf groups are controlled by different biological processes (i.e., limitations by carboxylation rate and electron transport rate for sunlit and shaded leaves, respectively) and have different light use efficiencies. The structural parameters LAI and CI as well as solar zenith angle are needed to separate sunlit and shaded LAI, while LCC can be used to determine the maximum carboxylation rate (Vcmax) in process-based models (e.g. Farquhar et al., 1980). As sunlit leaves have much larger photosynthesis rates than shaded leaves, the sunlit LAI has larger contributions than shaded LAI to the total canopy GPP, globally about 60% (Chen et al., 2012). However, as LAI increases, sunlit LAI reaches an asymptote not exceeding 2, so at high LAI, Vcmax, that can be determined by LCC (Croft et al., 2017; Lu et al., 2020), plays a critical role in determining the level of canopy GPP. Therefore, the increase in LAI would not compensate for the decrease in LCC in GPP estimation, especially at high LAI values (see the arrows at LAI = 5–6, Fig. 23.2a). Conversely, at low LAI values, most leaves are sunlit, so the amount of sunlit LAI is critical in determining the canopy-level GPP, and the increase in LCC would compensate slightly for the decrease in LAI in GPP estimation (see also the arrows at LAI = 1–2), as the radiation absorption has the first order importance while the carboxylation rate that determines the utilization of absorbed radiation may be regarded as the second order.

The above simple analysis qualitatively answers the questions related to mutual compensation of LAI and LCC in GPP estimation, leading to the conclusion that for accurate GPP estimation it is necessary to use both LAI and LCC inputs in ecosystem models because they have different functions: LAI for radiation absorption and distribution in the canopy and LCC for leaf carboxylation rate. The same principle would apply to ET estimation (Fig. 23.2b) because LAI would similarly determine the available energy absorbed by the canopy for vaporization and LCC would have effects on stomatal conductance that is proportional to the photosynthesis rate (Ball et al., 1987) that depends on LCC (Luo et al., 2019). We therefore suggest that synergistic retrievals of LAI and LCC be a next step of vegetation remote sensing for improving terrestrial carbon and water cycle estimation.

23.4 Summary

In this chapter, we show how canopy structural parameter LAI and leaf biochemical parameter LCC influence canopy reflectance and the usefulness to retrieve these two parameters synergetically using multi-spectral remote sensing data. Synergetic retrievals of both LAI and LCC can avoid the issues in current LAI and LCC algorithms that do not consider the influence of one parameter on canopy reflectance used to retrieve the other parameters and can be a next step in vegetation remote sensing for the purpose of improving regional and global carbon and water cycle estimation.

Acknowledgements J. M. Chen acknowledges a discovery grant (RGPIN-2020-05163) and Research Tools and Instrument grant (RTI-2021-00606) from Natural Science and Engineering Council of Canada in support of this research.

References

Ali, A. M., Darvishzadeh, R., Skidmore, A., Gara, T. W., & Heurich, M. (2021). Machine learning methods' performance in radiative transfer model inversion to retrieve plant traits from Sentinel-2 data of a mixed mountain forest. *International Journal of Digital Earth, 14*, 106–120.

Ball, J. T., Woodrow, I. E., & Berry, J. A. (1987). A model predicting Stomatal conductance and its contribution to the control of photosynthesis under different environmental conditions. In: Biggins, J. (Ed.), *Progress in Photosynthesis Research: Volume 4 Proceedings of the VIIth International Congress on Photosynthesis, Providence, Rhode Island, USA* (pp. 221–224), August 10–15, 1986. Springer Netherlands, Dordrecht.

Baret, F., Jacquemoud, S., Guyot, G., & Leprieur, C. (1992). Modeled analysis of the biophysical nature of spectral shifts and comparison with information content of broad bands. *Remote Sensing of Environment, 41*, 133–142.

Brown, L., Chen, J. M., Leblanc, S. G., & Cihlar, J. (2000). A shortwave infrared modification to the simple ratio for LAI retrieval in boreal forests: An image and model analysis. *Remote Sensing of Environment, 71*, 16–25.

Camacho, F., Cernicharo, J., Lacaze, R., Baret, F., & Weiss, M. (2013). GEOV1: LAI, FAPAR essential climate variables and FCOVER global time series capitalizing over existing products. Part 2: Validation and intercomparison with reference products. *Remote Sensing of Environment, 137*, 310–329.

Chen, J. M., & Black, T. A. (1992). Defining leaf area index for non-flat leaves. *Plant, Cell and Environment, 15*, 421–429.

Chen, J. M., Ju, W., Ciais, P., Viovy, N., Liu, R., Liu, Y., & Lu, X. (2019). Vegetation structural change since 1981 significantly enhanced the terrestrial carbon sink. *Nature Communications, 10*, 1–7.

Chen, J. M., & Leblanc, S. G. (1997). A four-scale bidirectional reflectance model based on canopy architecture. *IEEE Transactions on Geoscience and Remote Sensing, 35*, 1316–1337.

Chen, J. M., & Leblanc, S. G. (2001). Multiple-scattering scheme useful for geometric optical modeling. *IEEE Transactions on Geoscience and Remote Sensing, 39*, 1061–1071.

Chen, J. M., Liu, J., Cihlar, J., & Goulden, M. (1999). Daily canopy photosynthesis model through temporal and spatial scaling for remote sensing applications. *Ecological Modelling, 124*, 99–119.

Chen, J. M., Menges, C. H., & Leblanc, S. G. (2005). Global mapping of foliage clumping index using multi-angular satellite data. *Remote Sensing of Environment, 97*, 447–457.

Chen, J. M., Mo, G., Pisek, J., Liu, J., Deng, F., Ishizawa, M., & Chan, D. (2012). Effects of foliage clumping on the estimation of global terrestrial gross primary productivity. *Global Biogeochemical Cycles, 26*(1). https://doi.org/10.1029/2010GB003996

Chen, J. M., Pavlic, G., Brown, L., Cihlar, J., Leblanc, S. G., White, H. P., et al. (2002). Derivation and validation of Canada-wide coarse-resolution leaf area index maps using high-resolution satellite imagery and ground measurements. *Remote Sensing of Environment, 80*, 165–184.

Cheng, T., Riaño, D., & Ustin, S. L. (2014). Detecting diurnal and seasonal variation in canopy water content of nut tree orchards from airborne imaging spectroscopy data using continuous wavelet analysis. *Remote Sensing of Environment, 143*, 39–53.

Croft, H., Chen, J. M., Luo, X., Bartlett, P., Chen, B., & Staebler, R. M. (2017). Leaf chlorophyll content as a proxy for leaf photosynthetic capacity. *Global Change Biology, 23*, 3513–3524.

Croft, H., Chen, J. M., Wang, R., Mo, G., Luo, S., Luo, X., et al. (2020). The global distribution of leaf chlorophyll content. *Remote Sensing of Environment, 236*, 111479.

Croft, H., Chen, J. M., & Zhang, Y. (2014). The applicability of empirical vegetation indices for determining leaf chlorophyll content over different leaf and canopy structures. *Ecological Complexity, 17*, 119–130.

Demmig-Adams, B., & Adams, W. W. (1992). Photoprotection and other responses of plants to high light stress. *Annual Review of Plant Physiology and Plant Molecular Biology, 43*, 599–626.

Deng, F., Chen, J. M., Plummer, S., Mingzhen, C., & Pisek, J. (2006). Algorithm for global leaf area index retrieval using satellite imagery. *IEEE Transactions on Geoscience and Remote Sensing, 44*, 2219–2229.

Fan, W., Chen, J. M., Ju, W., & Zhu, G. (2014). GOST: A geometric-optical model for sloping terrains. *IEEE Transactions on Geoscience and Remote Sensing, 52*, 5469–5482.

Farquhar, G. D., Caemmerer, S. V., & Berry, J. A. (1980). A biochemical model of photosynthetic CO_2 assimilation in leaves of C_3 species. *Planta, 149*, 78–90.

Féret, J.-B., Gitelson, A. A., Noble, S. D., & Jacquemoud, S. (2017). PROSPECT-D: Towards modeling leaf optical properties through a complete lifecycle. *Remote Sensing of Environment, 193*, 204–215.

Friedl, M. A., Davis, F. W., Michaelsen, J., & Moritz, M. A. (1995). Scaling and uncertainty in the relationship between the NDVI and land surface biophysical variables: An analysis using a scene simulation model and data from FIFE. *Remote Sensing of Environment, 54*, 233–246.

Garrigues, S, Lacaze, R, Baret, F, Morisette, J. T., Weiss, M., Nickeson, J. E., et al. (2008). Validation and intercomparison of global Leaf Area Index products derived from remote sensing data. *Journal of Geophysical Research: Biogeosciences, 113*, G02028.

Gastellu-Etchegorry, J. P., Martin, E., & Gascon, F. (2004). DART: A 3D model for simulating satellite images and studying surface radiation budget. *International Journal of Remote Sensing, 25*, 73–96.

Gitelson, A. A. (2019). Remote estimation of fraction of radiation absorbed by photosynthetically active vegetation: Generic algorithm for maize and soybean. *Remote Sensing Letters, 10*, 283–291.

Gitelson, A. A., Viña, A., Ciganda, V., Rundquist, D. C., & Arkebauer, T. J. (2005). Remote estimation of canopy chlorophyll content in crops. *Geophysical Research Letters, 32*, L08403.

Gobron, N., Pinty, B., Mélin, F., Taberner, M., Verstraete, M. M., Robustelli, M., & Widlowski, J. L. (2007). Evaluation of the MERIS/ENVISAT FAPAR product. *Advances in Space Research, 39*, 105–115.

Gobron, N., Pinty, B., & Verstraete, M. M. (1997). Theoretical limits to the estimation of the leaf area index on the basis of visible and near-infrared remote sensing data. *IEEE Transactions on Geoscience and Remote Sensing, 35*, 1438–1445.

Guan, X., Chen, J. M., Shen, H., & Xie, X. (2021). A modified two-leaf light use efficiency model for improving the simulation of GPP using a radiation scalar. *Agricultural and Forest Meteorology, 307*, 108546.

He, L., Chen, J. M., Pisek, J., Schaaf, C. B., & Strahler, A. H. (2012). Global clumping index map derived from the MODIS BRDF product. *Remote Sensing of Environment, 119*, 118–130.

Houborg, R., Anderson, M., & Daughtry, C. (2009). Utility of an image-based canopy reflectance modeling tool for remote estimation of LAI and leaf chlorophyll content at the field scale. *Remote Sensing of Environment, 113*, 259–274.

Houborg, R., McCabe, M., Cescatti, A., Gao, F., Schull, M., & Gitelson, A. (2015). Joint leaf chlorophyll content and leaf area index retrieval from Landsat data using a regularized model inversion system (REGFLEC). *Remote Sensing of Environment, 159*, 203–221.

Hunt, E. R., Li, L., Yilmaz, M. T., & Jackson, T. J. (2011). Comparison of vegetation water contents derived from shortwave-infrared and passive-microwave sensors over central Iowa. *Remote Sensing of Environment, 115*, 2376–2383.

Inoue, Y., Guerif, M., Baret, F., Skidmore, A., Gitelson, A., Schlerf, M., et al. (2016). Simple and robust methods for remote sensing of canopy chlorophyll content: A comparative analysis of hyperspectral data for different types of vegetation. *Plant, Cell and Environment, 39*, 2609–2623.

Jacquemoud, S. (1993). Inversion of the PROSPECT + SAIL canopy reflectance model from AVIRIS equivalent spectra: Theoretical study. *Remote Sensing of Environment, 44*, 281–292.

Jiang, J., Comar, A., Burger, P., Bancal, P., Weiss, M., & Baret, F. (2018). Estimation of leaf traits from reflectance measurements: Comparison between methods based on vegetation indices and several versions of the PROSPECT model. *Plant Methods, 14*. https://doi.org/10.1186/s13007-018-0291-x

Jiao, T., Liu, R., Liu, Y., Pisek, J., & Chen, J. M. (2014). Mapping global seasonal forest background reflectivity with Multi-angle Imaging Spectroradiometer data. *Journal of Geophysical Research: Biogeosciences, 119*, 1063–1077.

Jiao, Z., Schaaf, C. B., Dong, Y., Román, M., Hill, M. J., Chen, J. M., et al. (2016). A method for improving hotspot directional signatures in BRDF models used for MODIS. *Remote Sensing of Environment, 186*, 135–151.

Ju, W., Chen, J. M., Black, T. A., Barr, A. G., Liu, J., & Chen, B. (2006). Modelling multi-year coupled carbon and water fluxes in a boreal aspen forest. *Agricultural and Forest Meteorology, 140*, 136–151.

Knyazikhin, Y., Martonchik, J. V., Myneni, R. B., Diner, D. J., & Running, S. W. (1998). Synergistic algorithm for estimating vegetation canopy leaf area index and fraction of absorbed photosynthetically active radiation from MODIS and MISR data. *Journal of Geophysical Research: Atmospheres, 103*, 32257–32275.

le Maire, G., François, C., Soudani, K., Berveiller, D., Pontailler, J.-Y., Bréda, N., et al. (2008). Calibration and validation of hyperspectral indices for the estimation of broadleaved forest leaf chlorophyll content, leaf mass per area, leaf area index and leaf canopy biomass. In *Remote Sensing of Environment* (pp. 3846–3864)

Li, D., Chen, J. M., Zhang, X., Yan, Y., Zhu, J., Zheng, H., et al. (2020). Improved estimation of leaf chlorophyll content of row crops from canopy reflectance spectra through minimizing canopy structural effects and optimizing off-noon observation time. *Remote Sensing of Environment, 248*, 111985.

Li, D., Cheng, T., Zhou, K., Zheng, H., Yao, X., Tian, Y., et al. (2017). WREP: A wavelet-based technique for extracting the red edge position from reflectance spectra for estimating leaf and canopy chlorophyll contents of cereal crops. *ISPRS Journal of Photogrammetry and Remote Sensing, 129*, 103–117.

Liu, Y, Liu, R, & Chen, J M. 2012. Retrospective retrieval of long-term consistent global leaf area index (1981–2011) from combined AVHRR and MODIS data. *Journal of Geophysical Research: Biogeosciences, 117*, G04003.

Lu, X., Ju, W., Li, J., Croft, H., Chen, J. M., Luo, Y., et al. (2020). Maximum carboxylation rate estimation with chlorophyll content as a proxy of Rubisco content. *Journal of Geophysical Research: Biogeosciences, 125*, e2020JG005748.

Luo, X., Chen, J. M., Liu, J., Black, T. A., Croft, H., Staebler, R., et al. (2018). Comparison of big-leaf, two-big-leaf, and two-leaf upscaling schemes for evapotranspiration estimation using coupled carbon-water modeling. *Journal of Geophysical Research: Biogeosciences, 123*, 207–225.

Luo, X., Croft, H., Chen, J. M., He, L., & Keenan, T. F. (2019). Improved estimates of global terrestrial photosynthesis using information on leaf chlorophyll content. *Global Change Biology, 25*, 2499–2514.

Manninen, T., Stenberg, P., Rautiainen, M., & Voipio, P. (2013). Leaf area index estimation of boreal and subarctic forests using VV/HH ENVISAT/ASAR data of various swaths. *IEEE Transactions on Geoscience and Remote Sensing, 51*(7), 3899–3909. https://doi.org/10.1109/TGRS.2012.2227327

Myneni, R. B., Hoffman, S., Knyazikhin, Y., Privette, J. L., Glassy, J., Tian, Y., et al. (2002). Global products of vegetation leaf area and fraction absorbed PAR from year one of MODIS data. *Remote Sensing of Environment, 83*, 214–231.

Nemani, R., Pierce, L., Running, S., & Band, L. (1993). Forest ecosystem processes at the watershed scale: Sensitivity to remotely-sensed leaf area index estimates. *International Journal of Remote Sensing, 14*, 2519–2534.

Pisek, J, Chen, J M, Alikas, K, & Deng, F. 2010. Impacts of including forest understory brightness and foliage clumping information from multiangular measurements on leaf area index mapping over North America. *Journal of Geophysical Research: Biogeosciences, 115*. https://doi.org/10.1029/2009jg001138.

Prince, S. D., & Goward, S. N. (1995). Global primary production: A remote sensing approach. *Journal of Biogeography, 22*, 815–835.

Running, S. W., Nemani, R. R., Peterson, D. L., Band, L. E., Potts, D. F., Pierce, L. L., & Spanner, M. A. (1989). Mapping regional forest evapotranspiration and photosynthesis by coupling satellite data with ecosystem simulation. *Ecology, 70*, 1090–1101.

Sprintsin, M., Chen, J. M., Desai, A., & Gough, C. M. (2012). Evaluation of leaf-to-canopy upscaling methodologies against carbon flux data in North America. *Journal of Geophysical Research: Biogeosciences, 117*. https://doi.org/10.1029/2010JG001407.

Tang, H., Dubayah, R., Brolly, M., Ganguly, S., & Zhang, G. (2014). Large-scale retrieval of leaf area index and vertical foliage profile from the spaceborne waveform lidar (GLAS/ICESat). *Remote Sensing of Environment, 154*, 8–18. https://doi.org/10.1016/j.rse.2014.08.007

Verger, A., Baret, F., Weiss, M., Filella, I., & Peñuelas, J. (2015). GEOCLIM: A global climatology of LAI, FAPAR, and FCOVER from VEGETATION observations for 1999–2010. *Remote Sensing of Environment, 166*, 126–137.

Verhoef, W. (1984). Light scattering by leaf layers with application to canopy reflectance modeling: The SAIL model. *Remote Sensing of Environment, 16*, 125–141.

Wang, R., Chen, J. M., Liu, Z., & Arain, A. (2017). Evaluation of seasonal variations of remotely sensed leaf area index over five evergreen coniferous forests. *ISPRS Journal of Photogrammetry and Remote Sensing, 130*, 187–201.

Wei, S., Fang, H., Schaaf, C. B., He, L., & Chen, J. M. (2019). Global 500 m clumping index product derived from MODIS BRDF data (2001–2017). *Remote Sensing of Environment, 232*, 111296.

Wu, C., Niu, Z., Tang, Q., & Huang, W. (2008). Estimating chlorophyll content from hyperspectral vegetation indices: Modeling and validation. *Agricultural and Forest Meteorology, 148*, 1230–1241.

Xiao, J., Chevallier, F., Gomez, C., Guanter, L., Hicke, J. A., Huete, A. R., et al. (2019). Remote sensing of the terrestrial carbon cycle: A review of advances over 50 years. *Remote Sensing of Environment, 233*, 111383.

Xiao, Z., Liang, S., Wang, J., Xiang, Y., Zhao, X., & Song, J. (2016). Long-time-series global land surface satellite leaf area index product derived from MODIS and AVHRR surface reflectance. *IEEE Transactions on Geoscience and Remote Sensing, 54*, 5301–5318.

Xu, C., Fisher, R., Wullschleger, S. D., Wilson, C. J., Cai, M., & McDowell, N. G. (2012). Toward a mechanistic modeling of nitrogen limitation on vegetation dynamics. *PLoS ONE, 7*, e37914.

Xu, M., Liu, R., Chen, J. M., Liu, Y., Shang, R., Ju, W., et al. (2019). Retrieving leaf chlorophyll content using a matrix-based vegetation index combination approach. *Remote Sensing of Environment, 224*, 60–73.

Yan, H., Wang, S. Q., Billesbach, D., Oechel, W., Zhang, J. H., Meyers, T., et al. (2012). Global estimation of evapotranspiration using a leaf area index-based surface energy and water balance model. *Remote Sensing of Environment, 124*, 581–595.

Zhao, M., Heinsch, F. A., Nemani, R. R., & Running, S. W. (2005). Improvements of the MODIS terrestrial gross and net primary production global data set. *Remote Sensing of Environment, 95*, 164–176.

Chapter 24
LiDAR Remote Sensing of Forest Ecosystems: Applications and Prospects

Qinghua Guo, Xinlian Liang, Wenkai Li, Shichao Jin, Hongcan Guan, Kai Cheng, Yanjun Su, and Shengli Tao

Abstract The three-dimensional (3D) structure of forests has long been recognized to have profound effects on forest ecosystems. However, the use of spectral and radar remotely sensed data for forest structure quantification is insensitive to changes in forest vertical structure. LiDAR has emerged as a robust means to measure forest structures. Numerous studies have been devoted to accurately quantifying forest structures from LiDAR data at various scales (from tree branches level to global level) and revolutionized the way we consider forest structure in ecosystem studies.

Q. Guo (✉)
Institute of Remote Sensing and Geographic Information System, School of Earth and Space Sciences, Institute of Ecology, Peking University, Beijing 100871, China
e-mail: guo.qinghua@pku.edu.cn

X. Liang
State Key Laboratory of Information Engineering in Surveying, Mapping and Remote Sensing, Wuhan University, Wuhan 430072, China

W. Li
Guangdong Provincial Key Laboratory of Urbanization and Geo-simulation, School of Geography and Planning, Sun Yat-Sen University, Guangzhou 100029, China
e-mail: liwenk3@mail.sysu.edu.cn

S. Jin
Plant Phenomics Research Centre, Collaborative Innovation Centre for Modern Crop Production Co-sponsored By Province and Ministry, Academy for Advanced Interdisciplinary Studies, Nanjing Agricultural University, Nanjing 210095, China

H. Guan · K. Cheng
School of Urban and Environmental Sciences, Peking University, Beijing 100871, China
e-mail: chengk@lreis.ac.cn

Y. Su
State Key Laboratory of Vegetation and Environmental Change, Institute of Botany, Chinese Academy of Sciences, Beijing 100093, China
e-mail: ysu@ibcas.ac.cn

S. Tao
Institute of Ecology, College of Urban and Environmental Sciences, Key Laboratory for Earth Surface Processes of the Ministry of Education, Peking University, Beijing 100871, China
e-mail: sltao@pku.edu.cn

B. Li et al. (eds.), *New Thinking in GIScience*,
https://doi.org/10.1007/978-981-19-3816-0_24

In this chapter, we outline how LiDAR sheds light on forest ecosystem studies and discuss current challenges and perspectives of LiDAR applications.

Keywords LiDAR · Forest · Monitoring · Radiative transfer model · Ecological processes

24.1 Introduction

Forest structure has a major impact on ecosystem processes including forest growth, energy flux, carbon and nitrogen cycling, vulnerability to droughts and fires, and biodiversity. Over the last decades, capacities of forest structure mapping are continually growing to meet the ever-increasing needs of forest management. Previously, mapping of forest structures relied on multispectral, hyperspectral, and radar remote sensing. However, these sensors have limited ability to penetrate canopies and often encounter the well-known saturation problem (i.e., reflectance values are not sensitive to the increase of biomass in dense forests), thus cannot effectively measure vertical forest structure (Guo, Jin, et al., 2020a; Guo, Su, et al., 2020b).

LiDAR is an active remote sensing technology that emits laser beams for measuring distances to objects and uses a positioning system to convert distance measurements into three-dimensional (3D) coordinates (x, y, and z) in space (Guo et al., 2017). The emitted pulses from LiDAR sensors can penetrate canopies through gaps in the foliage. Thus, LiDAR enables measurements of both vertical and horizontal forest structures, revolutionizing the way we investigate canopy structure in forest ecosystem studies (Hu et al., 2016). During the past decades, numerous studies have demonstrated the advantages of LiDAR for quantifying forest structures (Guo, Jin, et al., 2020a; Guo, Su, et al., 2020b).

In this chapter, we review the advancement of LiDAR technologies and their applications in 3D forest observation. We also discuss current challenges and perspectives of such applications.

24.2 Evolution of 3D Forest Observation

LiDAR sensors can be mounted on various types of platforms, offering 3D forest observations at a wide range of spatial scales (Fig. 24.1). Static terrestrial laser scanning (TLS) captures the 3D structure of a forest understory at the millimeter level. It is suitable for scanning individual trees and small plots (<1 ha in size). However, the relatively high cost of TLS equipment (e.g., the cost of Rigel VZ-400i laser scanner\RIEGL Laser Measurement Systems GmbH is more than 100 K USD) puts this method out of reach for many potential users. Recent lightweight systems, e.g., Lipod (GreenValley Technology Co., Ltd), make TLS easier and more affordable than ever before. Most recent research successfully scanned 12-ha tropical rainforests

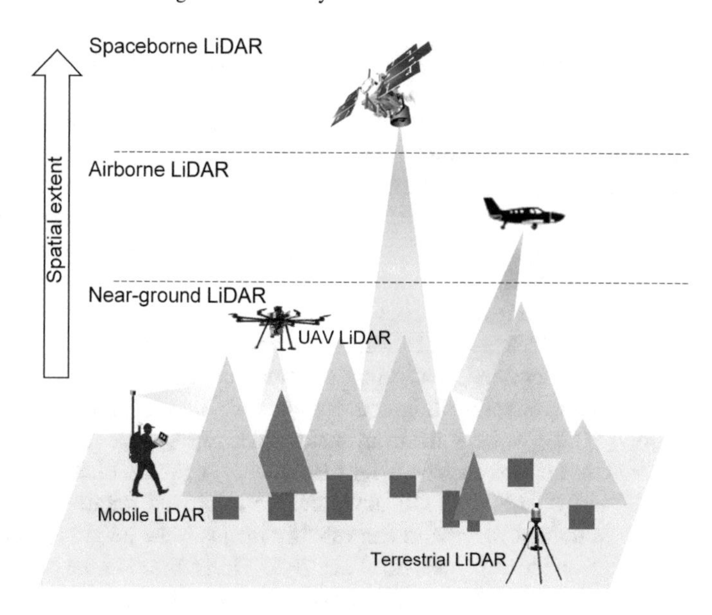

Fig. 24.1 LiDAR platforms used in forest studies

with a lightweight terrestrial laser scanner (Tao et al., 2021). Mobile systems (e.g., backpack LiDAR) are estimated to be at least 10-time faster than static techniques and are suitable for collecting data in large plots. Recent studies have shown that tree height and diameter at breast height (DBH) extracted by some commercial mobile systems (e.g., LiBackpack DGC50 system developed by GreenValley Technology Co., Ltd.) achieved equivalent accuracy to TLS (Su et al., 2021). Unmanned Aerial Vehicles (UAVs) outperform terrestrial mobile platforms in efficiency since their movement is unobstructed by ground obstacles (Liang et al., 2019). The current estimates (e.g., tree height and DBH) from high-end mobile platforms are comparable with those from static platforms in homogeneous forests with simple structures but do not yet meet the requirements of practical applications in heterogeneous forests (Liang, Hyyppä, et al., 2018a; Liang, Kukko, et al., 2018b).

Airborne laser scanning (ALS) systems, deployed on airplanes or helicopters, are often used for large-area forest observations. As ALS data are becoming increasingly available, it is possible to produce accurate wall-to-wall maps of forest structural and functional variables at regional and country-wide scales. For example, the U.S. Geological Survey (USGS) 3D Elevation Program (3DEP) was recently established to provide ALS data for the conterminous United States, Hawaii, and U.S. territories.

Spaceborne LiDAR systems have the longest ranging capability and can map forest structures globally. A series of space missions have been successfully implemented in recent years. For example, the Advanced Topographic Laser Altimeter System (ATLAS) on ICESat-2 and the Global Ecosystem Dynamics Investigation (GEDI) installed on the International Space Station (ISS) were both launched in 2018. With the continuous growth of spaceborne LiDAR data, it is possible to scale up local forest studies to regional and global scales by combining spaceborne, airborne, and UAV-borne LiDAR (Liu et al., 2022).

Each LiDAR platform has its own advantages and disadvantages. When choosing LiDAR techniques for forest ecology, one should consider the size of the studied area, the data quality expected, and the time and labor costs. For example, the low-cost UAV-borne laser scanning (ULS) is suitable for tree height and crown diameter measurements in forests with simple stand structures. In dense stands (ca. 2000 stems/ha or higher), TLS and ULS suit better for the estimation of tree DBH, height, volume, and biomass (Liang et al., 2016).

During the last decade, low-cost LiDAR systems have been the primary drivers of the boom of LiDAR systems in forestry applications. Challenges associated with low-cost LiDAR have been gradually mitigated. For example, the early handheld Personal Laser Scanning (PLS_{hh}) suffers from an insufficient measurement range (around 20 m) and introduces a bias in tree height estimates. A recent PLS_{hh} instrument, the ZEB-HORIZON (GeoSlam Ltd), integrates an improved ranging sensor with up to 100-m measurement range and can reliably measure the height of deciduous trees in a homogeneous forest (Jurjević et al., 2020). In addition to the performance improvement, the costs of LiDAR instruments have dropped dramatically. Recently, Da-Jiang Innovations (DJI) has released a LiDAR sensor called Livox MID40 priced at only 599 USD. Livox MID40 has been installed on the UAV for forest investigation (Hu et al., 2021).

Alternatively, point clouds can also be obtained with sensors such as cameras or smartphone cameras, but their applications in real forests are still rare. Pioneer studies used consumer cameras. More recently, systems such as Tango by Google and Azure Kinect by Microsoft were used to capture point clouds. They use optical and depth sensors to reconstruct surrounding environments into point clouds. For DBH estimation in forests, the accuracy (root mean square error, RMSE) was around 2 cm using Tango (Hyyppä et al., 2018). Azure Kinect was reported to have an RMSE of 8.43 cm for 51 urban trees (McGlade et al., 2020). ARKit and ARCore augmented reality (AR) projects were introduced by Apple and Google, respectively. Compared to Tango, both ARKit and ARCore do not require depth sensors, allowing a greater number of devices to use such AR solutions, with lower accuracy than Tango. The challenge with the AR solution is that direct sunlight can cause failures in data acquisitions, especially for infra-red sensors. Dense forests with full crown coverage are therefore more suited for data acquisition, reducing the likelihood of direct sunlight exposure of the camera. Another significant limitation is the short functional distance. The newest version of ARKit supports the LiDAR that is implemented on iPad Pro 2020/iPhone 12 pro (max)/iPhone 13 pro(max). LiDAR provides depth information; this may solve the problem associated with the short functional distance. The practical advantage is that the operator can see the resulting point cloud right away in the field and adjust the data acquisition according to realtime environmental conditions. In the future, these techniques may compete with both TLS and mobile LiDAR systems in forest inventories.

Another notable advancement in hardware in the last decade is the rapid development of multi- and hyperspectral LiDAR in 3D forest observations. It combines 3D and spectral information at two or more wavelengths. While the multi- and hyperspectral LiDAR prototypes are still being developed in laboratories, they are expected to open up new opportunities for, e.g., the separation of tree species and monitoring pest and disease. Currently, the utilization of multi- and hyperspectral LiDAR in forests is still at a very early stage. Two main challenges must be solved before they can be widely utilized. The first is the selection of optimal wavelengths for practical applications (e.g., their number and required spectral resolution). The second is the lack of applicable calibration solutions, especially for incidence angles and range effects.

24.3 Beyond 3D: New Spectrum of LiDAR Applications in Forest Ecosystem Studies

24.3.1 Application of LiDAR Structural, Temporal, and Spectral Information in Forest Ecosystem Studies

LiDAR transforms how we map forest ecosystems, thanks to its ability to provide accurate structural information of forests and their surrounding terrains. Researchers relate forest structural information to processes such as the evolutionary strategy of tree architecture (Su et al., 2020), the relationship between forest growth and stand structure (Forrester, 2019), and the interconnectivity of tree size and stand dynamics (West et al., 2009). Such studies typically use LiDAR-based forest traits like canopy height, canopy cover, vertical foliage profile, and undercanopy topography to achieve a quantitative depiction of forest 3D structure. It's particularly worth mentioning that LiDAR-based forest traits are excellent predictors of forest biomass. With statistical techniques such as linear regression and machine learning, LiDAR-based forest traits have enabled accurate mapping of forest biomass across a range of forest types (Coops et al., 2021).

The complexity of forest structure determines the quality of habitat available for species to occupy. Structural information from LiDAR has also contributed to substantial advances in wildlife behaviors, habitat modeling and biodiversity prediction. Most of these studies have been focused on flying animals (e.g., birds and bats) (Davies & Asner, 2014). Recently, the importance of topographical metrics is gradually being noticed by researchers for nonflying animals. For example, airborne LiDAR data revealed that fishers in the Southern Sierra Nevada Mountains in California prefer nesting sites on steeper slopes (Zhao et al., 2012).

A growing number of studies have used temporal and multi-band LiDAR to understand and monitor forest ecosystems. For instance, multi-temporal LiDAR metrics have opened up new opportunities for mapping dynamics of forest biomass and carbon stocks, which deepened the understanding of how forests are responding to rapid global changes. In phenology studies, repeated TLS data acquisition has provided an unprecedented opportunity to explore the phenological differences between understory and canopy trees, among trees with different functions, and between canopy dominants and emergent trees. Moreover, spectral information provided by multi-band LiDAR enables much more comprehensive and accurate descriptions of the physiochemical properties of leaves, organs, and trees, providing a better quantification of the distribution of matter and energy within trees, and a better understanding of the interaction between forest structure and function.

24.3.2 Linking the Forest Structure Information with Radiative Transfer Models and Ecological Processes

In addition to deriving metrics for forest ecosystem studies, LiDAR data can also be fed to radiative transfer (RT) models to study the interaction between light and forest ecosystems. RT models are useful tools for understanding many ecological processes, such as exchanges of energy, carbon, and water between the biosphere and atmosphere. Examples of RT models for such study purposes include: testing the possible biophysical mechanisms for the green-up of Amazon forests (Morton et al., 2014); simulation of the transport of energy, absorption of photosynthetically active radiation, and gross primary productivity in a plantation forest (van Leeuwen et al., 2015); understanding the relationship between heterogeneous vegetation structure, radiation, and snowmelt (Ni-Meister & Gao, 2011); revealing patterns of photosynthetic partitioning in an Arctic shrub (Magney et al., 2016). A key step to parameterize RT models is to reconstruct the 3D structures of forest scenes. Three typical approaches for LiDAR-parameterized RT models are:

1. Approximations of individual trees with simple geometric primitives. By applying individual tree segmentation in LiDAR data, the boundaries of individual trees can be delineated and the structural parameters of individual trees (e.g., locations, heights, crown radii, leaf area index, and leaf angle distribution) can be estimated. Tree crowns are abstracted by simple geometric shapes (e.g., cone, sphere, cylinder, and ellipsoid), which can be defined by the structural parameters of individual trees derived from LiDAR data. Models such as FLiES (Forest Light Environmental Simulator), RAPID (Radiosity Applicable to Porous IndiviDual objects), and DART (Discrete Anisotropic Radiative Transfer) can be parameterized using this approach.
2. Voxel-based approximations of canopies. The whole forest scene is subdivided into a set of voxels filled with specific elements. The voxel size is usually set to be large (e.g., 2 m), and vegetation voxels are assumed to be turbid media. The turbid medium voxels make statistical simplifications of tree architectures, and they are parameterized by LiDAR-derived attributes such as plant area

index or plant area density. A typical example of this approach is the voxel grid parametrization of the DART model in Schneider et al. (Schneider et al., 2014).

3. Reconstructions of realistic trees. It is also possible to reconstruct realistic 3D trees to approximate the fine-scale tree architectures from LiDAR data and provide more realistic RT simulations. The reconstructed 3D trees can be represented by triangular meshes, so we name this approach as triangular-based parametric reconstruction. Alternatively, tree architectures can be defined by voxelizing the point cloud data without requiring extra parameters or making certain assumptions on the tree architectures, but this approach requires a high density of point cloud data and uses high resolution (e.g., 0.01 m) solid voxels to ensures the accuracy of reconstruction, which is named voxel-based nonparametric reconstruction. The triangular-based parametric reconstruction is suitable for RT models like PBRT (physically based ray tracer), LESS (large-scale emulation system), DART, whereas the voxel-based nonparametric reconstruction is suitable for the voxel-based Monte Carlo radiative transfer model (VBRT) (Li et al., 2018).

The point density is relatively low in ALS data, and hence the first two approaches can be used to approximate the forest structures at coarse scales. For high-density point cloud data such as TLS, the last approach can be used for fine-scale approximations. With a defined 3D forest scene, numerical methods such as Monte Carlo ray tracing can be used to solve the RT equation.

24.4 Prospects for LiDAR Remote Sensing of Forest Ecosystems

With various types of LiDAR systems being used for forest observations in temporal, spatial, and spectral dimensions, LiDAR remote sensing of forest ecosystems enters the era of big data (Guo, Jin, et al., 2020a; Guo, Su, et al., 2020b), and is facing unprecedented challenges and opportunities.

Establishing and sharing LiDAR datasets is a prerequisite for translating LiDAR data into ecological knowledge. There have been several LiDAR datasets for developing and validating data-processing algorithms (Dong et al., 2020; Liang, Hyyppä, et al., 2018a; Liang, Kukko, et al., 2018b), but LiDAR datasets complemented by ecological observations (e.g., phenology and carbon flux) are rare. It remains challenging yet important to construct such LiDAR datasets following the principles of findable, accessible, interoperable, and reusable.

The LiDAR revolution is a part of the big data revolution. Deep learning has permeated into big data mining and contributed to LiDAR data processing and ecological cognition due to its ability to extract spatial–temporal features from massive data (Guo, Jin, et al., 2020a; Guo, Su, et al., 2020b). Deep learning has been used for processing LiDAR data in various aspects, such as registration, upsampling, completion, detection, classification and segmentation. These algorithms have contributed to the processing of LiDAR data in forested environments, including but not limited to, ground point filtering, tree species classification, individual tree segmentation, and wood and leaf separation (Jin et al., 2020). Thus, the combination of LiDAR data, environmental data, and deep learning has been advancing forest studies tremendously. For instance, a simple neural network has recently been used to map country-scale canopy height from spaceborne and UAV LiDAR data, aided by other ancillary datasets (Liu et al., 2022). Future research could endeavor to better fuse LiDAR data and other kinds of big data such as satellite radar data. Efforts should also be given to clarify why deep learning can achieve satisfying results, and to develop hybrid models by coupling data- and physical-process-driven models, for the ultimate purpose of improving ecosystem applications dominated by long-range spatial connections across multiple timescales.

Ecological modelling could also benefit from the rapid evolution of LiDAR remote sensing. As stated earlier, LiDAR provides accurate forest structural and topographical information, which can be fed into solar radiation models, vegetation dynamic models, biodiversity models, and ecological niche models, and are expected to increase the accuracies of these models. Moreover, LiDAR sheds new lights on classical ecological questions that could not be answered easily by traditional approaches. For example, TLS unlocks the possibility of characterizing 3D tree branch architecture, providing an excellent opportunity to quantify the branching system of trees accurately and efficiently without harvesting trees. LiDAR-derived vegetation and topographic structures have also led to a deeper understanding of animal behaviors such as nesting, hunting, and thermoregulation (Melin et al., 2014).

Despite its capability of providing accurate structural information, most LiDAR systems cannot obtain the spectral information of an object. As new, often more complex, questions emerge in forest ecology, future research should be encouraged to fuse LiDAR data with multispectral and hyperspectral data. This allows for the quantification of physiological properties of plants but in a 3D dimension, with the possibility of revolutionizing our understanding of plant structures and functions.

Another important direction for future research would be to merge LiDAR data acquired from different platforms, as LiDAR platforms differ hugely in point density, laser range, and data accuracy (Guan et al., 2019). For example, ULS measures top-canopy parameters (such as tree height and crown diameter) more accurately than terrestrial platforms. Terrestrial platforms, on the other hand, digitize targets under a forest canopy, providing e.g., stem and branch parameters (such as DBH and stem curve). Multi-platform collaborative observation for a forest ecosystem is the only way to obtain a comprehensive and accurate quantification of forest structural features.

24.5 Conclusions

LiDAR provides a powerful tool for forest structure quantification. Various types of LiDAR platforms have been applied to extract accurate forest structural traits across scales (from tree branch scales to global scales). These structural traits provide new sources for understanding and modelling forest ecosystem processes. Up to now, most studies focus on static structural information from LiDAR data, yet the full potential of LiDAR for forest ecosystem studies has not been realized. As LiDAR applications are moving into the era of big data, it opens new opportunities for linking both the structural, temporal, and spectral information to elucidate ecological processes at larger spatial and temporal scales. Moreover, with current LiDAR hardware equipment and processing algorithms are still undergoing a period of rapid development, we expect that LiDAR remote sensing will continue to be an exciting field in forest ecosystem studies for the near future.

References

Coops, N. C., Tompalski, P., Goodbody, T. R. H., Queinnec, M., Luther, J. E., Bolton, D. K., White, J. C., Wulder, M. A., van Lier, O. R., & Hermosilla, T. (2021). Modelling lidar-derived estimates of forest attributes over space and time: A review of approaches and future trends. *Remote Sensing of Environment, 260*, 112477.

Davies, A. B., & Asner, G. P. (2014). Advances in animal ecology from 3D-LiDAR ecosystem mapping. *Trends in Ecology & Evolution, 29*(12), 681–691.

Dong, Z., Liang, F., Yang, B., Xu, Y., Zang, Y., Li, J., Wang, Y., Dai, W., Fan, H., Hyyppä, J., & Stilla, U. (2020). Registration of large-scale terrestrial laser scanner point clouds: A review and benchmark. *ISPRS Journal of Photogrammetry and Remote Sensing, 163*, 327–342.

Forrester, D. I. (2019). Linking forest growth with stand structure: Tree size inequality, tree growth or resource partitioning and the asymmetry of competition. *Forest Ecology and Management, 447*, 139–157.

Guan, H., Su, Y., Hu, T., Wang, R., Ma, Q., Yang, Q., Sun, X., Li, Y., Jin, S., & Zhang, J. (2019). A novel framework to automatically fuse multiplatform LiDAR data in forest environments based on tree locations. *IEEE Transactions on Geoscience and Remote Sensing, 58*(3), 2165–2177.

Guo, Q., Jin, S., Li, M., Yang, Q., Xu, K., Ju, Y., Zhang, J., Xuan, J., Liu, J., Su, Y., Xu, Q., & Liu, Y. (2020). Application of deep learning in ecological resource research: Theories, methods, and challenges. *Science China Earth Sciences, 63*(10), 1457–1474.

Guo, Q., Su, Y., Hu, T., Guan, H., Jin, S., Zhang, J., Zhao, X., Xu, K., Wei, D., & Kelly, M. (2020). Lidar boosts 3D ecological observations and modelings: A review and perspective. *IEEE Geoscience and Remote Sensing Magazine, 99*(1), 232–257.

Guo, Q., Su, Y., Hu, T., Zhao, X., Wu, F., Li, Y., Liu, J., Chen, L., Xu, G., Lin, G., Zheng, Y., Lin, Y., Mi, X., Fei, L., & Wang, X. (2017). An integrated UAV-borne lidar system for 3D habitat mapping in three forest ecosystems across China. *International Journal of Remote Sensing, 38*(8–10), 2954–2972.

Hu, T., Su, Y., Xue, B., Liu, J., Zhao, X., Fang, J., & Guo, Q. (2016). Mapping global forest aboveground biomass with spaceborne LiDAR, optical imagery, and forest inventory data. *Remote Sensing, 8*(7), 565.

Hu, T, Sun, X, Su, Y, Guan, H, Sun, Q, Kelly, M, & Guo, Q. (2021). Development and performance evaluation of a very low-cost UAV-LiDAR system for forestry applications. *Remote Sensing, 13*(1).

Hyyppä, J, Virtanen, J.-P., Jaakkola, A., Yu, X., Hyyppä, H., & Liang, X. (2018). Feasibility of google tango and kinect for crowdsourcing forestry information. *Forests, 9*(1).

Jin, S., Su, Y., Zhao, X., Hu, T., & Guo, Q. (2020). A point-based fully convolutional neural network for airborne lidar ground point filtering in forested environments. *IEEE Journal of Selected Topics in Applied Earth Observations and Remote Sensing, 13*, 3958–3974.

Jurjević, L., Liang, X., Gašparović, M., & Balenović, I. (2020). Is field-measured tree height as reliable as believed—Part II, A comparison study of tree height estimates from conventional field measurement and low-cost close-range remote sensing in a deciduous forest. *ISPRS Journal of Photogrammetry and Remote Sensing, 169*, 227–241.

Li, W., Guo, Q., Tao, S., & Su, Y. (2018). VBRT: A novel voxel-based radiative transfer model for heterogeneous three-dimensional forest scenes. *Remote Sensing of Environment, 206*, 318–335.

Liang, X., Hyyppä, J., Kaartinen, H., Lehtomäki, M., Pyörälä, J., Pfeifer, N., Holopainen, M., Brolly, G., Francesco, P., Hackenberg, J., Huang, H., Jo, H.-W., Katoh, M., Liu, L., Mokroš, M., Morel, J., Olofsson, K., Poveda-Lopez, J., Trochta, J., & Wang, Y. (2018a). International benchmarking of terrestrial laser scanning approaches for forest inventories. *ISPRS Journal of Photogrammetry and Remote Sensing, 144*, 137–179.

Liang, X., Kankare, V., Hyyppä, J., Wang, Y., Kukko, A., Haggrén, H., Yu, X., Kaartinen, H., Jaakkola, A., Guan, F., Holopainen, M., & Vastaranta, M. (2016). Terrestrial laser scanning in forest inventories. *ISPRS Journal of Photogrammetry and Remote Sensing, 115*, 63–77.

Liang, X., Kukko, A., Hyyppä, J., Lehtomäki, M., Pyörälä, J., Yu, X., Kaartinen, H., Jaakkola, A., & Wang, Y. (2018b). In-situ measurements from mobile platforms: An emerging approach to address the old challenges associated with forest inventories. *ISPRS Journal of Photogrammetry and Remote Sensing, 143*, 97–107.

Liang, X., Wang, Y., Pyörälä, J., Lehtomäki, M., Yu, X., Kaartinen, H., Kukko, A., Honkavaara, E., Issaoui, A. E. I., Nevalainen, O., Vaaja, M., Virtanen, J.-P., Katoh, M., & Deng, S. (2019). Forest in situ observations using unmanned aerial vehicle as an alternative of terrestrial measurements. *Forest Ecosystems, 6*(1), 20.

Liu, X., Su, Y., Hu, T., Yang, Q., Liu, B., Deng, Y., Tang, H., Tang, Z., Fang, J., & Guo, Q. (2022). Neural network guided interpolation for mapping canopy height of China's forests by integrating GEDI and ICESat-2 data. *Remote Sensing of Environment, 269*, 112844.

Magney, T. S., Eitel, J. U. H., Griffin, K. L., Boelman, N. T., Greaves, H. E., Prager, C. M., Logan, B. A., Zheng, G., Ma, L., Fortin, E. A., Oliver, R. Y., & Vierling, L. A. (2016). LiDAR canopy radiation model reveals patterns of photosynthetic partitioning in an Arctic shrub. *Agricultural and Forest Meteorology, 221*, 78–93.

McGlade, J., Wallace, L., Hally, B., White, A., Reinke, K., & Jones, S. (2020). An early exploration of the use of the Microsoft Azure Kinect for estimation of urban tree Diameter at Breast Height. *Remote Sensing Letters, 11*(11), 963–972.

Melin, M., Matala, J., Mehtätalo, L., Tiilikainen, R., Tikkanen, O., Maltamo, M., Pusenius, J., & Packalen, P. (2014). Moose (A lces alces) reacts to high summer temperatures by utilizing thermal shelters in boreal forests–an analysis based on airborne laser scanning of the canopy structure at moose locations. *Global Change Biology, 20*(4), 1115–1125.

Morton, D. C., Nagol, J., Carabajal, C. C., Rosette, J., Palace, M., Cook, B. D., Vermote, E. F., Harding, D. J., & North, P. R. J. (2014). Amazon forests maintain consistent canopy structure and greenness during the dry season. *Nature, 506*(7487), 221–224.

Ni-Meister, W., & Gao, H. (2011). Assessing the impacts of vegetation heterogeneity on energy fluxes and snowmelt in boreal forests. *Journal of Plant Ecology, 4*(1–2), 37–47.

Schneider, F. D., Leiterer, R., Morsdorf, F., Gastellu-Etchegorry, J.-P., Lauret, N., Pfeifer, N., & Schaepman, M. E. (2014). Simulating imaging spectrometer data: 3D forest modeling based on LiDAR and in situ data. *Remote Sensing of Environment, 152*, 235–250.

Su, Y, Hu, T, Wang, Y, Li, Y, Dai, J, Liu, H, Jin, S, Ma, Q, Wu, J, Liu, L, Fang, J, & Guo, Q. (2020). Large-scale geographical variations and climatic controls on crown architecture traits. *Journal of Geophysical Research: Biogeosciences, 125*(2), e2019JG005306.

Su, Y., Guo, Q., Jin, S., Guan, H., Sun, X., Ma, Q., Hu, T., Wang, R., & Li, Y. (2021). The development and evaluation of a backpack lidar system for accurate and efficient forest inventory. *IEEE Geoscience and Remote Sensing Letters, 18*(9), 1660–1664.

Tao, S., Labrière, N., Calders, K., Fischer, F. J., Rau, E.-P., Plaisance, L., & Chave, J. (2021). Mapping tropical forest trees across large areas with lightweight cost-effective terrestrial laser scanning. *Annals of Forest Science, 78*(4), 103.

van Leeuwen, M, Coops, N C, & Black, T. A. (2015). Using stochastic ray tracing to simulate a dense time series of gross primary productivity. *Remote Sensing*, *7*(12). https://doi.org/10.3390/rs71215875.

West, G. B., Enquist, B. J., & Brown, J. H. (2009). A general quantitative theory of forest structure and dynamics. *Proceedings of the National Academy of Sciences, 106*(17), 7040.

Zhao, F., Sweitzer, R. A., Guo, Q., & Kelly, M. (2012). Characterizing habitats associated with fisher den structures in the Southern Sierra Nevada, California using discrete return lidar. *Forest Ecology and Management, 280*, 112–119.

Chapter 25
Dense Satellite Image Time Series Analysis: Opportunities, Challenges, and Future Directions

Desheng Liu and Xiaolin Zhu

Abstract Earth observation satellites provide important data for monitoring land surface dynamics. In recent years, with the development of new satellite constellations, supercomputing, artificial intelligence, and cloud computing, remote sensing studies of land surface changes have been gradually shifted from sparse time series analysis to dense time series anslysis. Dense satellite image time series dramatically improve our capability for capturing frequent changes in the land surface. It has changed the research questions, data processing techniques, and applications compared with the traditional sparse time series analysis. This chapter discussed the opportunities, challenges, and future directions of dense satellite time series data analysis. It can help researchers from the remote sensing community or other disciplines apply dense satellite time series analysis to solve real-world problems.

Keywords Dense satellite image time series · Remote sensing · Time series analysis

25.1 Introduction

Since the 1970s, earth observation satellites have opened a new dimension in our ability to observe and study the Earth system (Hansen & Loveland, 2012). The advantages of satellite images include the large spatial coverage and repeated observations, which are better than in-situ data to study changes of our Earth environment with high efficiency and low cost over large areas (Zhu, 2017). Due to their capabilities to monitor land cover and land use change, satellite image time series (SITS) have been widely used to detect land surface changes and model the human–environment

D. Liu (✉)
Department of Geography, The Ohio State University, Columbus, OH 43210, USA
e-mail: liu.738@osu.edu

X. Zhu
Department of Land Surveying and Geo-informatics, The Hong Kong Polytechnic University, Hong Kong 999077, China
e-mail: xlzhu@polyu.edu.hk

B. Li et al. (eds.), *New Thinking in GIScience*,
https://doi.org/10.1007/978-981-19-3816-0_25

interactions under the pressure of accelerated urbanization and global climate change (Liu & Cai, 2012; Zhu & Liu, 2019).

In the early stage of remote sensing development, studies mainly use two or more satellite images over multiple years with a relatively long time interval (e.g., every five years) to investigate land surface changes (Fig. 25.1). The multi-temporal images used in these studies can be named as sparse SITS. The main reason for using sparse SITS was the high cost of satellite images and the limited number of available satellites. In particular, Landsat satellite series provide the only available SITS over the past several decades at medium spatial resolutions to capture detailed land surface changes induced by human activities. However, Landsat images were not free to the public before 2008, so studies using Landsat images over large areas had a high cost. Traditional studies compared the classification results of individual images in the sparse SITS (i.e., post-classification methods) or directly calculated the difference of spectral signals between two images in the sparse SITS (i.e., change-vector based methods) to quantify the land cover and land use changes (Liu & Cai, 2012).

Although sparse SITS analyses provided us with new insights on the land surface changes at different scales and in various ecosystems, they have several drawbacks. First, the impact of seasonality brings uncertainties to change detection. The seasonality of land surface, such as vegetation phenology, crop phenology, and wetland dynamics, is affected by both human activities and climate. Sparse SITS cannot well differentiate between the land surface changes caused by human activities and seasonality, which leads to large uncertainties in the change detection results. Second, the snapshot mode (e.g., one image per year) of using sparse SITS cannot well differentiate land surface objects sharing high similarity in spectral signals in the period of which the image is acquired (Zhu & Liu, 2014). For example, many studies select summer images to compose the sparse SITS as the summer images can best show the distinct differences between vegetation and non-vegetation objects, but different

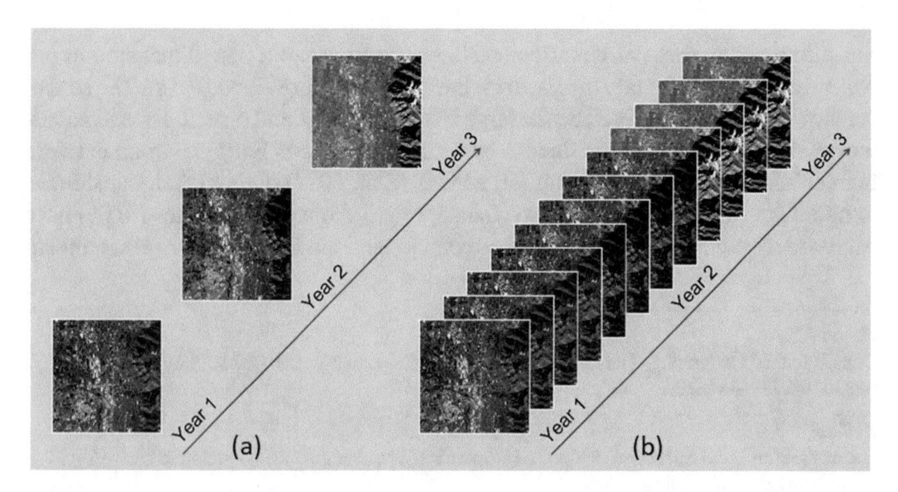

Fig. 25.1 Diagram of sparse time series (**a**) and dense time series (**b**) based studies

vegetation types (e.g., forests, shrub, grass, and crops) cannot be well classified only using images in summer due to their similar spectral responses.

In recent years, remote sensing studies have gradually shifted from sparse SITS analysis to dense SITS analysis. Dense SITS is defined as the images in the time series that contain high-frequency information, such as the seasonality and monthly dynamics (Fig. 25.1). Since 2008, dense SITS data from Landsat and other medium-resolution satellites, such as Sentinel, have been made available at no charge to end users. It has greatly promoted the studies of land surface dynamics using dense SITS because these data have a spatial resolution appropriate for heterogonous land surfaces, such as urban areas (Schneider, 2012). Meanwhile, a free cloud computing platform, Google Earth Engine (GEE), currently facilitates the use of dense SITS for large area monitoring of land and water dynamics, because it has a super ability to process massive satellite images (Gorelick et al., 2017). Studies based on dense SITS can overcome the drawbacks of sparse SITS mentioned above. However, analyzing dense SITS faces new challenges, which require new data analytics methods, such as new techniques in image processing, time series analysis, and data integration, as summarized in Fig. 25.2. This chapter aims to discuss the opportunities, challenges, and future directions of dense SITS studies.

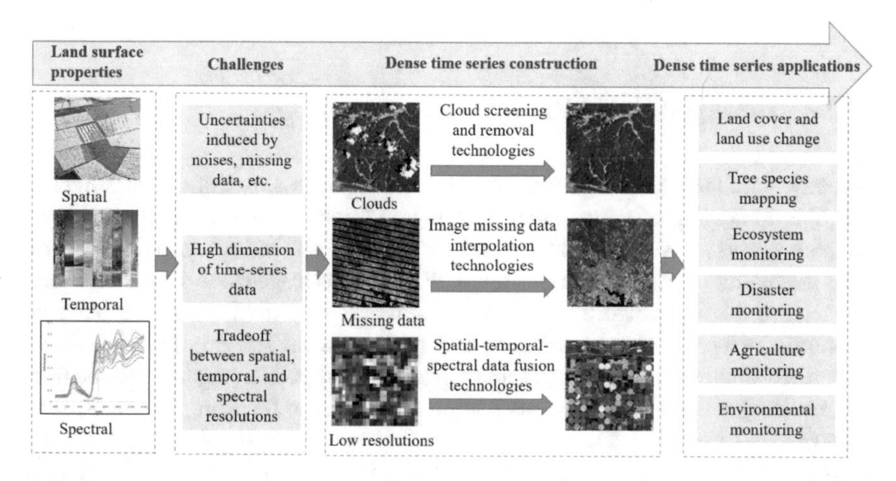

Fig. 25.2 Challenges, key technologies, and applications of dense time series based remote sensing studies

25.2 Opportunities for Developing Dense Time-Series Remote Sensing

25.2.1 New Data Sources

In recent years, to increase the frequency of satellite observations, many Earth observation missions adopted a constellation strategy, i.e., two or more satellites working together as a system to collect images with similar configurations. In particular, Sentinel-2 has two identical satellites (Sentinel-2A and -2B) flying on a single orbit plane but phased at 180°, which can improve the revisit cycle of a single satellite from 10 to 5 days. Besides, more microsatellites forming a constellation make SITS even denser. For example, Planet operates with two constellations. The PlanetScope (PS) consists of more than 180 CubeSats, and SkySat (SS) has 21 CubeSats. They provide daily and sub-daily coverage of the entire land surface of the Earth, respectively. As more Earth observation satellites expect to be launched owing to the development of new sensors and low-cost satellite platforms, more satellite data with high temporal frequencies will become available in the near future, providing great opportunities for dense SITS analysis.

25.2.2 Stronger Capability of Data Processing

Recently, with the rapid development of supercomputers and artificial intelligence (AI), the processing of dense SITS is facing an unprecedented revolution. Supercomputing cluster is the hardware foundation and computing support of dense SITS processing. Based on the existing CPU-GPU heterogeneous supercomputing architecture, big data algorithms such as statistical algorithms, machine learning algorithms, deep learning algorithms, and reinforcement learning algorithms can be further optimized by strengthening parallel computing, signal communication, memory or video memory, instruction set, and even field-programmable gate array (FPGA) hardware accelerating. It dramatically improves the efficiency and performance of SITS processing. The cloud-computing environment with big data management and analytics capabilities, e.g., GEE, has already opened opportunities for continuous and dynamic monitoring of our changing environment (Kennedy et al., 2018). Additionally, state-of-the-art geoscience algorithms are integrated and constantly upgraded in GEE-like cloud platforms, facilitating the long-term and large-scale research under a uniform framework (Gomes et al., 2020).

25.2.3 New Applications

Currently, dense SITS has been highly desired by applications that require timely spatial–temporal information, including emergency response, ecosystem monitoring, smart agriculture, and smart cities. First, with global climate changes and increases of extreme weather events, natural disasters such as floods and wildfires could occur more often; dense SITS can provide timely assessment of disasters to support prompt rescue missions (Xu et al., 2021). Second, ecosystems are threatened by climate change and human activities. Dense SITS can study vegetation-climate interactions (Zhu, 2017), including phenology (Tian et al., 2021) and carbon monitoring (Zhu & Liu, 2015). Third, modern agriculture needs timely information on crop growth and stress from water and nutrients. Dense SITS can provide such information up to daily frequency (Deng et al., 2018). Last, smart city is a new concept in current city planning and management, which requires spatial data with very high frequency to solve problems related to energy, water, food, waste, urban heat island, and air pollution (Zhu et al., 2019). These new applications will greatly promote the wide applications of dense SITS.

25.3 Challenges of Dense SITS Analysis

25.3.1 Data Quality Control

Data quality has a direct effect on the reliability of dense SITS analysis. Conventionally, masking clouds is a common method to control the quality of satellite data. In general, cloud masks can be generated for single images. For example, the Fmask algorithm has been used to generate the quality assessment (QA) band of the Landsat dataset (Zhu & Woodcock, 2012). However, significant errors of the Fmask algorithm have been widely reported (Bolton et al., 2020). As a result, time-series-based methods were developed to improve the accuracy of cloud masks, e.g., the Automatic Time-Series Analysis (ATSA) method (Zhu & Helmer, 2018), which adds temporal information into the screening of clouds. Moreover, inconsistency among different satellites, in terms of sensor configuration, viewing and illumination angles, and atmosphere, is a tough factor for data quality control when combining multi-source data to compose dense SITS. Several studies have tried to solve this problem. For example, the Harmonized Landsat Sentinel-2 (HLS) datasets normalized Sentinel-2 to match Landsat data using a linear correction (Claverie et al., 2017).

25.3.2 Data Analysis Techniques

Compared with sparse SITS, dense SITS brings more temporal information, but it also brings more challenges for data analysis and makes traditional techniques unsuitable. Dense SITS has a much higher dimension than sparse SITS. For example, there will be 23 Landsat-7 images if a study uses all the available Landsat images in a year to classify land cover types in a region covered by one Landsat scene. With six non-thermal bands in each image, there will be $23 \times 6 = 138$ input features for the classification, which could lead to the Hughes Problem, i.e., the classification accuracy decreases with number of features if the training samples are limited for supervised classifiers (Foody & Arora, 1997). Therefore, how to reduce the dimension of dense SITS or select the most informative features is a challenging issue to be addressed. This cannot be simply done by dimension reduction techniques (e.g., principal components analysis) or the feature searching algorithms (e.g., recursive feature elimination) (Zhu & Liu, 2014) since temporal information is complex and may be removed during the process. Moreover, pixels in dense SITS are spatially correlated especially for high spatial resolution satellite imagery. How to effectively model spatial–temporal information in dense time series analysis is another challenging issue to be addressed.

25.3.3 Cloud Impact

Frequent cloud contamination is an inevitable challenge for optical SITS. A previous study discovered that nearly one-third of the places in the world have less than six cloud-free Landsat images each year, which is far less than the regular 16-day satellite revisit (Ju & Roy, 2008). Therefore, some studies have explored the potential of time-series image reconstruction using partially cloudy images (Qiu et al., 2021). With the launch of new satellites (e.g., Landsat-9) and constellations (e.g., PlanetScope), combining data from multiple sources may alleviate the influence of cloud coverage. During the past decade, spaceborne SAR remote sensing, which can penetrate clouds, has become a significant data source for earth observation. The optical-SAR fusion may be a possible way for mitigating the cloud impact on dense SITS analysis (Zhu & Helmer, 2018).

25.4 Future Directions

25.4.1 Data Fusion to Reconstruct High-Quality Time Series

It is foreseen that a single satellite sensor cannot provide data with both high spatial and high temporal resolutions in the near future. For optical satellites, the data availability is further limited by clouds. Therefore, integrating data from multiple satellite sensors will be a feasible and cost-effective way to produce high-quality dense SITS. Three types of data fusion techniques can be developed for different situations. The first type is spatio-temporal-spectral fusion techniques which can integrate optical images from satellite sensors with different spatial, temporal, and spectral resolutions, such as Landsat, Sentinel-2, and MODIS. Although a substantial number of spatiotemporal fusion algorithms have been developed, most of them cannot fuse data of different spectral bands and their accuracies may not be good enough to support real-world applications (Zhu et al., 2018). The second type is cross-sensor fusion methods that can normalize the difference of data collected by the sensors onboard satellite constellations, such as PlanetScope constellation of more than 180 sensors. The cross-sensor fusion needs to normalize the differences of data caused by illumination and viewing geometries, radiometric range, atmospheric effect, and pointing accuracies (Wang et al., 2020). The third type is optical-radar fusion techniques that combine radar and optical images to solve the cloud problem since radar signals can penetrate clouds. However, these two types of satellites collect ground information with different principles so the optical-radar fusion algorithms should be designed to integrate the consistent features extracted from both satellites (Ahmad et al., 2020).

25.4.2 Modeling Spatial–Temporal Information

Dense SITS can be viewed as a collection of pixels in a spatially correlated time-series, a time-series of images, or a space–time image cube. New techniques for analyzing dense SITS need to consider both spatial and temporal dimensions. Existing methods for dense SITS analysis mainly focus on modeling the temporal information of individual pixels. For example, some studies have explored the potential of phenology features extracted from dense SITS for land cover mapping and showed promising results (e.g., Weisberg et al., 2021). Another strategy is to segment the dense time series into numerous periods with stable land cover types and then generate land cover maps by classifying time series images in each stable period, which can produce accurate land cover trajectories (Cai & Liu, 2015, 2018; Zhu & Woodcock, 2014). Future research can focus on adding spatial information in phenology modeling and segmentation-based approaches, extending Markov random field (MRF) based spatial–temporal classification model (Liu & Cai, 2012) to higher temporal frequency, and developing a fully integrated spatial–temporal model for dense SITS.

25.4.3 Development of Analysis-Ready Data and User-Friendly Tools

Analysis-ready data and user-friendly tools can facilitate the wide applications of dense SITS. First, analysis-ready SITS data are more convenient for researchers out of the remote sensing field. For example, the HLS dataset, a dense SITS product, harmonized three groups of fine-scale datasets, i.e., Landsat 8 and Sentinel-2A and Sentinel-2B. The HLS datasets have 30-m spatial resolution and frequent temporal resolution around 1–4 days, depending on the location (Li & Roy, 2017). Until now, the HLS products can cover the entire North America and partial Europe, and they can be freely downloaded. Therefore, future studies should put more effort in producing analysis-ready time series data with global coverage. Second, user-friendly analysis tools can remove the barrier of dense SITS applications since most time-series analysis tools are relatively more complicated than conventional ones. There are new demands of technologies to store, process, disseminate, analyze, and visualize dense SITS (Gomes et al., 2020). Cloud-based platforms such as GEE, can store analysis-ready time series datasets and provide high-performance and intrinsically parallel computation services (Gorelick et al., 2017). However, not everyone in the geospatial community or other disciplines is familiar with programming in cloud platforms. Many scholars developed open-source packages to perform geospatial analysis directly within the user interface without writing a single line of code (e.g., Aybar et al., 2020; Wu, 2021). With further development and application of backend analysis tools, it would be time- and cost-saving to monitor and predict land surface changes using dense SITS.

25.5 Conclusion

This chapter discussed the opportunities, challenges, and future directions of dense SITS analysis. This is a new paradigm of remote sensing studies, with the help of increasing free remote sensing data sources and cloud computing platforms. It has changed the research questions, data processing techniques, and applications compared with the traditional sparse SITS analysis. Free earth observation satellite data and new microsatellite constellations provide opportunities to capture the land surface dynamics at high frequency and from multiple dimensions. The development of computer hardware and software provides a more powerful capacity to process dense SITS with big volumes. At the same time, dense SITS brings new challenges, including data quality control to reduce the noises, maintaining a balance between dimension and information abundance, modeling spatial–temporal information, and generating reliable SITS in cloudy regions. To advance dense SITS analysis techniques and make dense SITS more accessible to researchers in broad disciplines, future studies should develop new data fusion and spatial–temporal modeling methods tailored for dense SITS, develop analysis-ready data sets with high quality and scientific standards, and integrate such analysis-ready data with powerful cloud computing platforms, which can significantly remove the barrier for users to directly apply dense SITS to solve real-world problems.

References

Ahmad, S. K., Hossain, F., Eldardiry, H., & Pavelsky, T. M. (2020). A fusion approach for water area classification using visible, near infrared and synthetic aperture radar for south asian conditions. *IEEE Transactions on Geoscience and Remote Sensing, 58*, 2471–2480.

Aybar, C., Wu, Q., Bautista, L., Yali, R., & Barja, A. (2020). rgee: An R package for interacting with Google Earth Engine. *Journal of Open Source Software, 5*, 2272.

Bolton, D. K., Gray, J. M., Melaas, E. K., Moon, M., Eklundh, L., & Friedl, M. A. (2020). Continental-scale land surface phenology from harmonized Landsat 8 and Sentinel-2 imagery. *Remote Sensing of Environment, 240*, 111685.

Cai, S., & Liu, D. (2018). Mapping Land cover trajectories using monthly MODIS time series from 2001 to 2010. In Q. Weng (Ed.), *Remote Sensing time series image processing* (pp. 137–155). CRC Press.

Cai, S., & Liu, D. (2015). Detecting change dates from dense satellite time series using a sub-annual change detection algorithm. *Remote Sensing, 7*(7), 8705–8727.

Claverie, M., Masek, J. G., Junchang, J., & Dungan, J. L. (2017). Harmonized Landsat-8 Sentinel-2 (HLS) Product User's Guide 2, pp. 1–17.

Deng, L., Mao, Z., Li, X., Hu, Z., Duan, F., & Yan, Y. (2018). UAV-based multispectral remote sensing for precision agriculture: A comparison between different cameras. *ISPRS Journal of Photogrammetry and Remote Sensing, 146*, 124–136.

Foody, G. M., & Arora, M. K. (1997). An evaluation of some factors affecting the accuracy of classification by an artificial neural network. *International Journal of Remote Sensing, 18*, 799–810.

Gomes, V. C. F., Queiroz, G. R., & Ferreira, K. R. (2020). An overview of platforms for big earth observation data management and analysis. *Remote Sensing, 12*, 1253.

Gorelick, N., Hancher, M., Dixon, M., Ilyushchenko, S., Thau, D., & Moore, R. (2017). Google Earth Engine: Planetary-scale geospatial analysis for everyone. *Remote Sensing of Environment, 202*, 18–27.

Hansen, M. C., & Loveland, T. R. (2012). A review of large area monitoring of land cover change using Landsat data. *Remote Sensing of Environment, 122*, 66–74.

Ju, J., & Roy, D. P. (2008). The availability of cloud-free Landsat ETM+ data over the conterminous United States and globally. *Remote Sensing of Environment, 112*, 1196–1211.

Kennedy, R. E., Yang, Z., Gorelick, N., Cohen, W. B., & Healey, S. (2018). Implementation of the LandTrendr Algorithm on Google Earth Engine. *Remote Sensing, 10*(5), 691.

Li, J., & Roy, D. P. (2017). A Global Analysis of Sentinel-2A, Sentinel-2B and Landsat-8 data revisit intervals and implications for terrestrial monitoring. *Remote Sensing, 9*(9), 902.

Liu, D., & Cai, S. (2012). A Spatial-Temporal modeling approach to reconstructing land-cover change trajectories from multi-temporal satellite imagery. *Annals of the Association of American Geographers, 102*, 1329–1347.

Qiu, Y., Zhou, J., Chen, J., & Chen, X. (2021). Spatiotemporal fusion method to simultaneously generate full-length normalized difference vegetation index time series (SSFIT). *International Journal of Applied Earth Observations and Geoinformation, 100*, 102333.

Schneider, A. (2012). Monitoring land cover change in urban and peri-urban areas using dense time stacks of Landsat satellite data and a data mining approach. *Remote Sensing of Environment, 124*, 689–704.

Tian, J., Zhu, X., Chen, J., Wang, C., Shen, M., Yang, W., Tan, X., Xu, S., & Li, Z. (2021). Improving the accuracy of spring phenology detection by optimally smoothing satellite vegetation index time series based on local cloud frequency. *ISPRS Journal of Photogrammetry and Remote Sensing, 180*, 29–44.

Wang, J., Yang, D., Detto, M., Nelson, B. W., Chen, M., Guan, K., Wu, S., Yan, Z., & Wu, J. (2020). Multi-scale integration of satellite remote sensing improves characterization of dry-season green-up in an Amazon tropical evergreen forest. *Remote Sensing of Environment, 246*, 111865.

Weisberg, P. J., Dilts, T. E., Greenberg, J. A., Johnson, K. N., Pai, H., Sladek, C., Kratt, C., Tyler, S. W., & Ready, A. (2021). Phenology-based classification of invasive annual grasses to the species level. *Remote Sensing of Environment, 263*, 112568.

Wu, Q. (2021). Leafmap: A Python package for interactive mapping and geospatial analysis with minimal coding in a Jupyter environment. *Journal of Open Source Software, 6*, 3414.

Xu, S., Zhu, X., Helmer, E. H., Tan, X., Tian, J., & Chen, X. (2021). The damage of urban vegetation from super typhoon is associated with landscape factors: Evidence from Sentinel-2 imagery. *International Journal of Applied Earth Observations and Geoinformation, 104*, 102536.

Zhu, X., Cai, F., Tian, J., & Williams, T. (2018). Spatiotemporal fusion of multisource remote sensing data: Literature survey, taxonomy, principles, applications, and future directions. *Remote Sensing, 10*, 527.

Zhu, X., & Helmer, E. H. (2018). An automatic method for screening clouds and cloud shadows in optical satellite image time series in cloudy regions. *Remote Sensing of Environment, 214*, 135–153.

Zhu, X., & Liu, D. (2019). Investigating the impact of land parcelization on forest composition and structure in southeastern Ohio using multi-source remotely sensed data. *Remote Sensing, 11*(19), 2195.

Zhu, X., & Liu, D. (2015). Improving forest aboveground biomass estimation using seasonal Landsat NDVI time-series. *ISPRS Journal of Photogrammetry and Remote Sensing, 102*, 222–231.

Zhu, X., & Liu, D. (2014). Accurate mapping of forest types using dense seasonal landsat time-series. *ISPRS Journal of Photogrammetry and Remote Sensing, 96*, 1–11.

Zhu, Z. (2017). Change detection using landsat time series: A review of frequencies, preprocessing, algorithms, and applications. *ISPRS Journal of Photogrammetry and Remote Sensing, 130*, 370–384.

Zhu, Z., & Woodcock, C. E. (2012). Object-based cloud and cloud shadow detection in Landsat imagery. *Remote Sensing of Environment, 118*, 83–94.

Zhu, Z., & Woodcock, C. E. (2014). Continuous change detection and classification of land cover using all available Landsat data. *Remote Sensing of Environment, 144*, 152–171.

Zhu, Z., Zhou, Y., Seto, K. C., Stokes, E. C., Deng, C., Pickett, S. T. A., & Taubenböck, H. (2019). Understanding an urbanizing planet: Strategic directions for remote sensing. *Remote Sensing of Environment, 228*, 164–182.

Chapter 26
Digital Earth: From Earth Observations to Analytical Solutions

Cuizhen Wang

Abstract Remote sensing collects the primary data for Earth observations. Social sensing especially citizen science offers crowdsourced volunteered geographic information (VGI) as patchworks of geospatial data infrastructure. Digital Earth integrates remote sensing and social sensing by employing Big Earth Data approaches. Via integration, geospatial information can be improved in four domains: spatial (coverage vs. details), temporal (timeliness), social (contextual), and data (credibility). While facing significant challenges in harnessing the soaring amount of spatial and social data, Digital Earth holds great opportunities for geospatial analytics to assist sustainable decision making.

Keywords Digital Earth · Remote sensing · Social sensing · Geospatial analytics · Big Earth data

26.1 Introduction

The first-generation Digital Earth was initially introduced by former U.S. Vice-President Al Gore to represent the Earth planet in a multi-resolution, three-dimensional system that allowed users to navigate through space and time (Gore, 1999). With rapid technological advances in Big Data era, it is now much closer to reality by utilizing vast amount of geographic information in both physical and social dimensions. Since 1992, more than 20,000 publications relevant to Digital Earth have been google-scholar indexed, and the numbers are steadily increasing (van Genderen et al., 2020).

Earth observations have heavily relied on remote sensing to collect imagery in a synoptic view. In 1858, the first aerial photograph was taken near Paris, France from a balloon tethered at 80 m high (Paine & Kiser, 2012). The real field of aerial remote sensing emerged in 1909 after airplane was invented by the Wright Brothers in 1903. Historical Earth observations were primarily based on analytical interpretation of

C. Wang (✉)
Department of Geography, University of South Carolina, Columbia, SC 29208, USA
e-mail: cwang@mailbox.sc.edu

B. Li et al. (eds.), *New Thinking in GIScience*,
https://doi.org/10.1007/978-981-19-3816-0_26

film-developed photographs, for example aerial reconnaissance during World War I and II. The first man-made satellites, Sputnik launched in 1957 by the former Soviet Union and Corona in 1960 by the U.S., represented Earth sensing transiting from airborne to spaceborne.

Digital remote sensing started in 1972 when the first Landsat satellite was launched by the U.S. National Aeronautics and Space Administration (NASA) to collect digital imagery instead of films for photo prints. Since then, multispectral, multi-sensor and multi-platform satellite remote sensing has become flourishing, providing countless digital imagery all over the world. Around 150 Earth observation satellites were in orbit in 2008 with daily acquisition of 10 terabytes (Tatem et al., 2008). By the end of 2020, it has increased to 906 Earth observation satellites amongst 3372 actively operating ones (Mohanta, 2021). More satellites are expected to be launched, providing an incredibly mass amount of spatial data for synoptic observations of our planet Earth.

Earth remote sensing, however, is restricted by various systematic and environmental constraints such as limited spatial resolutions, long revisit cycles, heavy cloud cover and atmospheric contamination. It is therefore often less useful in application cases such as rapid response of disasters like earthquakes and extreme weathers.

Social sensing emerges as a new, open source of spatial information. It falls in a broad category of sensing and data collection from humans or devices on their behalf (Dong et al., 2015). This chapter only studies social sensing in a perspective of spatial data collection from human sensors. In a sense of citizen science (Goodchild, 2007), volunteered geographic information (VGI) collected and distributed by non-authoritative individuals provides an opportunity to timely tackle the societal problems. The most common examples are VGI from social media and crowdsourcing, for example geo-tagged Twitter data (Palen et al., 2010), OpenStreetMap and Google Street View images. Although the reliability and validity of VGI are still heavily criticized (Schnebele & Waters, 2014), social sensing has become increasingly utilized in a variety of fields where spatial information is needed.

Digital Earth enables the integration of Earth remote sensing and social sensing for improved analytical solutions. This chapter explores the recent advancement of each category, elaborates the integrated approaches, and discusses the challenges and opportunities of Digital Earth for geospatial analytics in support of a sustainable society.

26.2 Remote Sensing: A Long Path of Earth Observations

Earth remote sensing has long been the primary source of spatial data and information. Early efforts in the 1950s–1960s were made primarily via analytical interpretation of film-developed photos. One famous example is the Cuban Missile Crisis (Len & Gerald, 2015). In 1962, with a set of photos collected by the U-2 spy aircraft, analysts detected the evidence of the Soviet Union installing medium-range missiles in Cuba, just 90 miles away from U.S. shores. After thirteen days (October 16–28) of confrontation between the two countries, the missiles were removed. Remote sensing played a major role in preventing a nuclear war.

Digital remote sensing since the 1970s dramatically expanded the capacities of Earth observations and applications. The electromagnetic spectrum it can sense spans from ultraviolet, visible, infrared (including thermal) to microwave regions. Beyond linear and area array cameras, image sensors could also be optical scanners, laser altimeters, synthetic aperture radar (SAR), and radiometers, etc. Depending on the energy source, it could be passive sensing that records the reflected sunlight or Earth surface emission, or active sensing that emits laser or radar signals and records the returns. Longer-wavelength sensors allow day-and-night observations such as land surface temperature or brightness temperature. Active sensors like LiDAR and SAR are also weather-free to be operated in cloudy environments.

Superior to visual interpretation of film-based photos, digital remote sensing employs numerical approaches for image analysis and interpretation. Digital imagery is built on its unit pixels in each band. While a pixel's size varies from sub-meter to tens of kilometers for different sensors, its digital number values provide numeric representation of the unit area on Earth surface, e.g., greenness, wetness, temperature, and terrain. Various models were developed to automatically identify the land use/cover types from the image and to determine their changes such as deforestation, urbanization, and agricultural expansion. Beyond this type of per-pixel analysis, object-based image analysis (OBIA) became popular in the 2000s to segment the imagery for improved feature extraction and classification (Blaschke, 2010). A large number of spectral indices were extracted to quantitatively measure land properties such as vegetation healthiness, impervious surface percentage, snow cover and soil wetness. More recently, machine learning and deep learning are increasingly applied in remote sensing to mine the mass amount of spatial data for improved classification and object detection. Deep learning, especially, mimics the function of human brain in neural networks to perform artificial intelligence for advanced image classification and object detection (Li et al., 2020; Ma et al., 2019).

Earth observations are more than 2-dimensional (2D) image view and mapping. Photogrammetry, the science and technology of measuring objects on photographic imagery, were intensively utilized to extract the 2D/3D information of an object such as distance, direction, area, and height on historical aerial photos. Digital photogrammetry software packages were later evolved to extract 3D products such as digital elevation models and ortho-rectified images based on the stereoscopic image pairs. The 30-m ASTER Global Digital Elevation Model (GDEM) product released in 2009, for example, was developed with the stereo image pairs collected from two TERRA/VNIR telescopes, one nadir and the other looking backward (Abrams et al., 2010). Similarly, radar remote sensing collects the paired SAR images in slightly different view angles to extract the interferogram and calculate elevation based on their phase differences. In 2000, NASA launched the Shuttle Radar Topographic Mission (SRTM) to develop its global DEM database (Farr & Kobrick, 2000). The interferogram of two SAR image scenes acquired at different times could also extract earth deformation such as subsidence and earthquake in mm-scale accuracies. Although not an imaging sensor, LiDAR emits discrete laser pulses and records the time ranges of the returned laser pulses. Both digital bare Earth model and digital surface model atop landscape are extracted.

Ready-to-use global products derived from satellite imagery have been popular. Earlier efforts include the Advanced Very High Resolution Radiometer (AVHRR) Global Inventory Modeling and Mapping Studies (GIMMS) NDVI3g product that is globally available at 8-km grid size and 15-day interval in 1981–2015 (Pinzón & Tucker, 2014). The most highly utilized ones are the NASA Moderate Resolution Imaging Spectroradiometer (MODIS) products from TERRA launched in 1999 and AQUA in 2002, which are providing optical imagery and products at 250–1000 m and global coverage at daily to monthly intervals. Both satellites are still in good operating conditions. A wealth of MODIS products is freely available, e.g., radiation budget variables such as land surface temperature; ecosystem variables such as leaf area index and net primary production; and land cover characteristics such as fire and snow.

Lastly, fine-scale Earth observations are under rapid development. In 1994, the Clinton Administration of the U.S. government announced its policy to allow civil commercial companies to market high spatial resolution imagery (Sneifer, 1996). The first commercial satellite, IKONOS, was launched in 1999 by SpaceImaging, an American company that was later merged to GeoEye and now Maxar Technologies. The IKONOS collected the multispectral (4 m) and panchromatic (1 m) imagery per custom orders in 1999–2015. Since then, a number of meter-scale commercial satellites such as Quickbird and WordView series have been launched, providing a wealth of high-resolution satellite imagery although the cost is also high. In more recent years, the concept of SmallSat—a constellation of small satellites—is well accepted internationally. Even smaller, nanosatellites and CubeSat—a large constellation of boxlike, lightweight satellites as low as 1 kg—are also under rapid growth. The most successful example is the Planet Lab Inc., which holds the world record of deploying a fleet of "Dove" CubeSats (10 × 10 × 30 cm in size). The 180+ Doves by 2021 have the imaging capacity of up to 350 million km^2 per day at 3–5 m resolution, covering nearly the entire landmass on Earth at a daily cadence. These new achievements may potentially meet our need of low-cost Earth imaging in fine-grain details.

In short, with more than one century's development, remote sensing for Earth observations has expanded from airborne to spaceborne, form film-based photography to digital imagery, from cameras to wide-spectral sensors, from narrow orbit swath to global observations, and from coarse-medium to meter-scale observations. Especially in the past 50 years, the long series of satellite observations enables us to collect spatially and temporally continuous information as essential geospatial support.

26.3 Social Sensing: VGI Collection and Dissemination

With the advancement of crowdsourcing, social sensing becomes increasingly utilized in tackling environmental problems. The concept of "Citizens as Sensors", or citizen science defined in Goodchild (2007), allows the collection and web-dissemination of VGI in forms of text, pictures, and videos. In this sense, social sensing is also a means of Earth observations, although the data collected heavily weighs in a social dimension. Superior to remotely sensed data that has fixed spatial resolutions and temporal revisit cycles, the real-time VGI provides timely information for situation awareness and rapid response during a specific event. Among some popular VGI examples, OpenStreetMap is an international effort to create open sourced map data by citizens anywhere of the world, and Google Earth allows the general public to reach and share their own data on a global web interface. Although not homogeneously available around the world, these volunteering efforts potentially fill in the gap of digital geographic information where official sources cannot be available.

Citizen science defines a network of citizens who serve as observers of an event in a spatial domain. While some large citizen science projects provide a fair degree of training, such projects are designed to open to any person who can contribute to the network without much expertise. A recent example is the Ghost of the Coast project by the Gedan Lab at the George Washington University. It engages citizen science to help document ghost forests along the U.S. east coast. Any person who spots the sign of dead or dying trees in coastal forest could post the photos with attributes including date, forest type and location (conveniently identified from phone apps such as Google Maps). The identified locations and their tagged attributes are visually available on satellite maps.

With more crowdsourcing platforms in social media (e.g., Twitter, Flickr, Baidu) available, there is an upsurging interest in VGI involvement. Information posted in social media are geotagged, i.e., the location of the post is asserted. Therefore, the posts related to a specific event such as a hurricane and the induced flooding (Huang et al., 2018a), could be effectively manipulated in geospatial analysis. For example, Huang et al. (2018b) established a multi-level flooding reconstruction model by integrating stream gauge gage data, satellite imagery and points of tweets with verified flooding.

The geotagged social media posts may contain both texts and visual (photo) contexts. Early efforts of verifying the VGI were through text-matching and manual verification of each post. However, social media posts have always been massive especially during an event. For example, during South Carolina's 1000-year flood on 1–18 October 2015, a total of 1.28 million georeferenced tweets were collected using Twitter Stream API and REST API (Li et al., 2017). It was extremely time consuming and practically infeasible to manually identify the valid tweets. Taking advantage of deep learning in pattern recognition, some textual and visual-textual fused Convolutional Neural Network architectures have been utilized for automated tagging of the tweeted texts and photos (Huang et al., 2020).

Social media relies on participant populations and access to internet. Therefore, VGI distributions are spatially heterogeneous, mostly clustered with human settlements with access to stable wired or wireless internet connections. VGI in vast natural environments or countries without low-cost internet access remains limited. It thus contributes to patchworks of spatial data infrastructure as guided by the National Spatial Data Infrastructure (NSDI) authorized by President Clinton administration in 1994 (Goodchild, 2007).

Social sensing and its applications are still debated. First, VGI holds high uncertainties collected by untrained, non-professional volunteers. Second, VGI is abundant in some areas but less frequent in other areas due to its patchwork nature. The abundance of VGI in an area should not be simply counted as the abundance or severity of a specific event. There is also debate that VGI may threaten individual privacy and homeland security when fine-detailed spatial and social information is posted online without authorization. Nevertheless, VGI has become more commonly utilized to augment conventional data sources in solving real-world problems.

26.4 Digital Earth: An Integrated Analytical Solution

The emerging Big Earth Data implements geospatial information from both remote sensing and social sensing. Remote sensing deploys a traditional top-down approach to collecting and disseminating geographic information. On the contrast, social sensing follows a reversal, bottom-up approach in a framework of citizens as sensors.

Both sensing approaches have their own advantages and disadvantages. Figure 26.1 summarizes their primary differences in aspects of spatial coverage and details, timeliness, imbedded information, and credibility. The integration of remote sensing and social sensing may achieve the full potential of Digital Earth. The improvement can be reflected in four domains as described below.

1. **Spatial domain**: Remote sensing provides global coverage in a continuous field. Social sensing is local, contributing to patchworks of geospatial infrastructure. By integrating the pixelized imagery with the mass amount of VGI in a specific "patch", we are able to gain more detailed, continuous spatial information in this area.
2. **Temporal domain**: Satellite imagery for timely Earth observations is restricted by its revisit cycle and spatial resolutions. VGI could be real-time. During a specific event, the integration of two data sources helps to gain better awareness of its spatiotemporal development of the event.
3. **Social domain**: Remote sensing imagery records land covers and the implicated uses on Earth surfaces. It does not contain any contextual information such as the names of places. We may imbed the imagery with the relevant social, cultural, and economic information asserted in VGI to assist social analytics.
4. **Data domain**: Remote sensing data is affirmative, i.e., the imagery fairly represents the land characteristics. Yet errors in image processing and interpretation

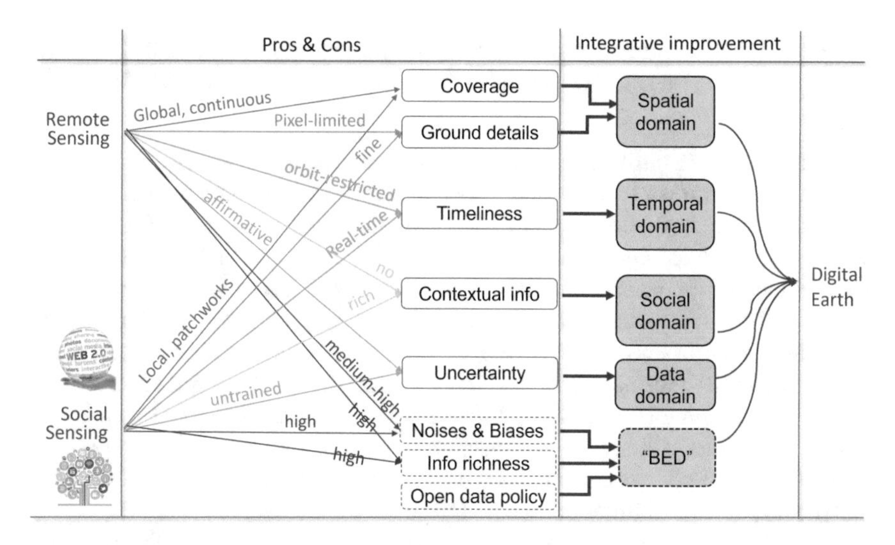

Fig. 26.1 Comparison of remote sensing and social sensing and their integrative usage in Digital Earth. The "BED" represents Big Earth Data approaches

may be introduced even by professional analysts. VGI holds high uncertainties through crowdsourcing of the untrained volunteering inputs. Digital Earth benefits from cross validation of the two information pieces.

Digital Earth does not simply treat remote sensing and social sensing as two discrete data sources. Big Earth Data (BED) complies with the five V's of Big Data: velocity, volume, value, variety, and veracity. Information richness and redundancy are intimately blended. The BED approaches thus face great challenges and opportunities toward the development of Digital Earth.

1. **Harnessing Big Earth Data**: All observational data contains noises and biases. Remote sensing is affirmative, but atmospheric interaction attributes to data loss or noises to Earth observations. In case of cloud cover, Earth information cannot be collected with optical remote sensing. Noises in social sensing is intrinsic. VGI is also heavily biased by the observer's inclination. The BED approaches to fusing remote sensing and social sensing require new and creative solutions in a well-designed architecture and system (van Genderen et al., 2020). Deep learning and artificial intelligence have been offering great opportunities for us to harness Big Earth Data. Yet many issues are still upfront to be solved, for example data standardization and strategies to maintain data quality and up-do-datedness in the rapidly changing geospatial landscape.
2. **Spatial data openness and privacy**: The high resolution satellite imagery and citizen as sensors allow the public to access the extremely detailed information in an unprecedented breadth and speed. This brings in the confrontation between data openness and social privacy. Homeland security has always been a concern with the release of spatial data with fine-grained details. How open should all

data in Digital Earth systems be? There are no answers yet. It is still challenging to clarify the transitional shift from data sharing to information and knowledge dissemination.

Digital Earth is built for adaptive human and societal analytics. It embraces geospatial data infrastructure, social networks, citizen science and human processes on Earth (Wang et al., 2015). Geospatial analytics is in an increasing demand for workforce in service employment sectors. Within the scope of Digital Earth, one active example is geospatial intelligence (GEOINT) that has been under rapid development in both high education and geospatial industry. Guided by a complete set of essential body of knowledge (BoK), the competency-based GEOINT certificates and certifications have been widely offered to promote Digital Earth education and professional training (Wang et al., 2015). Rooted on the fundamental components of GIScience, image analysis, data management and geo-visualization, the geospatial intelligence BoK has also included cross-field competencies such as qualitative analytical skills and abilities across the full scope of practice.

Digital Earth and geospatial intelligence are interactively evolving. The high resolution satellite imagery has allowed the emerging business of Earth Intelligence under the partnership between geospatial industry and government agencies and sometimes academia. Maxar Technologies in 2019, for example, won a significant contract to provide US Government users with on-demand access to mission-ready, high-resolution satellite imagery in multiple classification levels within 2–4 h after image acquisition. Other major geospatial companies such as Google and Planet have also built their own web API for user-friendly image services globally, including raw image and analytical products at various levels. Regarding the new development of citizen as sensors, small Unmanned Aircraft Systems (sUAS) have been treated as "personal remote sensing" for consumer-oriented 3D landscape monitoring (Wang et al., 2021). The rapid development of wireless sensor networks and Earth observation sensor webs (Chen et al., 2014) are becoming timely and pertinent data acquisition, processing, and service networks for Digital Earth.

Integrating geospatial information in physical and social dimensions, Digital Earth allows a deeper understanding of our natural and social environments and the change-response mechanisms in local to global scales. It enables better schemes of decision making on sustainable development and conservation of our Earth planet.

26.5 Conclusion

Remote sensing has long history of Earth observations, yet the imagery is restricted by spatial and temporal resolutions. Social sensing collects crowdsourced information via citizens as sensors although the applicability of VGI is often criticized. Digital Earth takes advantage of the merits of both sensing categories and filling their gaps utilizing Big Earth Data approaches. It expands our abilities from Earth observations to geospatial analytical solutions, preparing for a full capability of observing, interpreting, measuring, and decision making in tackling societal problems. Information in Big Earth Data is abundant and redundant in nature. Digital Earth faces the challenges of an effective Earth architecture to harness the five V's and a balanced system to maintain spatial data openness, human privacy, and social security. These could be better answered with the ever-changing digital technologies and internationally adopted data policies.

References

Abrams, M., Bailey, B., Tsu, H., & Hato, M. (2010). The ASTER global DEM. *Journal of American Society for Photogrammetry and Remote Sensing, 20*, 344–348.

Blaschke, T. (2010). Object based image analysis for remote sensing. *ISPRS Journal of Photogrammetry and Remote Sensing, 65*, 2–16.

Chen, N., Chen, X., Wang, K., & Niu, X. (2014). Progress and challenges in the architecture and service pattern of Earth observation sensor web for Digital Earth. *International Journal of Digital Earth, 7*, 935–951.

Farr, T. G., & Kobrick, M. (2000). Shuttle radar topography mission produces a wealth of data. *American Geophysical Union EOS, 81*, 583–585.

Goodchild, M. F. (2007). Citizens as sensors: The world of volunteered geography. *GeoJournal, 69*(4), 211–221.

Gore, A. (1999). The digital earth: Understanding our planet in the 21st century. *Photogrammetric Engineering and Remote Sensing, 65*(5), 528–530.

Huang, X., Li, Z., Wang, C., & Ning, H. (2020). Identifying disaster related social media for rapid response: A visual-textual fused CNN architecture. *International Journal of Digital Earth, 13*(9), 1017–1039.

Huang, X., Wang, C., & Li, Z. (2018a). A near real-time flood-mapping approach by integrating social media and post-event satellite imagery. *Annals of GIS, 24*(2), 113–123.

Huang, X., Wang, C., & Li, Z. (2018b). Reconstructing flood inundation probability by enhancing near real-time imagery with real-time gauges and tweets. *IEEE Transactions on Geoscience and Remote Sensing, 56*(8), 4691–4701.

Len, S., & Gerald, H. R. (2015). *The Cuban Missile crisis: A critical reappraisal.* Taylor and Francis. Archived at. Last accessed December 15, 2021.

Li, H., Wang, C., Ellis, J. T., Cui, Y., Miller, G., & Morris, J. T. (2020). Identifying marsh dieback events from Landsat image series (1998–2018) with an Autoencoder in the NIWB estuary, South Carolina. *International Journal of Digital Earth*, 1–17.

Li, Z., Wang, C., Emrich, T., & Guo, D. (2017). A novel approach to leveraging social media for rapid flood mapping: A case study of the 2015 South Carolina Floods. *Cartography and Geographic Information Sciences, 45*(2), 97–110.

Ma, L., Liu, Y., Zhang, X., Ye, Y., Yin, G., & Johnson, B. A. (2019). Deep learning in remote sensing applications: A meta-analysis and review. *ISPRS Journal of Photogrammetry and Remote Sensing, 152*, 166–177.

Mohanta N. 2021. How many satellites are orbiting the Earth in 2021? *Geospatial World*. Available at https://www.geospatialworld.net/blogs/how-many-satellites-are-orbiting-the-earth-in-2021/. Last accessed December 15, 2021.

Paine, D. P., Kiser, J. D. (2012). *Aerial photography and image interpretation* (3rd Ed.). Wiley.

Palen, L., Starbird, K., Vieweg, S., & Hughes, A. (2010). Twitter-based information distribution during the 2009 Red River Valley flood threat. *Bulletin of the American Society for Information Science and Technology, 36*(5), 13–17.

Pinzón, J. E., & Tucker, C. J. (2014). A non-stationary 1981–2012 AVHRR Ndvi3g time series. *Remote Sensing, 6*, 6929–6960.

Schnebele, E., & Waters, N. (2014). Road assessment after flood events using non-authoritative data. *Natural Hazards and Earth System Sciences, 14*(4), 1007–1015.

Sneifer, Y. (1996). The implications of national security safeguards on the commercialization of remote sensing imagery. *Seattle University Law Review, 19*, 539–572.

Tatem, A. J., Goetz, S. J., & Hay, S. I. (2008). Fifty years of earth-observation satellites. *American Scientist, 96*(5), 390.

van Genderen, J, Goodchild, M. F., Guo, H., Yang, C., Nativi, S., Wang, L., & Wang, C. (2020). Digital Earth challenges and future trends. In H. Guo, M. F. Goodchild, & A. Annoni (Eds.), *Manual of digital earth* (892p). SpringerOpen, Berlin.

Wang, C., Morgan, G., & Hodgson, M. E. (2021). sUAS for 3D tree surveying: Comparative experiments on a closed-canopy earthen dam. *Forests, 12*, 659.

Wang, D., Abdelzaher, T., & Kaplan, L. (2015). *Social sensing: Building reliable systems on unreliable data* (232p). Morgan Kaufmann. https://doi.org/10.1016/C2013-0-18808-3

Chapter 27
Spatial–Temporal Big Data Enables Social Governance

Jianya Gong and Gang Xu

Abstract The application of GIS technologies has extended from natural sciences to social sciences. The emerging spatial–temporal big data supported by GIS has broad applications in social governance. Through a unified time–space reference, multi-source big data from different departments can be linked and organized, forming a block data. A cloud platform based on the block data is developed for data processing, data fusion, data analysis, and data mining. This cloud platform can support the management of specific public affairs, such as natural resources management, urban and rural planning, and urban construction. In the future, we need to further explore and use spatial–temporal big data to constantly improve our spatial governance capabilities.

Keywords GIS · Social governance · Big data · Time–space reference

27.1 Introduction

Geographic Information Science (GIS) has long been applied beyond geography. GIS is not only widely applied in the fields of natural sciences such as geo-sciences, but also extended to social sciences such as history and public management (Liu, 2021). Following the development of GIS, its application includes at least the following four aspects: natural resources management, smart city, public services, and social governance (Fig. 27.1).

GIS has been applied to natural resource management early (Wright et al., 2009). It is well known that the earliest GIS system was the Canadian Geographic Information System (CGIS) developed for land-use management and resource monitoring in the early 1960s. The application of GIS to natural resource management is also constantly

J. Gong (✉)
School of Remote Sensing and Information Engineering, Wuhan University, Wuhan 430079, Hubei, China
e-mail: jygong@whu.edu.cn

G. Xu
School of Resource and Environmental Sciences, Wuhan University, Wuhan 430079, Hubei, China
e-mail: xugang@whu.edu.cn

B. Li et al. (eds.), *New Thinking in GIScience*,
https://doi.org/10.1007/978-981-19-3816-0_27

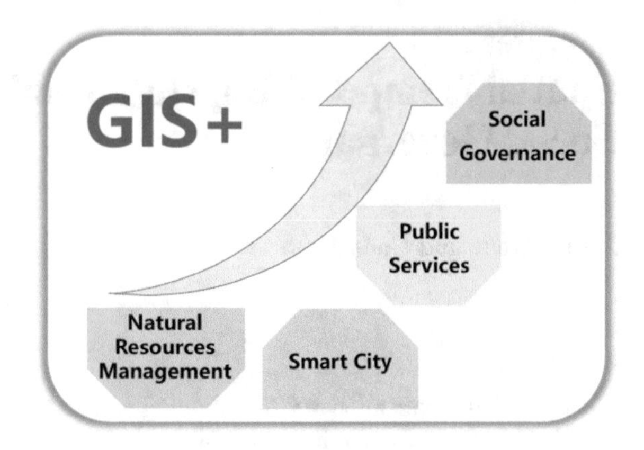

Fig. 27.1 The application of GIS extending from natural sciences to social sciences

evolving (Pei et al., 2021). At present, not only the management of land resources, but also water resources, forest and grass resources and other natural resources are inseparable from GIS, since natural resource data are all spatial data. The survey and monitoring of natural resources, ownership confirmation, and land use planning require the support of GIS technology. This type of application focuses on objects whose natural attributes change slowly.

Another field is the application of GIS to the construction of smart cities (Lv et al., 2018). Urbanization is the most important social transformation in the twenty-first century. By 2050, more than two-thirds of the world's population will live in cities (United Nations, 2015). At the turn of the new century, the field of geo-science proposed to build digital cities, which later developed into smart cities (Li et al., 2014). A city is composed of diverse types of networks, infrastructure and environmental systems that support the core functions of the city (Gong et al., 2020). The construction of smart cities requires the realization of comprehensive perception, ubiquitous interconnection, high-speed computing, and integrated applications (Gong & Wu, 2012). All of these applications are closely linked with cloud computing and other new-generation information technology (Trencher, 2019). GIS is indispensable in the construction of smart cities, and the typical representative is the building of "urban brain" of megacities.

The third aspect is the application of GIS to public services (Shi et al., 2018). Navigation and location-based services have been developed and improved in the past two decades; and they have quickly entered our daily lives. It is difficult to imagine how urban residents would find their destination in the complex road network without navigation. Location-based services have become inseparable for our daily lives (Ding et al., 2021). We use location-based services in all aspects of our daily travel, shopping, and other social activities. The difference from the previous two aspects is that public service is directly oriented towards people.

Today, application of GIS is extending to social governance (Lin, 2008, 2013). Social governance refers to the process by which the government, organizations, and citizens guide and regulate public affairs and social lives, and ultimately realize the maximization of public interests. China proposed to improve the modernization of the social governance system and governance capabilities. The improvement of spatial governance capabilities is an important step to achieve the above goals, which cannot be separated from GIS technologies. Different from public services with GIS, participants of social governance are broad, and social affairs are complex and diverse (Lazer et al., 2009, 2020). Therefore, the application of GIS to social governance faces unprecedented challenges (Zhou et al., 2020). This chapter attempts to explore the prospects, challenges and countermeasures of spatial–temporal big data supported by GIS in social governance.

27.2 Current Situation of Social Governance

The social structure is undergoing profound changes, and the needs of social participants have become more diverse. Therefore, social governance has become more challenging. The current social governance is transforming from digitization to intelligence, from fragmentation to integration, and from government-centered to citizen-centered. The emergence of spatial–temporal big data provides new ideas for social governance. Of course, this requires us to expand the functions of GIS to meet the needs of managing and analyzing spatial–temporal big data.

27.2.1 Why Social Governance Needs GIS?

We are now living in the new era of information. The advancement of information technology and the development of economy have made our lives more convenient. Great changes have taken place in our entire society and personal lifestyle, which are embodied in the following three aspects: (1) People's activity space expands from real geographic space to virtual space on the Internet. (2) Our social organization has changed from enterprises and public institutions to decentralized communities in a modern city. (3) We have transformed from a person constrained by conventional institutions to a person who can choose employment flexibly and independently or start a business on our own. In order to adapt to changes in social and organizational forms, social governance also needs to keep pace with the times (Mukherjee, 2018).

Our new life is fast-paced, and new technology research and development cycles are short. In addition, diverse social participants have very different demands and social governance involves numerous departments, objects, and events (Fig. 27.2). It is difficult for conventional top-down linear social governance to keep up with these changes (Mukherjee, 2020). At the same time, fast-paced social operation has produced massive and diverse spatial–temporal big data, which are often related to spatial location (Anselin et al., 2021). The effective organization and management of these massive spatial and temporal big data can resolve social problems and significantly improve the level of social governance. Therefore, social governance,

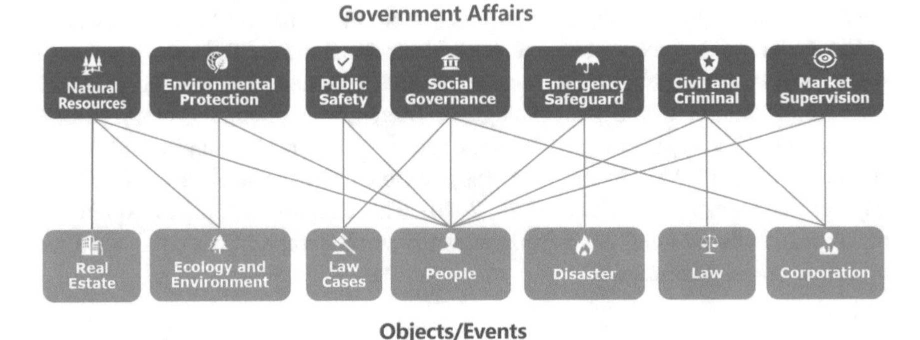

Fig. 27.2 Government affairs and social governance involving multiple departments, objects, and events

especially urban governance, urgently needs to introduce spatial–temporal big data supported by GIS technologies (Lin, 2008).

27.2.2 Problems and Challenges in Social Governance

There are many problems in current social governance (Fig. 27.3). First, at the community-level, staff personnel are overburdened. In order to fine-tune management, Chinese cities are divided into fine grids, which are usually finer than the community, and can be as fine as every building. Each grid has a dedicated person in charge, called a grid manager. Many specific management tasks of higher-level departments fall to the grid manager, which makes the community-level staff overburdened. Second, community-level departments produce data but do not have permission to use the data. Community-level departments are direct participants in social governance, and they have produced a lot of data in the management process;

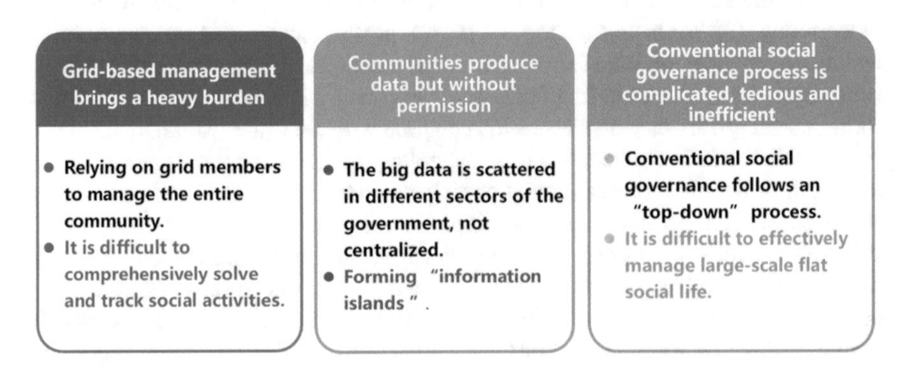

Fig. 27.3 Problems and challenges in current social governance

however, they do not have access to the data. Thus, these data cannot be directly used to support community-level management. Third, the conventional social governance process is overly complicated, tedious, and inefficient. The conventional top-down social governance is difficult to effectively manage the modern society that has a flat structure.

27.2.3 New Ways and Exploration of GIS for Social Governance

Conventional GIS has been mainly used to manage fixed objects on the Earth's surface, while social governance involves moving objects including people, vehicles, etc. How to extend the conventional functions of GIS to manage dynamic objects is a big challenge in social governance. It is necessary to build a new information system for social governance. The ideas and strategies for the improvement are refining subjects and objects, and integrating information from different departments, such as urban planning, urban construction, public security, etc. We need to build a unified platform with diversified participation in the social governance system to improve efficiency. This comprehensive system needs GIS technology.

Specific strategies are as follows:

- *Refining governance units.* We need to carry out refined community management and services in micro-units such as streets, courtyards, buildings, and grids.
- *Multiple co-governance.* We can encourage more organizations and individuals to participate in social governance, so as to achieve shared-governance and co-management.
- *New Information Technologies.* We need to introduce emerging information technologies such as big data and cloud computing to provide innovative ideas and strong support for social governance.

27.3 Spatial–Temporal Big Data in Social Governance

It is necessary to effectively manage spatial–temporal big data before using it in social governance, which relies on a unified time–space reference. The time–space reference is the infrastructure for urban informatization. Specific applications in social governance rely on cloud platforms, which establish the link between spatial–temporal big data and public affairs.

27.3.1 Time–Space Reference: Infrastructure for Urban Informatization

In order to process multi-source heterogeneous spatial–temporal big data in modern social governance, it is necessary to construct a unified time–space reference system to facilitate the organization and management of these big data (Fig. 27.4). A unified time–space reference system bridges data association and the cornerstone of intelligent social governance. In this system, each geographic entity is assigned a unique code that is a combination of its street address and geographic coordinates (Fig. 27.4). This step achieves the unification of the spatial reference. We can further add a time dimension in this framework so as to manage time-series big data. As a result, we build a spatial–temporal information database with a unified time–space reference.

Then, through the geographical identification of the management objects, the relationship between the social governance objects and the spatial–temporal information database is established. The objects of social management and services are mainly natural persons and legal persons, so population database and enterprise database are very important. We can associate them with the spatial–temporal information database through the residential address of the natural person and the registered address of the company. Next, we can develop specialized information systems for specific management affairs, such as urban planning systems and real estate management systems.

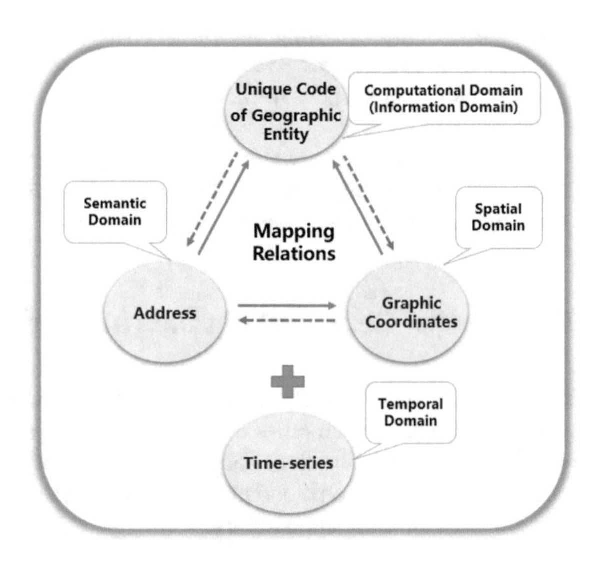

Fig. 27.4 Time–space reference system forming the infrastructure for urban informatization

27.3.2 Cloud Platform for Spatial–Temporal Big Data

Under the unified framework of time–space reference, we can build a spatial–temporal big data cloud platform by integrating multi-source heterogeneous spatial–temporal big data (Fig. 27.5). With the help of the spatial–temporal big data cloud platform, we can achieve the transformation from simple data publishing to comprehensive capability services, from data combination to multi-source data fusion, from static data to dynamic data, and from geographic positioning to geographic association. On this basis, we can realize geographic information fusion services, real-time data services, geographic event services, and space–time multidimensional analysis. The spatial–temporal big data cloud platform effectively helps us realize data processing, data fusion, data analysis and data mining. These technologies and methods can directly serve specific public affairs, such as natural resource management, urban and rural planning, urban construction, operation management, and public services.

27.4 Practice of Social Governance with GIS

This section introduces the practice of social governance with GIS in Shenzhen, China. We first explain the concept of block data, which consists of geographic data block, basic data block, and data block for specific affairs. Taking Shenzhen, China as an example, three typical applications are described to demonstrate how spatial–temporal big data play a role in specific public affairs.

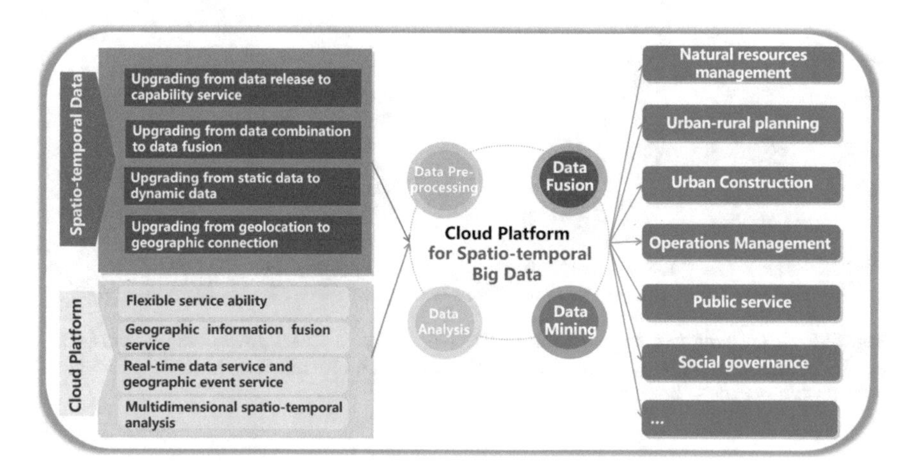

Fig. 27.5 Cloud platform for spatial–temporal big data supporting data pre-processing, data fusion, data analysis, and data mining

27.4.1 Building Data Blocks for Social Governance

Shenzhen, the youngest megacity in China, is at the forefront of social governance innovation. The government of Shenzhen proposed the concept of multiple governance based on block data. Block data encapsulates the social management elements (strip data, such as people, affairs, things, organizations) scattered in various departments into "blocks", which corresponds to provinces, cities, districts, streets, communities, and grids according to the management level. Through the establishment of block data models and relational graphs, we can perceive the social operation situation in an all-round way, so as to perform accurate analysis, conduct governance, provide services, and obtain feedback. This new system then provides scientific and technological support for improving the level of refinement of social governance.

Block data is composed of three parts: spatial geographic block, basic data block, and data block for specific affairs. Spatial geographic block mainly refers to spatial geographic information data, which is mainly divided into 8 levels of block units such as city block, district block, street block, community block, grid block, building block, apartment block and room block. The basic data block mainly refers to five major categories, including people, houses, corporations, events, and communications, and their respective subcategories. With the unified address code as the link, the five codes are associated to form a basic block data (Fig. 27.6). Data blocks for specific affairs are related to various departments, such as public security, fire protection, and market supervision.

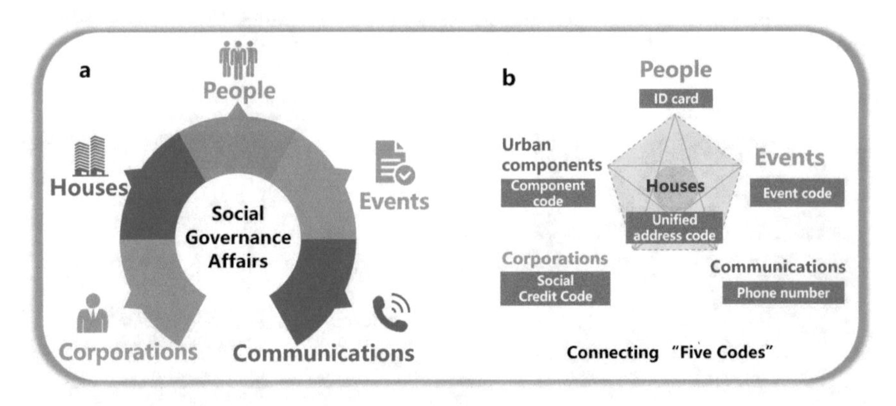

Fig. 27.6 Connecting five codes to form basic block data. **a** Social governance affairs involve people, houses, corporations, events, and communications. **b** Five elements are linked through the unified address code of houses

27.4.2 *Example Cases of Social Governance*

Through the technological innovation of block data, Shenzhen has achieved a number of social governance innovations. One is the change from grid management to block management. Grid management is a top-down refinement of administrative management, trying to carry out a comprehensive and refined management of the society. Grid management relies on a large number of people. The block management puts the elements of social governance into the eight-level blocks, making data management more refined and business governance more precise. The second is from ex-post disposal to pre-prediction. Often social governance problem can only be dealt with after the fact. Block data analysis helps to achieve early warning and prediction, more proactive discovery of social governance problems, and more humane push for social grassroots services. The third improvement is changing from top-down to point-to-point link. The conventional governance chain is hierarchical and cannot meet the needs of social governance which is flat in structure. Block data utilizes a point-to-point and end-to-end approach to realize the direct dissemination of information to specific responsible entities and optimize the chain of government management.

Example applications include the following 3 cases:

- **COVID-19 epidemic prevention and control**. Relying on the unified address and block data intelligent backplane, Shenzhen has enabled residents' self-declaration of itinerary information, which are accessible in the spatial database in real time, opening up the path to COVID-19 contact tracing and community investigation, improving the efficiency of epidemic prevention and control. During the epidemic, the self-declared itinerary data from more than 100 million people were matched to specific spatial locations (Fig. 27.7).
- **Integration of commercial management**. The registered addresses of enterprises have changed from filling in by oneself to selecting from the information system, which solves the problem of providing false registered addresses from the source (Fig. 27.8). Through the aggregation and linking of multi-department legal person data through the block data backplane, the "closed loop of law enforcement" in the entire process of registration and approval, service, supervision, and credit of

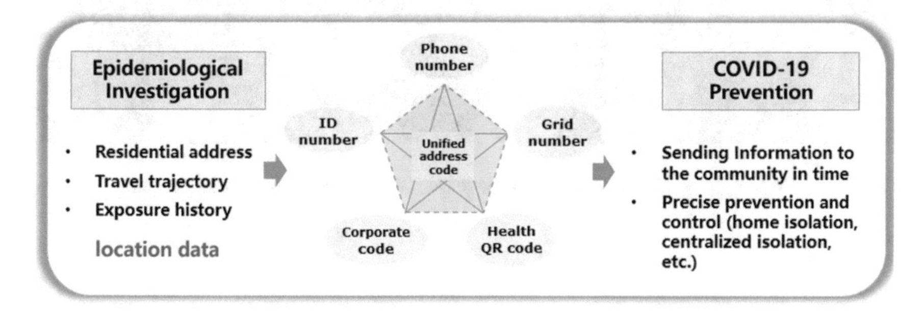

Fig. 27.7 Epidemiological investigation and COVID-19 prevention based on block data platform

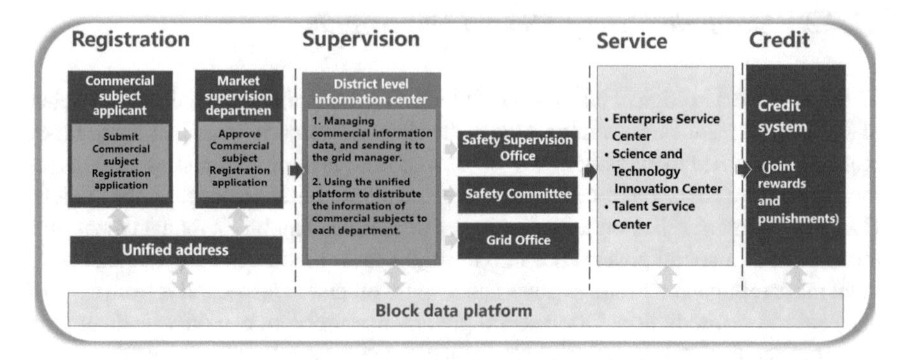

Fig. 27.8 Integration of commercial management and services based on block data platform

commercial entities has been realized, providing strong support for optimizing the business environment (Li et al., 2018).

- **Service for special patients**. Local government uses block data to accurately locate key personnel such as patients with mental health, drug addicts, and explores multiple co-treatment of key personnel. Based on this, a closed management loop is built for the discovery, visit, control, and treatment of special populations (Fig. 27.9).

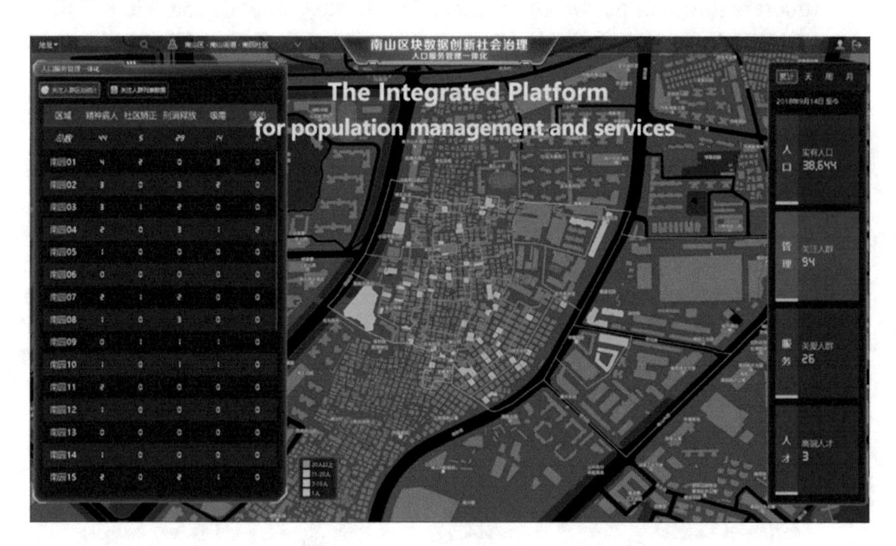

Fig. 27.9 The integrated platform for the management and services for special patients

27.5 Conclusions

The role of spatial–temporal big data in social governance is becoming increasingly important. Spatial–temporal big data can transform personal decision-making to intelligent decision-making based on data. With the support of muti-dimensional data, the needs of social governance are more precise and the description of social governance objects are clearer. We can build a new pattern of social governance by using data to extract information, to make decisions, to manage and to innovate, which can promote the governance system and governance capabilities.

From a single data source to multiple data datasets, we can realize the continuous aggregation and fusion of data across industries, departments, and fields. The highly correlated characteristics within the block data provide conditions for the continuous accumulation of data. Highly related data can be gathered together to reveal relationship patterns among people, things, in geo-space and through time. In the future, it is necessary to build a global social governance chain to serve public sector decision-making from a broader spatial data perspective.

Acknowledgements This work was supported by the National Natural Science Foundation of China (92038301).

References

Anselin, L., Li, X., & Koschinsky, J. (2021). GeoDa, From the Desktop to an Ecosystem for Exploring Spatial Data. *Geographical Analysis*, 1–28.

Ding, X., Fan, H., & Gong, J. (2021). Towards generating network of bikeways from Mapillary data. *Computers, Environment and Urban Systems, 88*, 101632.

Gong, J., & Wu, H. (2012). The geospatial service web: Ubiquitous connectivity with geospatial services. *Transactions in GIS, 16*(6), 741–743.

Gong, J., Liu, C., & Huang, X. (2020). Advances in urban information extraction from high-resolution remote sensing imagery. *Science China Earth Sciences, 63*(4), 463–475.

Lazer, D., Pentland, A., Adamic, L., Aral, S., Barabasi, A. L., Brewer, D., Christakis, N., Contractor, N., Fowler, J., Gutmann, M., Jebara, T., King, G., Macy, M., Roy, D., & Van Alstyne, M. (2009). Computational social science. *Science, 323*(5915), 721–723.

Lazer, D. M. J., Pentland, A., Watts, D. J., Aral, S., Athey, S., Contractor, N., Freelon, D., Gonzalez-Bailon, S., King, G., Margetts, H., Nelson, A., Salganik, M. J., Strohmaier, M., Vespignani, A., & Wagner, C. (2020). Computational social science: Obstacles and opportunities. *Science, 369*(6507), 1060–1062.

Li, D., Yao, Y., Shao, Z., & Wang, L. (2014). From digital Earth to smart Earth. *Chinese Science Bulletin, 59*(8), 722–733.

Li, F., Gui, Z., Wu, H., Gong, J., Wang, Y., Tian, S., & Zhang, J. (2018). Big enterprise registration data imputation: Supporting spatiotemporal analysis of industries in China. *Computers, Environment and Urban Systems, 70*, 9–23.

Lin, W. (2013). Digitizing the Dragon Head, geo-coding the urban Landscape: GIS and the transformation of China's Urban Governance. *Urban Geography, 34*(7), 901–922.

Lin, W. (2008). GIS development in China's urban governance: A case study of Shenzhen. *Transactions in GIS, 12*(4), 493–514.

Liu, Y. (2021). Core or edge? Revisiting GIScience from the geography-discipline perspective. *Science China-Earth Sciences, 64*, 1–4.
Lv, Z., Li, X., Wang, W., Zhang, B., Hu, J., & Feng, S. (2018). Government affairs service platform for smart city. *Future Generation Computer Systems, 81*, 443–451.
Mukherjee, F. (2018). GIS use by an urban local body as part of e-governance in India. *Cartography and Geographic Information Science, 45*(6), 556–569.
Mukherjee, F. (2020). Institutional networks of association for GIS use: The case of an urban local body in India. *Annals of the American Association of Geographers, 110*(5), 1445–1463.
Pei, T., Xu, J., Liu, Y., Huang, X., Zhang, L., Dong, W., Qin, C., Song, C., Gong, J., & Zhou, C. (2021). GIScience and remote sensing in natural resource and environmental research: Status quo and future perspectives. *Geography and Sustainability, 2*(3), 207–215.
Shi, Y., Deng, M., Yang, X., & Gong, J. (2018). Detecting anomalies in spatio-temporal flow data by constructing dynamic neighbourhoods. *Computers, Environment and Urban Systems, 67*, 80–96.
Trencher, G. (2019). Towards the smart city 2.0: Empirical evidence of using smartness as a tool for tackling social challenges. *Technological Forecasting and Social Change, 142*, 117–128.
United Nations, Department of Economic and Social Affairs, Population Division. (2015). World Urbanization Prospects: The 2014 Revision, (ST/ESA/SER.A/366).
Wright, D. J., Duncan, S. L., & Lach, D. (2009). Social power and GIS technology: A review and assessment of approaches for natural resource management. *Annals of the Association of American Geographers, 99*(2), 254–272.
Zhou, C., Su, F., Pei, T., Zhang, A., Du, Y., Luo, B., Cao, Z., Wang, J., Yuan, W., Zhu, Y., Song, C., Chen, J., Xu, J., Li, F., Ma, T., Jiang, L., Yan, F., Yi, J., & Xiao, H. (2020). COVID-19: Challenges to GIS with big data. *Geography and Sustainability, 1*(1), 77–87.

Chapter 28
Geo-computation for Humanities and Social Sciences

Kun Qin, Donghai Liu, Gang Xu, Yanqing Xu, Xuesong Yu, and Yang Zhou

Abstract Humanities and social sciences (HSS) are undergoing the transformation of quantification and spatialization. Geo-computation provides effective computational methods and tools for processing geographic information. Geo-computation for humanities and social sciences (GHSS) is a field coupling geo-computation with humanities and social sciences. This chapter introduces the concept of GHSS, and introduces the origin and development, the related theories and methods, and some applications of GHSS. At the end of the chapter, the future development directions of GHSS are discussed.

Keywords Geo-computation for humanities and social sciences · Computational social sciences · Geo-computation for social sciences · Spatially integrated humanities and social sciences

---

K. Qin (✉) · D. Liu · Y. Xu · X. Yu · Y. Zhou
School of Remote Sensing and Information Engineering, Wuhan University, Wuhan 430079, Hubei, China
e-mail: qink@whu.edu.cn

D. Liu
e-mail: ldhwhdx@whu.edu.cn

Y. Xu
e-mail: yanqing.xu@whu.edu.cn

X. Yu
e-mail: yxs95@whu.edu.cn

Y. Zhou
e-mail: zhouyang0612@whu.edu.cn

G. Xu
School of Resource and Environmental Sciences, Wuhan University, Wuhan 430079, China
e-mail: xugang@whu.edu.cn

B. Li et al. (eds.), *New Thinking in GIScience*,
https://doi.org/10.1007/978-981-19-3816-0_28

28.1 Introduction

Geo-computation is originally introduced in the first international conference on 'Geo-computation', hosted by the School of Geography at the University of Leeds in 1996 (Openshaw & Abrahart, 1996). Longley (1998) sees geo-computation as providing the framework for the execution of geo-computation science—in both advancing the state-of-the-art in the computation of geography as well as extending our understanding of geographical phenomena. Geo-computation is not simply about applying computational methods to explore geographical concepts, it offers an extensive toolkit for the examination and identification of new perspectives on spatial processes (Chen et al., 2012).

Humanities are the disciplines about the knowledge of human heart and feeling, including philosophy, history, literature, linguistics, journalism, art, and so on. Social sciences are the disciplines about the research of various social phenomena, including economy, politics, sociology, law, management, and so on.

In the big data era, humanities and social sciences are undergoing a transition from traditional methods to quantification and spatialization. Some research directions, such as spatially integrated humanities and social sciences (SIHSS) (Lin et al., 2006, 2010), geo-computation for social sciences (GCSS) (Li et al., 2020), are emerging in recent years. SIHSS refers to introducing spatial thinking, spatial concepts into the fields of humanities and social sciences (Lin et al., 2006, 2010; Qin, Lin, et al., 2020). GCSS is a discipline that employs remote sensing earth observations and is driven by spatiotemporal big data that reflects surface features and human activities. It senses, analyzes, and mines categories and intensities of human activities and their influences on natural and social environments in multiple spatial and temporal dimensions (Li et al., 2020). Both humanities and social sciences need geo-computation to provide theories, methods, and toolkits for them. Thus, we refer to this research direction as Geo-computation for Humanities and Social Sciences (GHSS).

The remainder of this chapter is organized as follows. In Sect. 28.2, we discuss the origin and development of GHSS. In Sect. 28.3, we discuss the related theories and methods of GHSS. In Sect. 28.4, we introduce some applications of GHSS. In the conclusion, the future development directions of GHSS are discussed.

28.2 The Origin and Development of GHSS

Geo-computation for humanities and social sciences (GHSS) stemmed from computational social sciences (CSS), geo-computation for social sciences (GCSS), spatially integrated humanities and social sciences (SIHSS), and so on. So, we firstly introduce the origin and development of CSS, GCSS, SIHSS, then discuss the origin and development of GHSS.

In the era of big data, scientific research enters the fourth paradigm: data-intensive scientific discovery (Hey et al., 2009). With the development and application of the internet, especially the internet of things, large quantities of behavior records about human production and life activities are recorded as data, which provide big data sources for social scientists to research problems in social sciences fields. Computational social sciences (CSS) mainly referred to applying computational simulation technologies into the research of social problems (Bankes, et al., 2002; Kuznar, 2006). The paper *Computational Social Science* is published in *Science* in 2009, which boomed large quantity research about this theme (Lazer et al., 2009).

Computational Social Sciences (CSS) apply computational models and algorithms into the big data analysis about human behaviors, and build models, simulate, and analyze different social phenomena. Compared with biology and physics, there are more obstacles for CSS, for privacy and privilege reasons. CSS in the big data era has the characteristics of complexity, self-adaptability, social interaction, and time–space. It is a new research direction to develop spatiotemporal analysis theories and methods of CSS from spatiotemporal viewpoints (Qin, Wang, et al., 2020). The research about the human-nature relationship becomes an essential section of CSS. The spatial information sciences and technologies including geographic information system (GIS), remote sensing (RS), global navigation satellite system (GNSS), provide new technologies for the research of the human-nature relationship. They research the human-nature relationship from different viewpoints and derive geo-computation for social sciences (GCSS), which is illustrated in Fig. 28.1.

Spatially integrated humanities and social sciences (SIHSS) is a new inter-discipline which introduces spatial thinking and spatial concepts into humanities and social sciences. Since the first forum on spatially integrated humanities and social sciences was held at the Chinese University of Hong Kong in 2009, this new inter-discipline has developed vigorously and achieved significant progress. Qin, Lin, et al. (2020) gave a review of SIHSS in the recent ten years. SIHSS is a typical multi-discipline, which combines "3S (GIS, RS, GNSS)" with humanities and social sciences.

Geo-computation for humanities and social sciences (GHSS) is a discipline which provides theories and methods of geo-computation for humanities and social sciences. The framework of GHSS is illustrated in Fig. 28.2.

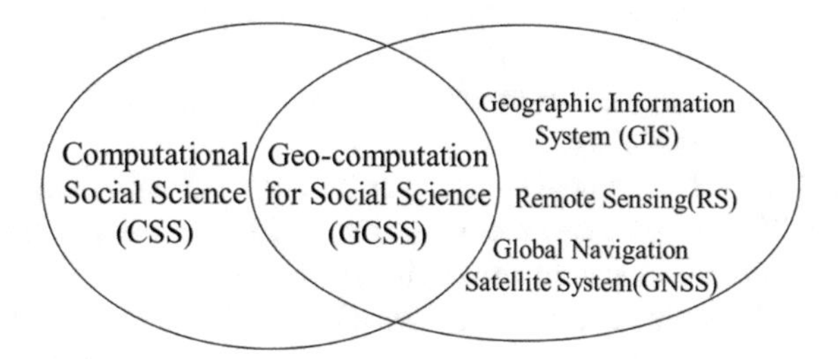

Fig. 28.1 Geo-computation for social sciences (GCSS)

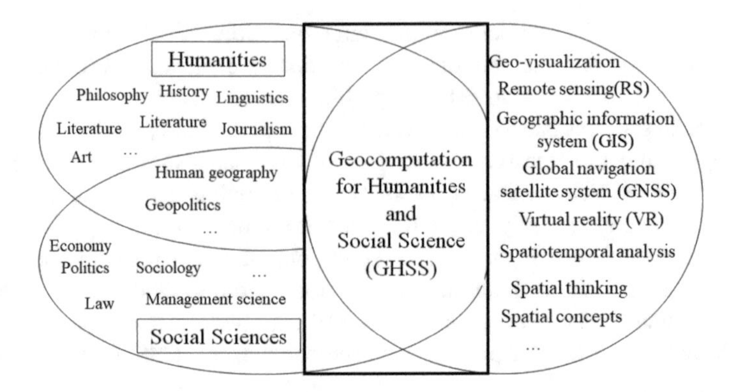

Fig. 28.2 Framework of GHSS

28.3 The Related Theories and Methods of GHSS

GHSS is developed based on the theories and methods of network sciences, social physics, 3S (GIS, RS, GNSS), spatiotemporal analysis, and so on. These related theories and methods of GHSS will be introduced in the following.

Network science, especially complex network science, is the basic theoretical foundation of GHSS. Different from regular network and random network, complex network includes the characteristics of small world, scale-free, hierarchical architecture, self-similarity, and self-organization. The interaction phenomena may be expressed by complex networks. The spatial interaction network is a kind of directed network which embeds population flow, commodity flow, and information flow into geographical space. Geographical multiple flow (GMF) is shaped when different types of geographical flows exist together (Pei et al., 2020). Communication, trade, population migration, population movement, transportation, international relations, and social contacts provide an environment for spatial interaction network.

Social physics firstly refers to applying physics concepts into the research of social science. Pentland (2015) pointed out that social physics aims to describe the flow of information and thoughts, and the reliable mathematical relationship among the flows with human behavior. Social physics is a discipline about thought flow. With the help of thought flow, we can improve collective intelligence, and promote the development of a smart society.

3S (RS, GIS, GNSS) provides technologies for GHSS. RS provides abundant data sources for GHSS, GIS provides methods of spatial data management, and spatial analysis, GNSS provides location and navigation for GHSS.

Spatiotemporal big data analysis provides advanced technologies for GHSS. For example, spatiotemporal big data analysis methods are utilized to explore hotspots based on taxi trajectory data (Zhao et al., 2017), detect abnormal trajectories (Wang et al., 2018) and abnormal behavior patterns, analyze congestion events (Qin, Xu, et al., 2019), and so on.

28.4 Applications of GHSS

There are many examples of application of GHSS including history integrated GIS, literature integrated GIS, linguistic integrated GIS, GIS and philosophy, human dynamics, human geography, politics GIS, GIS and management sciences, spatial econometrics, spatial social networks, spatial interaction networks, geography of crime, and geography of public health, etc. Some applications are introduced in the following section.

28.4.1 Geo-computation for Politics and International Relations

The research concerns of political science include public policies, authoritative distribution of benefits or power, and the behaviors of national subjects. Distribution of power and benefits involves complex political behavior.

Taking elections as an example, some scholars utilized GIS tools to research how geographic elements affect elections and the spatial distribution of election elements. For examples, Bowen (2014) studied the cost of different principles in the process of determining elections in the United State. Based on geo-tagged tweets, Liu et al. (2021) studied the relationship between voting results and county economic growth, approval rates and other factors, and established a prediction model.

Scholars have also used GIS tools to study satisfaction for public policies, the distribution of policy target groups, and so on. For examples, Wang (2020) explored the methodology of applying public health policies to geographic information systems. Pedro et al. (2019) used GIS for sustainability assessment to assist in the formulation of land-use policies.

In addition, international relations (IR) studies the behavior between countries, which is also a research priority of GIS in political science. Scholars use the tools of network science to conduct research on IR. For examples, Zhukov and Stewart (2013) researched how to determine the proximity of two countries (regions) in the IR network. Similarly, related scholars also combined geospatial relations with the social network of national relations to explain the phenomenon of international conflicts or alliances (Flint et al., 2009).

GDELT dataset is composed of real-time international publication news. Assuming that countries A and B often participate in cooperation, the relationship between them is positive, otherwise it would be negative. Therefore, GDELT can be applied to IR studies. An IR network $G = (V, E, W)$ can be constructed, where V represents nodes in the network, and E represents edges in the network (which means two entities appear together in the events), W represents the weights of the edges (the number of events that both A and B participate in for a certain period). The structural mining of the network shows that IR networks have the scale-free characteristic, that is, a small number of countries have a large number of interactions with other countries, and most countries have very little interaction with other countries (Qin, Luo, et al., 2019).

Geo-computation for politics and international relations is an important application of GHSS. Geo-computation may provide spatiotemporal analysis methods for politics and international relations.

28.4.2 Geo-computation for Human Mobility

The last decade has witnessed a trend in literature of combing GIS and human mobility to solve more sophisticated problems. Human geography has always been an important subject in geographic studies. Volunteered geographical information (Goodchild, 2007) and social sensing (Liu et al., 2015) provided new data sources and new thoughts for the research of human mobility. As our understanding of human mobility advances, models to explain the mechanisms governing human traveling behaviors become increasingly more refined. Tools and methods in GIS like layer overlapping and spatial buffering are required to support the application and visualization of human mobility theory (Barbosa et al., 2018; Pindolia et al., 2012).

Related works can be divided into two categories: (i) using geo-tagged data of human movements to solve GIS problems; (ii) using GIS approaches to explain the spatial pattern of human mobility. In the first category, varied data such as taxi trajectory data (Zhao et al., 2016), social media check-in data (Jia et al., 2019), and mobile phone data (Kang et al., 2013) are used, aiming to find hotspots in the city, delineate urban regions, or infer land uses from the perspective of real human interactions. In the second category, tools in GIS can be used in many studies. For example, when building parameterized gravity-like human mobility models, distances and travel time need to be estimated first, in this situation GIS resources like road networks and methods like route planning can help (Pindolia et al., 2012). For another example, POIs can be overlapped on street blocks as indications of different regional functions, therefore the travel pattern derived from GNSS trajectories among those blocks can be understood (Yuan et al., 2012). Both methods have proven to benefit from the combination of GIS and human mobility.

Geo-computation for human mobility is an important application of GHSS. Geo-computation may provide effective spatiotemporal analysis methods for human mobility.

28.4.3 Geo-computation for Public Health

Public health issues are closely related to location, including the natural environment, personal behavior, and health outcomes. Geographic information science (GIS) is an emerging science concerning spatial location. The application of GIS theories and methods in the field of public health is becoming more and more extensive. In short, it mainly includes two aspects. The first is revealing the spatiotemporal pattern of disease incidence and its spatial correlation with environmental elements; the second is evaluating the spatial distribution of medical service facilities and resources and their accessibility, effectiveness, and fairness.

Since the outbreak of the COVID-19 epidemic, the application of GIScience in the field of public health has been particularly emphasized. GIS has played an irreplaceable role in the prevention and control of the COVID-19 epidemic. For example, GIS technology can analyze the temporal and spatial evolution of the epidemic. Previous studies revealed that the outflow population of Wuhan determined the spatial pattern of the epidemic in other cities in China (Xu, Wang, et al., 2021). This also shows that the lockdown of Wuhan has prevented the further spread of the COVID-19 epidemic in China. GIS technologies can also reveal the influencing factors of the COVID-19 spatiotemporal pattern within the city, generally including urban population density, the demographic structure, population interaction frequency, and natural factors such as temperature and air quality (Xu et al., 2022). GIS can also quantify the social and economic impact of public measures such as epidemic prevention and control. For example, non-pharmaceutical intervention measures such as lockdowns were generally adopted to prevent and control the epidemic, which led to a general dimming of city lights at night, indicating that public measures have significantly affected social and economic activities (Xu, Xiu, et al., 2021).

Health will be a topic of increasing concern for urban residents, healthy cities are an important aspect of sustainable urban development. The application of GIS in the field of public health and health geography is bound to become more robust.

28.5 Summary

GHSS (Geo-computation for humanities and social sciences) couples geo-computation with HSS (humanities and social sciences), it is a multidisciplinary intersection research direction. Geo-computation can provide spatiotemporal analysis and computation methods and toolkits for humanities and social sciences. In the future, the research directions of GHSS include the research framework of GHSS, and GHSS's computational models, spatiotemporal analysis methods, geo-visualization methods, and the development of online platforms.

Acknowledgements The authors are grateful for the financial support for the National Natural Science Foundation of China (No. 42171448), and the constructive comments from professor A-Xing Zhu.

References

Bankes, S., Lempert, R., & Popper, S. (2002). Making computational social science effective: Epistemology, methodology, and technology. *Social Computer Review, 20*(4), 377–388.
Barbosa, H., Barthelemy, M., Ghoshal, G., James, C. R., Lenormand, M., Louail, T., Menezes, R., Ramasco, J. J., Simini, F., & Tomasini, M. (2018). Human mobility: Models and applications. *Physics Reports, 734*, 1–74.
Bowen, D. C. (2014). Boundaries, redistricting criteria, and representation in the US house of representatives. *American Politics Research, 42*(5), 856–895.
Chen, T., Haworth, J., & Manley, E. (2012). Advances in geocomputaion (1996–2011). *Computers, Environment and Urban Systems, 36*, 481–487.
Flint, C., Diehl, P., Scheffran, J., Vasquez, J., & Chi, S. H. (2009). Conceptualizing conflictspace: Toward a geography of relational power and embeddedness in the analysis of interstate conflict. *Annals of the Association of American Geographers, 99*(5), 827–835.
Goodchild, M. F. (2007). Citizens as sensors: The world of volunteered geography. *GeoJournal, 69*, 211–221.
Hey, T., Tansley, S., & Tolle, K. (2009). The fourth paradigm: data-intensive scientific discovery. *Microsoft Research.*
Jia, T., Yu, X., Shi, W., Liu, X., Li, X., & Xu, Y. (2019). Detecting the regional delineation from a network of social media user interactions with spatial constraint: A case study of Shenzhen, China. *Physica a: Statistical Mechanics and Its Applications, 531*, 121719.
Kang, C., Sobolevsky, S., Liu, Y., & Ratti, C. (2013). Exploring human movements in Singapore: a comparative analysis based on mobile phone and taxicab usages. In *Proceedings of the 2nd ACM SIGKDD International Workshop on Urban Computing. Association for Computing Machinery, New York.*
Kuznar, L. A. (2006). High fidelity computational social science in anthropology: Prospects for developing a comparative framework. *Social Science Computer Review, 24*(1), 15–29.
Liu, R., Yao, X., Guo, C., & Wei, X. (2021). Can we forecast presidential election using Twitter data? An integrative modelling approach. *Annals of GIS, 27*(1), 43–56.
Openshaw, S, & Abrahart, R. J. (1996). Geocomputation. In R. J. Abrahart (Ed.), *Proceedings of 1st international Conference on Geocomputation* (pp. 665–666). London: University of Leeds.
Pedro, J., Silva, C., & Pinheiro, M. D. (2019). Integrating GIS spatial dimension into BREEAM communities sustainability assessment to support urban planning policies, Lisbon case study. *Land Use Policy, 83*, 424–434.
Pei, T., Shu, H., Guo, S. H., et al. (2020). The concept and classification of spatial patterns of geographical flow. *Journal of Geo-Information Science, 22*(1), 30–40.
Pentland, A. (2015). Social physics: How social networks can make us smarter.
Pindolia, D. K., Garcia, A. J., Wesolowski, A., Smith, D. L., Buckee, C. O., Noor, A. M., Snow, R. W., & Tatem, A. J. (2012). Human movement data for malaria control and elimination strategic planning. *Malaria Journal, 11.*
Qin, K., Xu, Y. Q., Kang, C. G., Sobolevsky, S., & Kwan, M. P. (2019). Modeling spatio-temporal evolution of urban crowd flows. *ISPRS International Journal of Geo-Information, 8*(12), 570.
Qin, K., Luo, P., & Yao, B. (2019). Networked mining of GDELT and international relations analysis. *Journal of Geo-Information Science, 21*(1), 14–24.
Qin, K., Lin, H., Hu, D., Xu, G., Zhang, X. X., Lu, B. B., & Ye, X. Y. (2020). A review of spatially integrated humanities and social sciences. *Journal of Geo-Information Sciences, 22*(5), 912–928.
Qin, K., Wang, Q. X., Li, S., Luo, P., & Xu, Y. Q. (2020). Spatial interaction network analysis methods for computation social sciences and geocomputation for social sciences. *Social Journal, 3*, 64–67.
Lazer, D., Pentland, A., Adamic, L., Aral, S., Barabási, A. L., Brewer, D., Christakis, N., Contractor, N., Fowler, J., Gutmann, M., Jebara, T., King, G., Macy, M., Roy, D., & Alstyne, M. V. (2009). Computational social science. *Science, 323*(5915), 721–723.

Li, D. R., Guo, W., Chang, X. M., & Li, X. (2020). From earth observation to human observation: Geocomputation for social science. *Journal of Geographical Sciences, 30*(2), 233–250.

Lin, H., Zhang, J., Yang, P., et al. (2006). Research progress of spatially integrated humanities and social science. *Journal of Geo- Information Science, 8*(2), 30–37.

Lin, H., Lai, J. G., & Zhou, C. H. (2010). *Research on spatially integrated humanities and social sciences.* Science Press.

Liu, Y., Liu, X., Gao, S., Gong, L., Kang, C., Zhi, Y., Chi, G., & Shi, L. (2015). Social sensing: A new approach to understanding our socioeconomic environments. *Annals of the Association of American Geographers, 105*(3), 512–530.

Longley, P. A. (1998). Foundations. In P. A. Longley, S. M. Brooks, R. McDonnel, & B. Macmillan (Eds.), *Geocomputation: A primer* (pp. 1–15). Wiley.

Wang, F. (2020). Why public health needs GIS: A methodological overview. *Annals of GIS, 26*(1), 1–12.

Wang, Y. L., Qin, K., Chen, Y. X., & Zhao, P. X. (2018). Detecting anomalous trajectories and behavior patterns using hierarchical clustering from taxi GPS data. *ISPRS International Journal of Geo-Information, 7*(1), 25.

Xu, G., Wang, W., Lu, D., Lu, B., Qin, K., & Jiao, L. (2021). Geographically varying relationships between population flows from Wuhan and COVID-19 cases in Chinese cities. *Geo-Spatial Information Science.* https://doi.org/10.1080/10095020.2021.1977093

Xu, G., Jiang, Y., Wang, S., Qin, K., Ding, J., Liu, Y., & Lu, B. (2022). Spatial disparities of self-reported COVID-19 cases and influencing factors in Wuhan, China. *Sustainable Cities and Society, 76*, 1–9.

Xu, G., Xiu, T., Li, X., Liang, X., Jiao, L. (2021b). Lockdown induced night-time light dynamics during the COVID-19 epidemic in global megacities. *International Journal of Applied Earth Observation and Geoinformation*, 102421.

Yuan, J., Zheng, Y., & Xie, X. (2012). Discovering regions of different functions in a city using human mobility and POIs. In *Proceedings of the 18th ACM SIGKDD International Conference on Knowledge Discovery and Data Mining*. New York: Association for Computing Machinery.

Zhao, P. X., Qin, K., Ye, X. Y., Wang, Y. L., & Chen, Y. X. (2017). A trajectory clustering approach based on decision graph and data field for detecting hotspots. *International Journal of Geographical Information Science, 31*(6), 1101–1127.

Zhukov, Y. M., & Stewart, B. M. (2013). Choosing your neighbors: Networks of diffusion in international relations. *International Studies Quarterly, 57*(2), 271–287.

Chapter 29
Four Methodological Themes in Computational Spatial Social Science

Fahui Wang

Abstract This chapter outlines four methodological themes in spatial analytics with broad applications in social sciences and public policy, all grouped under the umbrella of "Computational Spatial Social Science". Spatial accessibility measures the relative ease by which the locations of activities or services can be reached, and serves as a major matric for location advantages. Regionalization constructs regions by merging small areas that are similar in attributes or are tightly connected. The former forms homogenous regions and the latter defines functional regions. Both can be scale flexible and thus produce a series of area units to support analysis, management, and planning. Spatial simulation imitates real-world social, economic, and human environments, behaviors and interactions in a lab setting, and empowers social scientists for discovery and cost-effective policy experiments. Finally, the maximal accessibility equality problem (MAEP) is proposed as a new location-allocation paradigm in spatial optimization to plan public resources and services.

Keywords Computational spatial social science · Spatial accessibility · Regionalization · Spatial simulation · Spatial optimization · Maximal accessibility equality problem (MAEP)

29.1 Introduction

In the last three decades or so, the advancement of social science can be characterized by three major trends, going scientific, pursuing public policy relevance, and making a spatial turn. Increasingly social science relies on data analysis that can be computationally intensive, and this is especially true when data are spatial. Harnessing discoveries of social science often means actionable public policy. Effective and efficient policy needs to be place sensitive, for example, many advocate precision public health. The convergence of these forces gives rise to the growth of Computational Spatial Social Science. In the U.S., the Center for Spatially Integrated

F. Wang (✉)
The Graduate School and Department of Geography and Anthropology, Louisiana State University, Baton Rouge, LA 70803, USA
e-mail: fwang@lsu.edu

B. Li et al. (eds.), *New Thinking in GIScience*,
https://doi.org/10.1007/978-981-19-3816-0_29

Social Science (CSISS) at the University of California Santa Barbara, funded by the National Science Foundation in 1999, has been an important force in promoting the usage of Geographic Information System (GIS) technologies, increasingly known as geographic information science (GIScience), in social science.

This short chapter discusses four exemplary methodological themes in spatial analytics and showcases some best practices in each. They are chosen as the author has been intimately involved in developing them, witnessed their broad applications in social sciences and public policy, and believes in their potentials for further advancing the field of computational spatial social science. Each theme begins with a brief review of methods and then discusses various applications. Due to space limit, only representative or most recent literature is cited.

29.2 Spatial Accessibility for 4As

Spatial accessibility refers to the relative convenience by which services can be reached from a given location, and thus captures the very essence of location advantage. Since its inception about two decades ago, the *2-step floating catchment area (2SFCA)* method has been a popular measure of spatial accessibility. It overcomes the shortcomings of preceding methods that focus on either proximity to the nearest facility or simply supply–demand ratios within fixed geographical or administrative boundaries. Later the generalized 2SFCA method was proposed to synthesize various refinements to the original 2SFCA method, such as:

$$A_i = \sum_{j=1}^{n} \left[S_j f(d_{ij}) / \sum_{k=1}^{m} (D_k f(d_{kj})) \right] \tag{29.1}$$

where A_i is accessibility at demand (population) location i, D_k is amount of demand at location k, S_j is the capacity of supply facility at location j, d is the distance or travel time between them, $f(d)$ is a distance decay function that can be continuous, discrete or hybrid between them, and n and m are the total numbers of facility locations and population locations, respectively. The popularity of 2SFCA method is aided by an intuitive interpretation of accessibility score (e.g., physicians per 1000 people) and its automation in an ArcGIS toolkit (Wang, 2015: 112–113).

By switching the notations for demand (D) and supply (S) in Eq. (29.1), it turns to the *inverted 2SFCA* (or *i2SFCA*) method that measures potential crowdedness for facilities (or scarcity of resource). The population-based accessibility and facility-oriented crowdedness are two sides of the same coin in examining the geographic variability of resource allocation but have their distinctive emphases for different purposes. A recent paper (Wang, 2021) provides the theoretical derivations for both methods and also validates them by empirical data. Another extension is the proposition of *2-step virtual catchment area* (*2SVCA*) method that measures spatial accessibility to telehealth by accounting for broadband availability (Alford-Teaster et al., 2021). Specifically, the 2SVCA method replaces the distance decay function f by a measurement of virtual connection strength between supply and demand, e.g., their broadband strengths.

The broad applications of spatial accessibility studies can be summarized as: accessibility matters for "4As":

1. for *anything* (e.g., healthcare, job, education, recreation, food),
2. to *anyone* (e.g., on disparity in access between socio-demographic groups),
3. by *any means* (e.g., via different transportation modes, challenges by the handicapped), and
4. at *any time* (e.g., accounting for temporal variability of supply, demand, and transportation between daytime versus nighttime, seasonally, normal vs. in the event of natural disasters) (Li et al., 2022).

Once the disparity of accessibility is quantified, public policy and planning strategy can be designed to mitigate the problem for promoting equal opportunities for all, an issue examined in Sect. 29.5.

29.3 Scale-Flexible Regionalization

Regionalization groups a large number of small areas to a relatively small number of regions while optimizing a given objective function and satisfying certain constraints. Aided by automation in a GIS environment, some regionalization methods generate a series of different numbers and thus different sizes of regions to enable a researcher to examine a study area at different scales. These methods are termed "*scale-flexible regionalization*", a classic task in geographic analysis modernized by GIS-based computational methods. This section covers two families of scale-flexible regionalization methods.

One type of regionalization is to derive *homogeneous regions.* In other words, it merges contiguous areas that are similar in attributes. For example, the *regionalization with dynamically constrained agglomerative clustering and partitioning* (*REDCAP*) method first (1) constructs a cluster hierarchical tree based on attribute similarities among small areas, and then (2) partitions the spatially contiguous cluster tree to generate a series of regions of different sizes while explicitly optimizing a homogeneity measure (Guo & Wang, 2011). The homogeneity measure is the total sum of squared deviations (SSD) defined as

$$SSD = \sum_{r=1}^{k} \sum_{i=1}^{n_r} \sum_{j=1}^{d} \left(x_{ij} - \overline{x}_j\right)^2$$

where k is the number of regions, n_r is the number of small areas in region r, d is the number of variables considered, x_{ij} is an attribute variable value and $\overline{x}_j$ is the regional mean for variable j. Each input variable should be normalized and be assigned a weight as its relative importance in contributing to the overall measure of homogeneity. The *mixed-level regionalization* (*MLR*) method decomposes areas of large population and merges areas of small population to derive regions with comparable population size, and thus the final regions are composed of different (mixed) areal units (Mu et al., 2015). The core algorithm of MLR remains merging neighboring areas that are most similar in attributes. In other words, similarity is defined by the attribute distance D_{ij} between them, such as

$$D_{ij} = \sum_t (x_{it} - x_{jt})^2$$

where an area i and its adjacent object j have their tth attributes standardized as x_{it} and x_{jt}, respectively.

There are several values for constructing homogeneous regions in various applications. For instance, crime analysis and health studies often encounter the *small population* (*numbers*) *problem* when the subject under investigation is a rare event (e.g., cancer, AIDS, homicide). Its rates in sparsely populated areas are unreliable or sensitive to missing data and other data errors, and the data can be suppressed for privacy protection. Regionalization mitigates the problem by constructing larger geographic areas to obtain more stable and reliable rates. Additional benefits include: newly-derived regions tend to be spatially independent from each other (since areas of similar attributes are already merged) and thus traditional statistical analysis methods (e.g., ordinary least square regression) may be applied without the need to control for spatial autocorrelation, and a series of regions of various sizes enable analysis at different geographic scales and affords a researcher an opportunity to examine the *modifiable area unit problem* (*MAUP*). A recent study uses regionalization to divide China into two regions of nearly identical area size and greatest contrast in population and shreds new light to the scientific foundation of classic "Hu Line" (Wang et al., 2019a). Another uses mobile app data to define a hierarchical urbanization "source-sink" regions in China in terms of intensity of labor force import or output (Y. Wang et al., 2019).

The other type of regionalization is to delineate *functional regions*. A functional region is coherent so that connection strengths in terms of service flow, passenger volume, financial linkage, or communication are stronger within a region than beyond. Functional region can be in various forms such as *catchment area* for a service facility, *trade area* for a store, and *hinterland* or *urban sphere of influence* for a central city. Here, *hospital service area (HSA)* is used as an example to illustrate the methods and its application in defining healthcare market areas (Wang & Wang, 2022). *Dartmouth method* pioneered the delineation of HSAs in the USA. It uses a simple plurality rule by assigning an area (e.g., a ZIP code area) to a hospital if its residents visit the hospital most often out of alternatives, and then collects the areas assigned to the same hospital to form an HSA. The Dartmouth method lacks a systematic perspective and cannot ensure the maximal total service volumes within derived HSAs. It only defines one set of HSAs and thus not scale flexible. Some recent studies use the *network community detection* approach to segment a network of patient-to-hospital OD flows into subnetworks (communities) so that the resulting subnetworks have the maximum connections within each and the minimum connections between them. The network community detection approach delineates a given number of HSAs corresponding to a resolution value defined by the analyst, and thus is scale flexible.

Defining HSAs helps provide a basic geographic unit for health care delivery assessment, management, and planning. The unit needs to pertain to the specific medical service (e.g., cancer service areas, pediatric surgical areas, primary care service areas), and be updated in a timely fashion and at a scale suitable for the purpose of research and public policy relevance. Two promising methods, namely "spatially constrained Louvain and Leiden algorithms", are automated in ArcGIS tools to meet these challenges. The tools can certainly benefit other applications that involve delineation of functional regions such as trade areas in market analysis and urban hinterlands in planning for a system of cities.

29.4 Spatial Simulation for Pursuit of Finest Scale in Individuals

Spatial simulation is a spatially explicit, bottom-up modeling approach to explore how spatial patterns emerge from simulated individuals and their interaction in space. It has rather a broad scope. This section covers two areas of spatial simulation the author has involved in. One focuses on *Monte Carlo simulation* of individuals in space, and by a simple extension, connections between locations (e.g., commuting). The other is *agent based modeling* (*ABM*) that is truly dynamic with individual agents moving spatio-temporally. Both can be considered a pursuit of modeling individuals, the finest scale in spatial analysis.

Spatial data often come as aggregated data in various areal units. Analysis of aggregated data incurs several problems such as the aforementioned MAUP, ecological fallacy, and loss of spatial accuracy in calibrating measures such as location precision, distance, and travel time. Monte Carlo simulation generates a set of random points according to a defined probability distribution function (PDF). In some studies, it is desirable to first disaggregate data in area units to individual points according to a spatial pattern revealed in observed data, then aggregate data back to an area unit of one's chosen. For example, Wang et al. (2019b) uses Monte Carlo simulation to generate individual residents that are consistent with land use data, and then aggregates population back to various uniform area units to examine the scale and zonal effects. The study indicates that the logarithmic function, instead of the popular exponential function, is the best fitting one for the urban population density pattern in Chicago. Similarly, Hu and Wang (2019) use it to simulate individual resident workers and individual jobs, and then their linkages in individual trips in order to improve the estimation of commute distance and time for journey-to-work trips.

Another popular application of Monte Carlo simulation is to test statistical hypotheses using randomization tests. For example, it is used to design statistical significance tests for global and local spatial autocorrelation indices such as *Moran's I* and *G statistic*. Wang et al. (2017) also use Monte Carlo simulation to calibrate the *global and local indicator of colocation quotients* with corresponding statistical significance tests to detect whether two types of points tend to cluster (collocate) or disperse (the opposite) from each other. The methods are now available in ArcGIS Pro.

More advanced spatial simulation techniques such as cellular automata (CA) and agent-based model (ABM) simulate multiple agents and their movements and interactions in a nearly-realistic environment. Here a recent agent-based crime simulation model (Zhu & Wang, 2021) is used to illustrate its basic features and functionalities and demonstrate its potentials. The model defines three types of agents such as motivated offenders, vulnerable targets, and police for their distinct roles. It then dynamically simulates agents' daily routines including (1) mandatory activities such as work and rest, and (2) flexible activities (e.g., shopping, dining, and recreation). Police may follow a hotspot policing, random patrol, or other strategy so the model can test the effect of each. Within a detailed representation of a study area with a road network and points of interest, crime opportunities and deterrence emerge from intersecting space–time trajectories of three types of agents. The model shows good fitness between predicted vs. reported crime hotspots.

29.5 The Maximal Accessibility Equality Problem (MAEP) in Spatial Optimization

Spatial optimization uses computational approaches to find the optimal solutions to decision variables, which are spatial, for objective function(s) under defined constraints. It is grouped under a broad umbrella "*operations research (OR)*", widely taught and practiced in business management, applied mathematics, engineering, planning, and geography. Here spatial optimization refers to its narrow definition related to *location-allocation problems* that seek the best decision on where to locate facilities and how large those should be.

Planning often faces two competing goals, efficiency vs. equality. It has been long debated on which should be prioritized among academics. Among traditional location-allocation models, the *p-median problem* minimizes total travel burden, the *location set covering problem* (*LSCP*) minimizes resource commitment, and the *maximum covering location problem* (*MCLP*) maximizes demand coverage. All are designed for a goal related to the principle of efficiency such as maximum gain or minimum cost for the whole system. An exception is the *minimax problem* that seeks to minimize the largest travel burden. However, the minimax problem only attempts to reduce the inconvenience of the least accessible user without accounting for the distribution of demands across the whole spectrum, and therefore is considered as marginally addressing the equality issue. These classic models and their variants have sustained extensive popularity, in part because of the convenience of solving them in ArcGIS.

Only modest progress has been made in research on location-allocation models addressing equality issues. One of such early efforts formulates the optimization objective as minimal inequality in accessibility of facilities across geographic areas (Wang & Tang, 2013). Specially, accessibility is measured by the 2SFCA method in Eq. (29.1), and inequality is defined as the variance (i.e., least squares) of the accessibility index. Therefore, the objective function is

$$\text{Min} \sum_i (A_i - a)^2 \quad \text{or} \quad \text{Min} \sum_i [D_i (A_i - a)^2] \tag{29.2}$$

where a is the weighted average accessibility score in a study area and a constant, the first formula minimizes the total diversions from that average, and the second minimizes the weighted total diversions by adding a weighting factor D_i (demand at each location i). The decision variables are supply capacities for facilities to be solved, and the problem is subject to the total supply constraint. Such a model is termed a *quadratic programming (QP) problem* and can be solved in various open-source programs or commercial optimization software.

The above QP only solves the capacities for facilities. Many location-allocation problems also need to decide where to site the facilities. Luo et al. (2017) argues that such a decision is often sequential by deciding on their sites first and then their capacities, and formulates a method termed "*two-step optimization for spatial accessibility improvement (2SO4SAI)*". The first step is to find the best locations to site new facilities by emphasizing accessibility as proximity to the nearest facilities. The second step adjusts the capacities of facilities for minimal inequality in accessibility measured by the 2SFCA method. The solution to the first step is to strike a balance among the solutions to the p-median, MCLP and minimax problems, and the second step solves the QP problem as defined in Eq. (29.2). This trade-off approach is further validated in a recent study on planning emergency medical services (EMS) in Shanghai (Li et al., 2022).

These spatial optimization models advocate a common objective of maximal equality (or minimal inequality) and emphasize that achieving such a goal begins with equal accessibility, not equal utilization nor equal outcome, a principle consistent with the consensus reached by Culyer and Wagstaff (1993). They are grouped under the term "*Maximal Accessibility Equality Problem (MAEP)*" (Wang & Dai, 2020). Many variant models can be derived from this broad framework. For instance, beyond the variance definition in Eq. (29.2), inequality can be formulated as maximum deviation, mean absolute deviation, coefficient of variation, Gini coefficient and others. Beyond the 2SFCA method or spatial proximity in distance, accessibility can also be measured in cumulative opportunities, gravity-based potential model and others

(Wang, 2015: 93–95). The applications are seen in planning health care services, schools, senior care facilities and EMS.

In summary, Computational Spatial Social Science (CSSS) is truly an interdisciplinary field that has benefited from the advancements in computational science, social sciences, and GISc, and more importantly their increasing intersections.

References

Alford-Teaster, J., Wang, F., Tosteson, A. N. A., & Onega, T. (2021). Incorporating broadband durability in measuring geographic access to healthcare in the era of telehealth: A case example of the Two-Step Virtual Catchment Area (2SVCA). *Journal of the American Medical Informatics Association, 28*, 2526–2530.

Culyer, A. J., & Wagstaff, A. (1993). Equity and equality in health and health care. *Journal of Health Economics, 12*(4), 431–457.

Guo, D., & Wang, H. (2011). Automatic region building for spatial analysis. *Transactions in GIS, 15*(s1), 29–45.

Hu, Y., & Wang, F. (2019). *GIS-based simulation and analysis of intraurban commuting*. CRC Press.

Li, M., Wang, F., Kwan, M.-P., Chen, J., & Wang, J. (2022). Equalizing the spatial accessibility of emergency medical services in Shanghai: A trade-off perspective. *Computers, Environment and Urban Systems, 92*, 101745.

Luo, J., Tian, L., Luo, L., Yi, H., & Wang, F. (2017). Two-step optimization for spatial accessibility improvement: a case study of health care planning in rural China. *BioMed Research International, 2017*, 2094654.

Mu, L., Wang, F., Chen, V. W., & Wu, X. (2015). A place-oriented, mixed-level regionalization method for constructing geographic areas in health data dissemination and analysis. *Annals of the Association of American Geographers, 105*, 48–66.

Wang, F. (2015). *Quantitative methods and socioeconomic applications in GIS* (2nd Ed.). CRC Press.

Wang, F. (2021). From 2SFCA to i2SFCA: Integration, derivation and validation. *International Journal of Geographical Information Science, 35*, 628–638.

Wang, F., & Dai, T. (2020). Spatial optimization and planning practice towards equal access of public services. *Journal of Urban and Regional Planning (Cheng-shi-yu-qu-yu-gui-hua-yan-jiu), 12*(2), 28–40 (in Chinese).

Wang, F., Liu, C., & Xu, Y. (2019a). Analyzing population density patterns in China with GIS-automated regionalization: the Hu Line revisited. *Chinese Geographical Science, 29*, 541–552.

Wang, F., Liu, C., & Xu, Y. (2019b). Mitigating the zonal effect in modeling urban population density functions by Monte Carlo simulation. *Environment and Planning B: Urban Analytics and City Science, 46*, 1061–1078.

Wang, F., & Tang, Q. (2013). Planning toward equal accessibility to services: A quadratic programming approach. *Environment and Planning B-Planning & Design, 40*, 195–212.

Wang, F., & Wang, C. (2022). *GIS-Automated Delineation of Hospital Service Areas*. CRC Press.

Wang, F., Hu, Y., Wang, S., & Li, X. (2017). Local Indicator of Colocation Quotient with a statistical significance test: Examining spatial association of crime and facilities. *Professional Geographer, 69*, 22–31.

Wang, Y., Wang, F., Zhang, Y., & Liu, Y. (2019). Delineating urbanization "source-sink" regions in China: Evidence from mobile app data. *Cities, 86*, 167–177.

Zhu, H., & Wang, F. (2021). An agent-based model for simulating urban crime with improved daily routines. *Computers, Environment and Urban Systems, 89*, 101680.

Chapter 30
Geosocial Analytics

Kai Cao, Yunting Qi, Mei-Po Kwan, and Xia Li

Abstract The adoption of spatially integrated approaches has become an increasing trend in social sciences. Concurring with the spatial turn in social sciences, there has been a social turn in geography. Inspired by the theoretical debates in the social turn in geography, particularly the debate around the concept of space, we critically reflect on existing studies of spatially integrated social sciences. Following that, we propose a geosocial analytical framework for a more comprehensive knowledge of our lived society. The geosocial analytical framework should lay geographical research pathways at its center, keeps open to both quantitative and qualitative methods as well as computational technologies, remains interested in any topics relevant to human societies and potentially engages with conventional social theories. Some challenges possibly faced by the implication of geosocial analytics have been identified as well, namely data sources, ethical concerns, and the difficulties in the combination of different research approaches. We take the chapter as an initiative to introduce the geosocial analytics and we encourage more researchers to further work on it.

Keywords Geosocial analytics · Spatially integrated social science · GIScience

K. Cao (✉) · X. Li
Key Laboratory of Geographic Information Science (Ministry of Education), School of Geographic Sciences, and Key Laboratory of Spatial-temporal Big Data Analysis and Application of Natural Resources in Megacities (Ministry of Natural Resources), Shanghai 200241, China
e-mail: kcao@geo.ecnu.edu.cn

X. Li
e-mail: lixia@geo.ecnu.edu.cn

Y. Qi
Department of Geography, Royal Holloway University of London, London, UK
e-mail: Yunting.Qi.2017@live.rhul.ac.uk

M.-P. Kwan
Department of Geography and Resource Management and Institute of Space and Earth Information Science, The Chinese University of Hong Kong, Shatin, Hong Kong, China
e-mail: mpkwan@cuhk.edu.hk

B. Li et al. (eds.), *New Thinking in GIScience*,
https://doi.org/10.1007/978-981-19-3816-0_30

30.1 Introduction

In early 2020, when people still possessed very little knowledge of COVID-19 per se and merely had limited ideas about its spread throughout the world, scholars at John Hopkins University publicly published a global map of COVID-19 that displays the number of confirmed cases and death by regions, tracks critical data in the last 28 days and in its later version, provides more explicit and diverse statistics on the pandemic (Dong et al., 2020). The COVID-19 Dashboard is a map but definitely is more than a geographical map. Through frankly and vividly presenting the distribution and spread of COVID-19 at different geographical scales, it has great significance in terms of public health, medical service and related research in sociology, medicine, social work, politics, other disciplines. The global map of COVID-19 can be taken as a splendid example of how spatial thinking and computational technology could benefit knowledge and studies in social sciences.

Since the late 1990s when they were launched, spatially integrated social sciences (which is part of the result of the so-called spatial turn in social science) have been proved as an insightful approach to examining social problems. In this chapter, we first briefly review related literature regarding the spatial turn in social sciences. We then suggest a framework of geosocial analytics taking both social and geographical thoughts into consideration. Finally, we reflect on future directions and potential challenges of conducting geosocial analytics.

30.2 Spatially Integrated Social Sciences

Space is one of the most fundamental concepts in the discipline of geography but has received rather limited attention in social sciences for a long time. The distinct attitudes towards space and related spatial thinking could be explained by research focus varying between geography and social sciences. To put it more specifically, geographers are interested in the embeddedness of human activities and relationships in specific places as well as human-place interactions, while social scientists are paying dominant attention to human activities per se. Considering that people's activities, subjectivities and various social relationships always occur in space, the ignorance of space in social research has been accused of uprooting human behaviours from relevant contexts and further potentially causing biased knowledge (Goodchild et al., 2000; Sui, 2010). In this regard, social scientists have increasingly adopted the concept of space and spatial analytic toolkits, like geographical information systems, remote sensing, spatial statistics, in their research (Wang, 2011), which is an exciting trend at least from our viewpoint as geographers. But the integration of spatial approaches into social sciences has been far from sufficient in both theoretical and methodological terms. In order to ground social research better in specific time–space, scholars call for more advocation to spatially integrated social sciences (Anselin et al., 2004; Goodchild & Janelle, 2003; Nyerges et al., 2011). Spatially integrated social sciences do not refer to specific disciplines and instead appear as an extensive term which "describe[s] the integration of space and place in social science research using Geographic Information Systems (GIS)" (Lechner et al., 2019, p. 1) and attempts to deal with conventional and emerging problems in social sciences through quantitative analysis of spatial (big) data and qualitative GIS approaches (Bainbridge, 1999; Lechner et al., 2019; Nyerges, 2009).

The spatially integrated approach could be witnessed in a range of social disciplines. The most famous example is found in Economics and the work by Krugman (1991), who re-introduces the importance and significance of place in economic activities, particularly in international trades. The award of the Nobel Prize in Economics to Kurgman in 2008 undoubtedly demonstrates the tremendous potential of spatial understanding and reasoning for social sciences. Closely relevant to economics and urban/regional studies, sub-disciplines like land market studies, real estate studies and others benefit greatly from GIS and spatial econometrics, which were introduced by Anselin et al. (2004) and many other scholars. Besides, due to the successful application of GIS in indicating the distribution and spread of disease and everyday health risks, health studies have become an important area in the spatial turn of social sciences (Cromley & McLafferty, 2002; Richardson et al., 2013) and the importance of GIS has been further proved in recent public health research regarding COVID-19 (Dong et al., 2020; Franch-Pardo et al., 2021; Kan et al., 2021). Work conducted by Kwan and her associates indicates the relationships between health conditions, activities, and urban environment (Hawthorne & Kwan, 2012; Huang et al., 2021; Kwan, 2013, 2018; Zhao et al., 2018). In their *Science* report, Richardson et al. (2013) comment that "Research agendas that systematically incorporate spatial data and analysis into global health research hold extraordinary potential for creating new discovery pathways in science" (p.1391).

Criminology is another area that benefits significantly from the increasing integration of spatial technologies due to its spatial characteristics. For instance, Cohen and Tita (1999) illustrated the successful application of GIS and spatial analysis to homicide patterns detection in Pittsburgh. More recent studies employ spatial analysis in criminology to challenge some long-standing assumptions regarding ethnicity, social segregation, crime rate (Wang & Minor, 2002). Anthropology is also pioneering in the application of GIS and geo-visualisation, as significant supplementary tools to the traditional ethnographic approaches. A rapidly growing number of studies using GIS and analytic mapping can also be found in anthropology in past decades (Aldenderfer & Maschner, 1996; Padilla, 2013; Roberts, 2016). Scholars also start employing GIS and other spatial analysis methods to examine conventional topics in sociology, such as the interactions and isolations between social groups and various social processes (like gentrification, segregation, cultural expansion, etc.), particularly those occurring in urban spaces (Goldfield & Schlichting, 2007; Sohoni & Saporito, 2009; Zambrano et al., 2021).

Compared to conventional approaches, the spatially integrated approaches have largely emerged outside the mainstream of social sciences. Based on a brief review of key works in selected disciplines in the last paragraph, we suggest that the spatially integrated methods, including but not limited to GIS and spatial statistical analysis, could promote social sciences from the three following perspectives. First, the spatially integrated approaches provide social scientists with a powerful analytical toolkit to identify, interrogate and find solutions to various social problems. Taking health research as an example, GIS data illustrates people's daily movement in urban built environments and helps to calculate the exposure possibility to (un)healthy urban facilities like fast food stores and gyms, which leads to the argument that a better urban design could lower urban residents' health risks like obesity (Cromley & McLafferty, 2011; Richardson et al., 2013). Both the research data and argument are distinct from usual health studies that mostly rely on longitudinal observations and surveys to gain knowledge of human daily activities while overlooking the circumstances where these activities happen. Similar cases could be witnessed in Criminology, GIS and spatial statistical analysis contribute to more reliable and feasible crime prediction modelling compared to those merely on the basis of existing crime records (Wang & Minor, 2002).

Second, the spatially integrated approaches equip social scientists with fresh thinking to conduct their research and opens the possibility of practical innovations and even theoretical advancement. Different from conventional approaches that largely exclude space with specific temporal attributes into the discussion, spatially integrated social sciences disclose the time–space situatedness of human activities, decisions, and relationships. Thus, some innovative research methods have been emerging in social sciences, like collecting data through wearable GPS devices and mobile sensing devices (Ho et al., 2021; Kwan & Ding, 2008; Wang et al., 2021). With the awareness of the spatial and temporal dimensions, "we are better able to inform decision making targeted at addressing the challenges facing people and places" (Corcoran et al., 2021, p. 445).

Third, the spatially integrated approaches further promote the integration of social sciences and other disciplines, such as computational sciences and engineering. With the development of computational and internet technology, computational social sciences (hereafter CSS) have become another notable emerging interdisciplinary field which "advances theories of human behaviour by applying computational techniques to large datasets from social media sites, the Internet, or other digitised archives such as administrative records" (Edelmann et al., 2020, p. 62). Largely driven by big data, CSS offers scholars great potential to conduct research involving massive samples and data with huge volume and surprising diversity which was once unrealistic (Hox, 2017). Some computational technology like machine learning even helps to overcome the obstacles encountered by quantitative social research examining the emotional world of human beings and mitigating some long-standing bias in social sciences like socio-culturally and ethnically unbalanced sample structures (Giles, 2012; Lazer et al., 2020). While not so many researchers adopting the approach of CSS have explicitly claimed that they were using spatial analysis, the relevant hints are rather obvious, such as the visualisation of data in the format of maps. We suggest that GIS and other spatial analytics should be taken as an essential part of CSS, because spatial analytics helps to ground digitised data in the real world and links the cyberspace with the material space humans actually live in.

As can be seen in existing literature, spatial analytics have played an essential role in recent social sciences. Admitting that the approach is far from being perfect, it has brought so much inspiration in both conceptual perspectives and practical applications. In the next section, we critically reflect on the spatial turn in social sciences and call for a more comprehensive integration between geographical and social thinking.

30.3 From Geospatial to Geosocial

The first step of critically reflecting on the spatial turn in social sciences should be to enquire about the meaning of space, one of the most vital geographical concepts. At least in the works we have reviewed in the last section, 'space' in spatially integrated social sciences largely remained a geometric container and a spatio-temporal context where human activities occur and where recorded matters exist. While admitting the material and geometrical aspect of space, geographers have significantly furthered the understanding of space in recent theoretical debates and empirical exploration. In the latest literature, space is ontologically cognitive (people need to perceive and establish their personal knowledge and emotions regarding the space), relational (various social relations are embedded in specific spaces and there exist mutual shaping forces between social relations and spaces) and temporally dynamic (spaces are changing across time, life stages and other temporal elements, and space is a becoming process) (Massey, 2005; Thrift & Crang, 2000). In short, the concept of space in geography has become ontologically dialectical. Considering that such conceptualization of space has spoken a lot about emplaced human activities, emotions, relations, which are the primary concerns of social sciences, the narrowed understanding of space as a geometric container may hinder the future progress of spatially integrated social sciences.

Second, spatially integrated social sciences remain geometrically topological and dismisses more or less a deep understanding of micro-level details of human activities. For example, although researchers can gain information on residents' exposure to (un)healthy facilities and ethnic diversities through GIS data of their daily commute (Kwan, 2018; Tan et al., 2022; Wang, 2011), we still have limited knowledge of how people interact with other individuals or social groups in specific sites and how they attach specific emotions to these interactive activities and spaces (Kwan, 2002, 2007), which could be explored widely and insightfully in conventional social sciences adopting interviews and observations. Third, following the first and second points of criticism, the subjective world of human beings is hardly fully represented in the current framework of spatially integrated social sciences. While there have been CSS works exploring human emotions on sites through collecting and analysing emotional words (Giles, 2012), we doubt whether the coded texts can sufficiently deliver emotional expressions given the unspeakable, perceivable nature of emotions (Bondi, 2005). As the critical aspect of everyday lives, the emotional world of humans should never be excluded from academic discussion.

Almost simultaneously with the spatial turn in social sciences, there happens the social turn in geography, in which geographers firmly reject positivism and eagerly embrace social thoughts like post-structuralism, post-modernity, humanism and feminism (Del Casino & Vincent, 2009). While the spatial turn in social sciences is looking into massive trends and phenomena in emplaced societies, the social turn in geography is interested in deep knowledge at the micro-level. The two seemingly contradictory directions in different fields propose an important question to researchers: Is there a better way to gain knowledge of our lived world? Or in another enquiring way, how should we academically investigate our lived world? Considering the (dis)advantages of different approaches, neither being purely spatial nor being purely social is sufficient and here we propose a geosocial analytical framework for a more comprehensive knowledge of our lived societies.

In this initiative, we suggest some fundamental guidelines for conducting geosocial analytics. *First, following both the conventions of geography and social sciences, spatial approaches should be a kind of fundamental, or even instinctive in exaggerated rhetoric thinking way, and more than just a simple format of data presentation and analysis.* To put it precisely, geosocial analytics should address both spatio-temporal situatedness and place embeddedness. The former refers to the venue of social activities/processes and is related to geo-visualisation, which has been widely acknowledged in existing spatially integrated social sciences. The latter means a range of place-specificity like social norms, governance systems and cultures which powerfully shape human activities and emotions but receive limited attention in existing geospatial analysis.

Second, geosocial analytics does not advocate the binary between quantitative and qualitative research in terms of methodology and seeks data diversity and innovative methods. Quantitative data and analysis are good at disclosing macro social and geographical trends, while qualitative data and analysis compensate the weaknesses of quantitative approaches with in-depth detailed evidence. Thus, both approaches are needed in geosocial analytic and deserve an efficient combination. *Third, geosocial analytical framework should be open to any topic relevant to human and human societies.* Particularly, we encourage scholars to investigate topics that are relatively unusual in existing spatially integrated social sciences and computational social sciences, like everyday encounters which seem to be unremarkable but deliver extraordinary meanings regarding individuals and the whole society. But we must admit that the investigation of specific topics might require methodological and conceptual innovations. *The last but not least, geosocial analytics should seek greater collaborations with social theories* which are usually explored, verified, and developed through conventional social approaches while undoubtedly having spatial implications, like feminism and post-colonialism. It is not easy but definitely deserves efforts, as it would be an important step to break down the barrier between spatially integrated social sciences and conventional geography/social sciences, and will help to integrate the emerging trend into the mainstream of social sciences.

Obviously, geosocial analytics should not be a closed term, but a broad term that is open to any further rigorous discussion and debates. The above four guidelines can be seen as initiatives to open up more possibilities of research under the geosocial analytical frameworks.

30.4 Potential Challenges

While we firmly believe geosocial analytics would be an inspired approach to further existing research in social sciences and geography, geosocial analytics still face a number of challenges. Some challenges have been indicated by social scientists adopting spatially integrated approaches and some challenges are newly appearing while the geosocial analytical framework is constructed.

The first challenge is regarding data. Many scholars have indicated that the problem of data accessibility and sharing might hinder the long-term development of spatially integrated social sciences (Lazer et al., 2020); in the framework of geosocial analytics potentially adopting various formats of data (e.g., qualitative interviews, observations, visual data, statistics, and so on), organisation and integration possibly bring extra challenge. The second challenge is ethical concerns. Existing research has reflected that big data may violate people's privacy and that relying on data collected from cyberspace likely causes a biased sampling structure (Giles, 2012; Kwan, 2016). While geosocial analytics hope to look into both macro trends and micro details, the ethical issues become emergent concerns for researchers. The third challenge is the combination of different research approaches and the integration between spatial methodology and seemingly abstract social theories, which is derived from the comprehensive nature of geosocial analytics. The challenge cannot be handled through one-off attempts, but through continuously careful research design.

In this chapter, we have critically reflected on existing literature regarding spatially integrated social sciences and computational social sciences. Following that, we took a step further towards geosocial analytics by providing fundamental guidelines and indicating potential challenges. While our ideas are far from mature, we hope that the chapter would encourage more geographers and social scientists to explore and improve the geosocial analytical framework.

Acknowledgements We would like to thank all the editors for their contribution to this chapter.

References

Aldenderfer, M., & Maschner, H. D. (1996). *Anthropology, space, and geographic information systems*. Oxford University Press.

Anselin, L., Florax, R., & Rey, S. J. (2004). *Advances in spatial econometrics: Methodology, tools and applications*. Springer.

Bainbridge, W. S. (1999). International network for integrated social science. *Social Science Computer Review, 17*(4), 405–420. https://doi.org/10.1177/089443939901700401

Bondi, L. (2005). Making connections and thinking through emotions: Between geography and psychotherapy. *Transactions of the Institute of British Geographers, 30*(4), 433–448. https://doi.org/10.1111/j.1475-5661.2005.00183.x

Cohen, J., & Tita, G. (1999). Diffusion in homicide: Exploring a general method for detecting spatial diffusion processes. *Journal of Quantitative Criminology, 15*(4), 451–493.

Corcoran, J., Lomax, N., & Lombard, J. (2021). Contemporary applications for spatially integrated social science. *Applied Spatial Analysis and Policy, 14*(3), 445–447. https://doi.org/10.1007/s12061-021-09425-z

Cromley, E., & McLafferty, S. (2002). GIS and public health New York: Guilford Press.

Cromley, E. K., & McLafferty, S. L. (2011). *GIS and public health*. Guilford Press.

Del Casino, J., & Vincent, J. (2009). *Social geography: a critical introduction*. Wiley.

Dong, E., Du, H., & Gardner, L. (2020). An interactive web-based dashboard to track COVID-19 in real time. *The Lancet Infectious Diseases, 20*(5), 533–534. https://doi.org/10.1016/S1473-3099(20)30120-1

Edelmann, A., Wolff, T., Montagne, D., & Bail, C. A. (2020). Computational Social Science and Sociology. *Annual Review of Sociology, 46*(1), 61–81. https://doi.org/10.1146/annurev-soc-121919-054621

Franch-Pardo, I., Desjardins, M. R., Barea-Navarro, I., & Cerdà, A. (2021). A review of GIS methodologies to analyze the dynamics of COVID-19 in the second half of 2020. *Transactions in GIS, 25*(5), 2191–2239.

Giles, J. (2012). Computational social science: Making the links. *Nature, 488*(7412), 448–450. https://doi.org/10.1038/488448a

Goldfield, J., & Schlichting, K. (2007). Foreign language and sociology: Exploring French society and culture. *Understanding place: GIS and Mapping Across the Curriculum*, 227–236.

Goodchild, M. F., Anselin, L., Appelbaum, R. P., & Harthorn, B. H. (2000). Toward Spatially Integrated Social Science. *International Regional Science Review, 23*(2), 139–159. https://doi.org/10.1177/016001760002300201

Goodchild, M. F., & Janelle, D. G. (2003). *Spatially integrated social science*. Oxford University Press.

Huang, J., Kwan, M.-P., & Kan, Z. (2021). The superspreading places of COVID-19 and the associated built-environment and socio-demographic features: A study using a spatial network framework and individual-level activity data. *Health & Place, 72*, 102694.

Hawthorne, T. L., & Kwan, M.-P. (2012). Using GIS and perceived distance to understand the unequal geographies of healthcare in lower-income urban neighbourhoods. *The Geographical Journal, 178*(1), 18–30. https://doi.org/10.1111/j.1475-4959.2011.00411.x

Ho, E.L.-E., Zhou, G., Liew, J. A., Chiu, T. Y., Huang, S., & Yeoh, B. S. A. (2021). Webs of care: Qualitative GIS research on aging, mobility, and care relations in Singapore. *Annals of the American Association of Geographers, 111*(5), 1462–1482. https://doi.org/10.1080/24694452.2020.1807900

Hox, J. J. (2017). Computational social science methodology, anyone? *Methodology, 13*(Supplement 1), 3–12. https://doi.org/10.1027/1614-2241/a000127

Kan, Z., Kwan, M.-P., Wong, M. S., Huang, J., & Liu, D. (2021). Identifying the space-time patterns of COVID-19 risk and their associations with different built environment features in Hong Kong. *Science of the Total Environment, 772*(10), 145379.

Krugman, P. R. (1991). *Geography and trade*. MIT press.

Kwan, M.-P. (2002). Feminist visualization: Re-envisioning GIS as a method in feminist geographic research. *Annals of the Association of American Geographers, 92*(4), 645–661.
Kwan, M.-P. (2007). Affecting geospatial technologies: Toward a feminist politics of emotion. *The Professional Geographer, 59*(1), 22–34.
Kwan, M.-P. (2013). Beyond space (as we knew it): Toward temporally integrated geographies of segregation, health, and accessibility. *Annals of the Association of American Geographers, 103*(5), 1078–1086. https://doi.org/10.1080/00045608.2013.792177
Kwan, M.-P. (2018). The Limits of the Neighborhood effect: Contextual uncertainties in geographic, environmental health, and social science research. *Annals of the American Association of Geographers, 108*(6), 1482–1490. https://doi.org/10.1080/24694452.2018.1453777
Kwan, M.-P. (2016). Algorithmic geographies: Big data, algorithmic uncertainty, and the production of geographic knowledge. *Annals of the American Association of Geographers, 106*(2), 274–282.
Kwan, M.-P., & Ding, G. (2008). Geo-narrative: Extending geographic information systems for narrative analysis in qualitative and mixed-method research. *The Professional Geographer, 60*(4), 443–465. https://doi.org/10.1080/00330120802211752
Lazer, D. M. J., Pentland, A., Watts, D. J., Freelon, D., Aral, S., Athey, S., Contractor, N., et al. (2020). Computational social science: Obstacles and opportunities. *Science, 369*(6507), 1060–1062. https://doi.org/10.1126/science.aaz8170
Lechner, A. M., Owen, J., Ang, M., & Kemp, D. (2019). Spatially integrated social sciences with qualitative GIS to support impact assessment in mining communities. *Resources, 47*(8), 1–12. https://doi.org/10.3390/resources8010047
Massey, D. (2005). *For space.* SAGE.
Nyerges, T. (2009). GIS and Society. In R. Kitchin & N. Thrift (Eds.), *International encyclopedia of human geography* (pp. 506–512). Elsevier.
Nyerges, T., Couclelis, H., & McMaster, R. (2011). *The SAGE handbook of GIS and society.* SAGE.
Padilla, S. G. (2013). Anthropology and GIS: Temporal and Spatial Distribution of the Philippine Negrito Groups. *Human Biology, 85*(1/3), 209–230, 222.
Richardson, D. B., Volkow, N. D., Kwan, M.-P., Kaplan, R. M., Goodchild, M. F., & Croyle, R. T. (2013). Spatial turn in health research. *Science, 339*(6126), 1390–1392.
Roberts, L. (2016). Deep mapping and spatial anthropology. *Humanities, 5*(1). https://doi.org/10.3390/h5010005
Sohoni, D., & Saporito, S. (2009). Mapping school segregation: Using GIS to explore racial segregation between schools and their corresponding attendance areas. *American Journal of Education, 115*(4), 569–600. https://doi.org/10.1086/599782
Sui, D. Z. (2010). GeoJournal: A new focus on spatially integrated social sciences and humanities. *GeoJournal, 75*(1), 1–2. https://doi.org/10.1007/s10708-010-9341-2
Tan, Y., Kwan, M.-P., & Chai, Y. (2022). How Chinese hukou system shapes ethnic dissimilarity in daily activities: A study of Xining, China. *Cities, 122*, 103520. https://doi.org/10.1016/j.cities.2021.103520
Thrift, N., & Crang, M. (2000). *Thinking space.* Routledge.
Wang, F. (2011). Spatially-integrated social sciences and GIS: A personal perspective. *Acta Geographica Sinica, 66*(8), 1011–1089.
Wang, F., & Minor, W. W. (2002). Where the jobs are: Employment access and crime patterns in Cleveland. *Annals of the Association of American Geographers, 92*(3), 435–450. https://doi.org/10.1111/1467-8306.00298
Wang, J., Kou, L., Kwan, M.-P., Shakespeare, R. M., Lee, K., & Park, Y. M. (2021). An integrated individual environmental exposure assessment system for real-time mobile sensing in environmental health studies. *Sensors, 21*(12), 4039. https://doi.org/10.3390/s21124039
Zambrano, J. P., Calle-Jimenez, T., & Orellana-Alvear, B. (2021). *GIS for decision-making on the gentrification phenomenon in strategic territorial planning.* Paper presented at the Conference on Information and Communication Technologies of Ecuador.
Zhao, P., Kwan, M.-P., & Zhou, S. (2018). The uncertain geographic context problem in the analysis of the relationships between obesity and the built environment in Guangzhou. *International Journal of Environmental Research and Public Health, 15*(2). https://doi.org/10.3390/ijerph15020308

Chapter 31
Defining Computational Urban Science

Xinyue Ye, Ling Wu, Michael Lemke, Pamela Valera, and Joachim Sackey

Abstract Uncovering the multi-dimensional interplay between computation and urban life's spatial-social aspects has both theoretical and practical implications for urban planning and public health science. Many analytical methods have been implemented and applied to deal with high-dimensional, heterogeneous, and unstructured location-based social data drawn from urban locales. Computational urban science has four interdependent layers: human dynamics-centered, platform-based, action-oriented, and convergence-driven. As a research paradigm based on computational thinking and spatiotemporal synthesis, computational urban science can provide a needed framework for addressing many pressing urban sustainability challenges from a systematic perspective.

Keywords Computational urban science · Human dynamics · Computational science · Urban science · Community engagement · Convergence

31.1 Introduction

World-wide, unprecedented urban growth has generated fascinating issues for interdisciplinary scholarly research. Urban communities in the twenty-first century have increasingly transitioned into complex systems and systems-of-systems, consisting of many dynamically interdependent human, environmental, and technical systems (Bibri, 2021). Simultaneously, advances in Information Communication Technology have been accompanied by social innovations, which continues to drive the techno-centric versus human-centric debates (Costales, 2022). In this context, new urban data involves the life qualities and connections that characterize individuals and/or

X. Ye (✉) · L. Wu
Texas A&M University, College Station, Texas 77840, USA
e-mail: xinyue.ye@tamu.edu

M. Lemke
University of Houston-Downtown, Houston, Texas 77002, USA

P. Valera · J. Sackey
Rutgers, The State University of New Jersey, New Brunswick, USA

B. Li et al. (eds.), *New Thinking in GIScience*,
https://doi.org/10.1007/978-981-19-3816-0_31

communities in urban areas. Rapid growth in urbanization and intensive urban–rural interaction enables massive flows of virtual and physical elements, including people. At the same time, an increasing number of extreme weather-driven urban hazards are a significant source of biodiversity loss, social disruption, and economic disparity, threatening progress towards the United Nations' 17 Sustainable Development Goals (Ye et al., 2021a, 2021b). In response, the United States and China have recalled the Paris Agreement's aim to address the challenges of the global climate crisis, pursuing efforts to limit the increase of global average temperature to below 1.5 °C (U.S. Department of State, 2021).

The U.S. National Science Foundation summarized the following key research questions to address continued and emerging challenges relevant to urban sustainability (NSF, 2020): (1) How can science improve forecasts and make predictions about the future states of rural, suburban, and urban systems? (2) What theories explain the structure and function of communities in the twenty-first century and what are the critical drivers to social change? (3) What aspects and intersections of social, built, and natural systems influence the resilience and sustainability of communities and the well-being of the people living in them? (4) How can successful innovations in one community be transferred to other communities? (5) How can integrative research along with community engagement improve the quality of life in those communities? To address these research questions and enable more resilient responses to persistent and emerging challenges associated with urban growth (Ye et al., 2021a, 2021b), a framework of computational urban science (CUS) is needed in the context of complexity systems. The advantages of incorporating elements of complex systems (e.g., dynamic complexity, interactions, etc.) shed light on the aforementioned questions raised. Incorporating computational science into urban science and vice versa has led to the increasing popularity of CUS (defined as urban computing in engineering schools or public informatics in public policy schools). At the same time, CUS should not be limited to either urban science or computational science; instead, it is a new research paradigm of studying urban phenomena based on computational thinking and spatiotemporal synthesis. CUS is composed of four layers: (1) Human dynamics-centered, (2) Platform-based, (3) Action-oriented, and (4) Convergence-driven. Together, these four layers that constitute CUS can contribute to urban sustainability in the twenty-first century through the integration of multidisciplinary and transdisciplinary theory, methodology, analysis, action planning, and community engagement.

31.2 Human Dynamics-Centered

Human dynamics-centered CUS aims to put humans at the center of technology driven advances to optimize the human–environment interaction in the built environment. Understanding human dynamics would facilitate the development of human-centered CUS towards a trans-disciplinary field of research and practice drawing on three elements: people, place, and space. To appreciate cities, we must view them as systems of networks and flows to better understand an individual by examining his/her social ties and interactions (Batty, 2013). The growing volume and diversity of human data in the urban communities further justify the need for human dynamics-centered CUS. The critical importance of centering human dynamics within CUS is grounded in historical and contemporary insights. Hägerstrand (1953) argued that human communication is mainly constrained by geographical social networks, indicating that social actions are spatially organized. Hence, Tuan (1977) pointed out that a place is a social construction of physical space, with special meaning for each individual across time. Nowadays, data logs from individual activities and interactions are typically collected at a fine-scale spatial and temporal level, thus bringing human-centered analytics to the frontier of urban science. Barabasi (2005) highlighted the role of human dynamics from a complex network perspective in revealing the mechanisms underlying many complex social, technological, and economic phenomena. Shaw et al. (2016) emphasized the geographical view of human dynamics research in the context of the emerging mobile and big data era.

Bringing a humanistic approach to the centre of urban computing helps us revisit the traditional urban theories and allows us to observe the role of computation shifting from lab-based academic use by research groups to urban daily use for individuals. User-generated data is by nature multi-scale and can be harnessed to gain insights into the urban activity structures (Huang et al., 2021; Steiger et al., 2016). With growing investments from government and business, the increasing affordability of technology and infrastructure can enable a revolution in urban management because decision-makers can more rapidly understand citizens' feedback and incorporate this information into policymaking. In this way, research insights derived from CUS can change our lives and neighbourhoods more instantly. Human dynamics-centered CUS can also re-orient urban management from the centralized and top-down that currently represents the status quo to a decentralized and bottom-up perspective, providing a more democratized decision- and policy-making environment in urban areas (Batty, 2013).

31.3 Platform-Based

A platform-based system is a process of developing, implementing, and refining a complex systems-based procedures that can serve as a critical enabler for the empirical study of space–time dynamics within urban environments. Platform-based CUS is expected to derive knowledge from heterogeneous streams of big urban data. Furthermore, a platform-based CUS has the added value of being a signaling term to attract more attention from industry. Computation is the essential methodological platform for urban data fusion, data mining, and simulation (Kontokosta, 2021). With fast-evolving disruptive digital technologies, such platforms will trigger new modes of analytics that could tackle the fine-scale data-rich urban environment and thus simultaneously advance both computational science and urban and social science. Gelernter (1993) predicted that software can go beyond codes and tools and towards "crystal balls" where we can observe and deeply understand the urban world. In addition, such "crystal balls" can facilitate computer-aided urban scenario experiments by leveraging a virtual urban environment workbench integrating "multi-dimensional visualization, dynamic phenomenon simulation, and public participation" (Lin et al., 2013; Wan et al., 2021).

Advances in artificial intelligence (AI) present additional opportunities to help address some of the most pressing and enduring urban challenges. Platform-based CUS can serve as common ground for the long-term collaborations between AI researchers and application-domain urban experts. For instance, cutting-edge sensing technologies are increasingly adept at recording urban mobility patterns in massive trajectory datasets, such as the movements among humans, taxis, buses, fleets, and cars. Visualizing and analysing such big dynamic data plays a critical role in knowledge discovery with an appropriate platform for trajectory data management and interactive visualizations facilitating user engagement (Ye et al., 2021a, 2021b). A large number of users, including students, teachers, researchers in various urban domains, and data analysts and software developers from industrial sectors need more open-source platforms to utilize big urban datasets better and transform these data into actionable knowledge. The revolution of computation as a platform for urban science has further blurred the boundaries of sub-disciplines in urban research. At the same time, the increasing affordability of computing and the flatter learning curve has removed long-standing barriers that had previously hindered the proliferation of computing platforms across disciplines.

One of the novelties involved with CUS is the creation of digital twins of existing cities for policymaking and design, which can weave the four layers mentioned above. A digital twin is a digital representation of a physical object or system, linked to real-time data inputs (Li et al., 2021). Such technologies have expanded to include large items such as buildings, factories, and cities. Digital twin technology has moved beyond manufacturing and into the merging worlds of AI and big data analytics. Through such an approach, computer programs take real-world data about a physical object or system as inputs and produce, as outputs, predications, or simulations of how that physical object or system will be affected by other inputs. The participants will also be afforded the ability to conduct experimental research in built environments through the use of mobile devices. By linking urban sensors with the developed digital twin with advanced virtual and augmented reality technologies, we can provide contextualized, real-time in-situ modelling of cities, which can be used to dynamically analyse real-time built environments and test scenarios for sustainable urban growth.

Through software and hardware integration, digital twin can be used for action to address the built environment challenges. It will enable ubiquitous networked immersion and virtual human teleportation to any location and scale of the built environment to assist humanity in solving current societal challenges and design needs. For example, the capabilities to accurately visualize and dynamically update the conditions of underserved and marginalized communities will afford such communities with the data to create real change in their neighbourhoods and solve existing and future racial inequity issues. Simultaneously, the ability to model climate change-based scenarios and test their impacts on the built environment offer unprecedented capabilities for altering the projected effects of issues such as sea-level rise. Such capabilities have tremendous possibilities for flood preparedness and recovery in coastal communities, which can be applied globally. It would enable a scalable simulation and interactive visualization platform that reveals the interactions and dynamics of urban flows, entities, and social phenomena from site/community to the regional/national scale. For example, a user can trigger a virtual fire, building collapse, or natural disaster (e.g., flood event, chemical plant explosion, etc.), and/or observe the predicted behaviour of pedestrians and other inhabitants. We can simulate how communities will be transformed by moving to autonomous vehicles. These platforms would also allow the understanding of existing urban infrastructure conditions and how to increase its efficacy.

31.4 Action-Oriented

CUS can inform high-impact, sustainable, and citizen-driven action planning, interrelationships, and policymaking, thereby empowering the decision-making of individuals, communities, and other key stakeholders. There are growing interests in utilizing big urban data for a broader audience in government agencies, practitioners, and citizens. Urban data contains abundant knowledge about a given city and its citizens. The extracted information through CUS can be utilized in many important and practical applications to optimize urban planning and improve built environments in urban settings. Both the Columbia Climate School (established in 2020) and the Climate and Sustainability School at Stanford University (established in 2022) have identified transdisciplinary initiatives to develop actionable, evidence-based, and realistic pathways for impacting communities, which is expected to be highly relevant to urban science. For example, the Envisioning the Neo-traditional Development by Embracing the Autonomous Vehicles Realm Institute (ENDEAVR Institute) at Texas A&M University aims to translate emerging urban computing technologies into action-oriented solutions for small communities (ENDEAVR, 2022). ENDEAVR Institute also serves as an interdisciplinary project-based learning platform that connects science, technology, engineering, art, and mathematics disciplines with industries and communities.

The field of public health is focused on improving health and addressing health inequities of populations, communities, and individuals through prevention, health promotion and implementing interventions. With recent advances in information technology, big data, and communication technology, we can collect, analyse, and store large volumes of real-time data at the population level. The field of information technology, urban/social science, and public health are converging into the study of complex environments calling for a comprehensive review of how to achieve better outcomes and interventions to address those contemporary issues effectively. An example is how during the COVID-19 pandemic, some testing and vaccination centers were overwhelmed whilst others were not. A better understanding and implementation of CUS could enable health authorities to plan better. Another example is using geolocation data to estimate physical activity and how people move to access food. This information could be used to help authorities improve the built environment and decide where to allocate resources. The Healthy People 2030 also has several action objectives that can benefit from better integration with CUS, such as "reduce the number of days people are exposed to unhealthy air", "increase trips to work made by mass transit", and "increase the proportion of adults who walk or bike to get places".

31.5 Convergence-Driven

Advancements in location-aware technology, information and communication technology, and mobile technology during the past two decades have transformed the focus and need of built environment research. It moves from mostly indoor-based, community-level, or metropolitan-scaled static assessments to spatial, temporal, and dynamic relationships that integrate human behaviours across multiple environments and scales (mixed environmental models now including natural, built, and virtual elements). Simultaneously, projections show that, globally, more people will live in areas designated as vulnerable or high-risk relative to contemporary and future urban issues (e.g., sea level rise, depopulation, natural disasters, etc..), which suggests that urban communities will experience increases in multi-hazard risks. Disasters cause significant sources of property loss, social disruption, and inequality. Communities can reduce vulnerability while increasing social and physical resilience through research-driven and evidence-based planning, design, and policy development. However, silos within the design, social, and engineering sciences, and gaps between research and practice have made sustainable and equitable development difficult.

Given the interconnected systems and systems-of-systems that are inextricably tied to human dynamics and are ubiquitous within urban environments, complementary approaches that can embody diverse theories, methodologies, and data types are necessary. Thus, themes of CUS are increasingly towards the convergence and synthesis of theories and methods across multiple disciplines, as well as big and open data, computing technology, and interactive and collaborative environments. Goodchild (2020) pointed out any single discipline cannot solve pressing social and environmental challenges, and hence integration across conceptualizations, analytical methods, and software environments is necessary. From the convergent perspective, CUS is a human-centric synthesis that is responsive to the rapidly changing data and computing environment and reflective of the multi-faceted nature of social complexity. To extract profound insights from the massive amount of urban data, users must conduct iterative and evolving information foraging and sense-making and guide the process using their domain knowledge. Hence, visualization and iterative visual exploration are important in this process that relies on the deep integration of computational science and urban science. Due to the lack of synthesis, despite myriad visualization techniques and systems developed from urban data, there remains a gap between the demand of urban researchers and the availability of free-accessor open-source software. Such tools should include interactive visualizations and provide data curation, management, and logistic functions.

31.6 Promising Outlook and Next Steps

We live in the Urban Millennium. With over three million people estimated to be moving into areas each week (Perry et al., 2021), the world's population is expected to become increasingly concentrated in urban areas, which are projected to grow from 55% to 68% of the world's population over the next 30 years (WHO, 2022). As these demographics manifest worldwide, endemic problems associated with increases in urbanization can be expected to worsen. For example, nearly half of urban growth occurs in informal settlements, which are noted for their lack of clean water and proper sanitation and are considered especially vulnerable to adverse health and safety outcomes (Vojnovic et al., 2019; WHO, 2022). Further, many of those individuals locating to urban areas have been forcibly displaced, often due to man-made forces such as climate change (Perry et al., 2021). Additionally, the growth of urban areas has gone hand-in-hand with social inequality (Loukaitou-Sideris, 2020). As a result, urbanization—and in particular, unplanned and rapid urbanization—is connected with a majority of the leading causes of death (WHO, 2022).

Meeting persistent and emerging challenges related to twenty-first century urbanization will require approaches that can meaningfully embody those complex systems and systems-of-systems that characterize urban environments. Accordingly, there is a need to create a paradigm shift to develop a systematic and theoretical framework to proliferate CUS. The CUS framework presented in this chapter can propel a paradigm shift in the state of urban science through its integration of multidisciplinary and transdisciplinary theory, methodology, analysis, action planning, and community engagement. Along these lines, the four layers that constitute the CUS framework delineated herein can contribute to urban sustainability in the twenty-first century. In particular, with its emphasis on community engagement, this CUS framework can also address longstanding issues in urban research and design, most notably historical failures of these fields to be fully participatory and inclusive and to account for justice (Loukaitou-Sideris, 2020). By collaborating with underrepresented and diverse communities and by increasing communication with these communities, the field of CUS can enable urban design that can reduce biases and create a shared language and set of norms to bring a greater understanding of underlying complex systems and systems-of-systems. As a result, CUS can provide a means for enabling authentic bottom-up input from marginalized urban communities that can enable sustainable, equitable, and just urban development moving forward.

References

Batty, M. (2013). *The new science of cities.* MIT press.

Batty, M. (2020). *Foreword I: Charting computational social science from a spatial perspective.* In *Spatial Synthesis* (pp. 3–5). Springer, Cham.

Barabasi, A. L. (2005). The origin of bursts and heavy tails in human dynamics. *Nature, 435*(7039), 207–211.

Bibri, S. E. (2021). The core academic and scientific disciplines underlying data-driven smart sustainable urbanism: An interdisciplinary and transdisciplinary framework. *Computational Urban Science, 1*(1), 1–32.

Costales, E. (2022). Identifying sources of innovation: Building a conceptual framework of the Smart City through a social innovation perspective. *Cities, 120*, 103459.

ENDEAVR. (2022). *Envisioning the neo-traditional development by embracing the autonomous vehicles realm.* http://endeavr.city/about-us/team-leaders

Gelernter, D. (1993). *Mirror worlds: Or the day software puts the universe in a shoebox... How it will happen and what it will mean.* Oxford University Press.

Goodchild, M. F. (2020). *Convergence and Synthesis.* In *Spatial Synthesis.* Springer.

Hägerstrand, T. (1953). *Innovationsförloppet ur korologisk synpunkt* (Vol. 25). Gleerupska Univ.-Bokhandeln.

Huang, X., Li, Z., Jiang, Y., Ye, X., Deng, C., Zhang, J., & Li, X. (2021). The characteristics of multi-source mobility datasets and how they reveal the luxury nature of social distancing in the US during the COVID-19 pandemic. *International Journal of Digital Earth, 14*(4), 424–442.

Kontokosta, C. E. (2021). Urban informatics in the science and practice of planning. *Journal of Planning Education and Research, 41*(4), 382–395.

Li, D., Yu, W., & Shao, Z. (2021). Smart city based on digital twins. *Computational Urban Science, 1*(1), 1–11.

Lin, H., Chen, M., Lu, G., Zhu, Q., Gong, J., You, X., Wen, Y., Xu, B., & Hu, M. (2013). Virtual geographic environments (VGEs): a new generation of geographic analysis tool. *Earth-Science Reviews, 126*, 74–84.

Loukaitou-Sideris, A. (2020). Responsibilities and challenges of urban design in the 21st century. *Journal of Urban Design, 25*(1), 22–24.

NSF. (2020). *Urban systems and communities in the 21st century.* https://www.nsf.gov/ere/ereweb/c21c/index.jsp. Released on September 16, 2020.

Perry, G., Upchurch, C., & Kline, L. (2021). *Displacement, migration, and urbanization in the 21st century.* Wilson Center.

Shaw, S., Tsou, M., & Ye, X. (2016). Human dynamics in the mobile and big data era. *International Journal of Geographical Information Science, 30*(9), 1687–1693.

Steiger, E., Resch, B., & Zipf, A. (2016). Exploration of spatiotemporal and semantic clusters of Twitter data using unsupervised neural networks. *International Journal of Geographical Information Science, 30*(9), 1694–1716. https://doi.org/10.1080/13658816.2015.1099658

Tuan, Y. F. (1977). *Space and place: The perspective of experience.* University of Minnesota Press.

U.S. Department of State. (2021). *U.S.-China joint statement addressing the climate crisis.* https://www.state.gov/u-s-china-joint-statement-addressing-the-climate-crisis/

Vojnovic, I., Pearson, A. L., Asiki, G., DeVerteuil, G., & Allen, A. (2019). Global urban health: Inequalities, vulnerabilities, and challenges in the 21st century. In I. Vojnovic, A. L. Pearson, G. Asiki, G. DeVerteuil, & A. Allen (Eds.), *Handbook of global urban health* (pp. 3–32). Routledge.

Wan, G., Lin, H., Zhu, Q., & Liu, Y. (2021). Virtual geographical environment. In *Advances in cartography and geographic information engineering* (pp. 443–477). Springer.

World Health Organization. (2022). *Urban health.* https://www.who.int/health-topics/urban-health#tab=tab_1.

Ye, X., Du, J., Gong, X., Zhao, Y., Shamal, A. D., & Kamw, F. (2021a). SparseTrajAnalytics: An interactive visual analytics system for sparse trajectory data. *Journal of Geovisualization and Spatial Analysis, 5*(1), 1–11.

Ye, X., Wang, S., Lu, Z., Song, Y., & Yu, S. (2021b). Towards an AI-driven framework for multi-scale urban flood resilience planning and design. *Computational Urban Science, 1*(11). https://doi.org/10.1007/s43762-021-00011-0

Chapter 32
What Can We Learn from "Deviations" in Urban Science?

Fan Zhang and Xiang Ye

Abstract "Deviation" is common in scientific research, referring to the phenomenon that the output of a process is different from the expected. Deviation may possess various appearances and definitions, e.g., deviation of an observation from the truth, the general trend, or the theoretical value under assumptions, etc. Although in many cases it is perceived by the researcher as unwanted, it may be an inspirer and facilitator, leading to new discoveries and insights from innovative pathways. This chapter initiates a discussion on what and how we can learn from deviations, particularly in urban science. We use several application examples featuring big data and deep learning to illustrate our points.

Keywords Deviation · Quantitative analysis · Urban science · Deep learning · Street view imagery

32.1 Introduction

"Deviation" is a very common phenomenon in scientific research. In this chapter we use this term to refer to the difference between what is obtained and what is referenced. In many cases, a deviation is not favored by the researcher, as it brings the feeling that something is "wrong". In this case, a common response is to mitigate, offset, or eradicate its appearance and effects, as if it was a nuisance to tidy data, accurate models, correct predictions, and unregretful decisions (Yuan et al., 2020).

However, treating the deviation as a nuisance is not necessarily the only or even the principal way of dealing with it in scientific research. Depending on objectives,

F. Zhang (✉)
Senseable City Lab, Massachusetts Institute of Technology, Cambridge, MA 02139, USA
e-mail: zhangfan@mit.edu

X. Ye
Research Institute for Smart Cities, Shenzhen University, Shenzhen 518060, Guangdong, China
e-mail: yexiang@szu.edu.cn

School of Architecture and Urban Planning, Shenzhen University, Shenzhen 518060, Guangdong, China

B. Li et al. (eds.), *New Thinking in GIScience*,
https://doi.org/10.1007/978-981-19-3816-0_32

viewpoints, contexts, and available options, there is a collection of alternative possibilities to harness the deviation as a tool, an information source, or an innovative entry point, to answer questions that have been previously left unattended.

With this chapter, we intend to initiate a discussion on deviation, exploring the diversity of its appearance, meaningfulness (Sect. 32.2), and statistical properties (Sect. 32.3), as well as a variety of strategies toward it (Sect. 32.4). A selection of application examples in urban science are demonstrated thereafter (Sect. 32.5), leading to the conclusion that how deviations can bring us more benefits than challenges if we take an alternative viewpoint (Sect. 32.6).

32.2 A Variety of Deviations

A general understanding of *deviation* is "the difference between what is obtained and what is referenced". Therefore, what is to be referenced can be used as the benchmark to define deviations in different contexts.

The most straightforward benchmark is *truth* (Fig. 32.1a). A deviation from truth is an *error*, originated from the limited accuracy and ability to describe the real world. Rarely do we appreciate this kind of deviation, albeit its existence is almost inevitable and omnipresent.

Another frequently adopted benchmark is the theoretical extremum of an indicator (Fig. 32.1b). In fitting and predicting, an indicator is chosen to assess the performance of the model. While we usually do not expect the indicator to reach the theoretical extremum, a smaller deviation is always preferred: We like to see a small RMSE, a large R^2, or a confusion matrix with its non-zero elements mostly along the main diagonal.

A related benchmark is the theoretical expectation under the null hypothesis (Fig. 32.1c). Contrary to the situation of theoretical extremum, the deviation from the null hypothesis is usually favored, because only when it is large enough, can we reject the null hypothesis and claim the usually more interesting alternative.

Meanwhile, a benchmark can be the theoretical output value from a model under the model's assumptions (Fig. 32.1d). Assumptions are pre-set conditions for a model to perform as expected. On the one hand, the model itself does not possess any ability to check if the desired assumptions are honored. On the other hand, failure to meet these assumptions leads to model misspecification (Hansen, 2021, pp. 213–214), and the variable of interest may deviate from its theoretical value. The deviation from that benchmark typically leads to a deeper exploration and refinement of the model: How should we determine if an assumption is violated? How would the model behave if a certain assumption is relaxed?

In some situations, existing observations are referenced as a benchmark (Fig. 32.1e). This is typically the case if the same model is applied to different data sets. When the model outcomes based on a newer set of data are available, they are naturally compared with the existing ones to see if any deviations appear. In social and environmental sciences, this deviation is a manifestation of the weak replicability

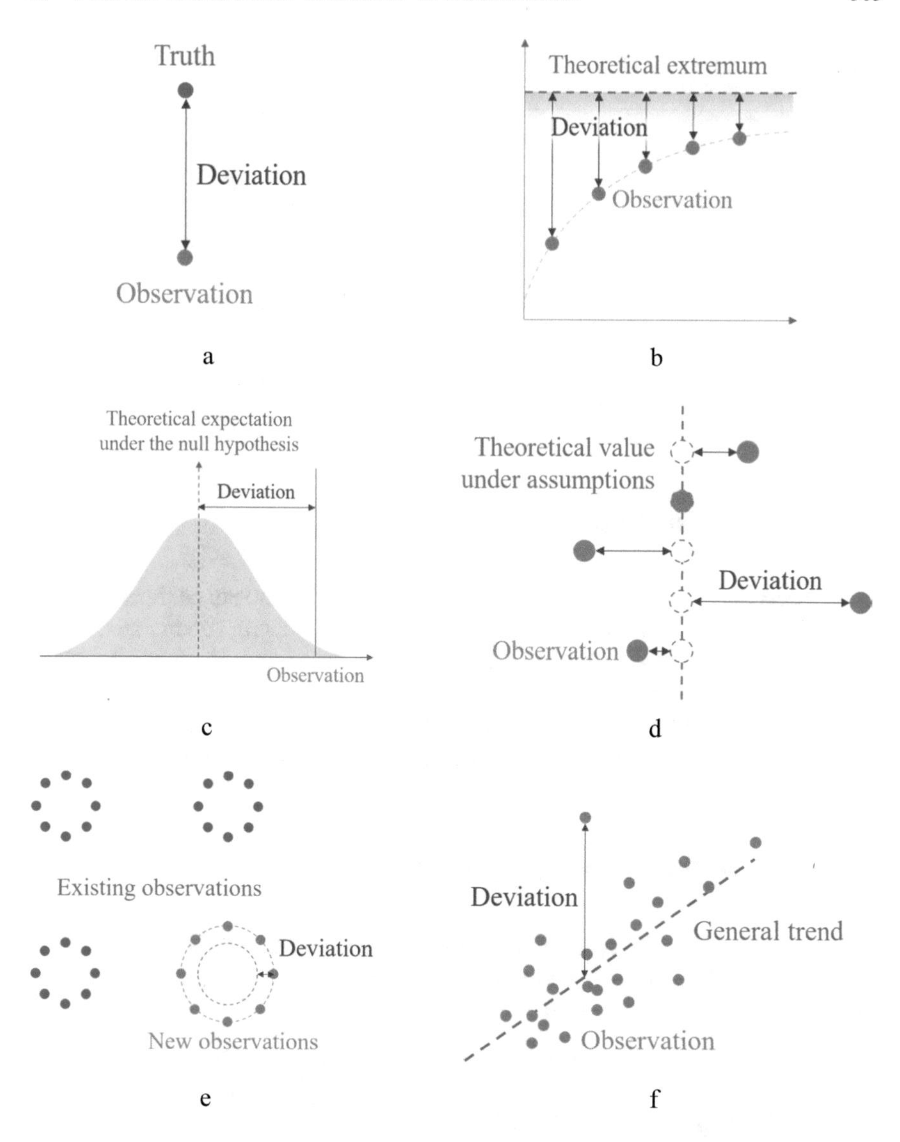

Fig. 32.1 Deviations categorized by the benchmark. **a** The deviation from truth. **b** The deviation from the theoretical extremum of an indicator. **c** The deviation from the theoretical expectation under null hypothesis. **d** The deviation from the theoretical value under assumptions. **e** The deviation from existing observations. **f** The deviation from general trend

(Goodchild & Li, 2021), suggesting a limited adaptability of a model established based on one dataset when applying to another dataset. Though considered negative at the first glance, it can help understanding heterogeneity when properly adopted.

Last but not least, the benchmark can be general trend (Fig. 32.1f). General trend serves as a background for investigating individual cases, during which we care more

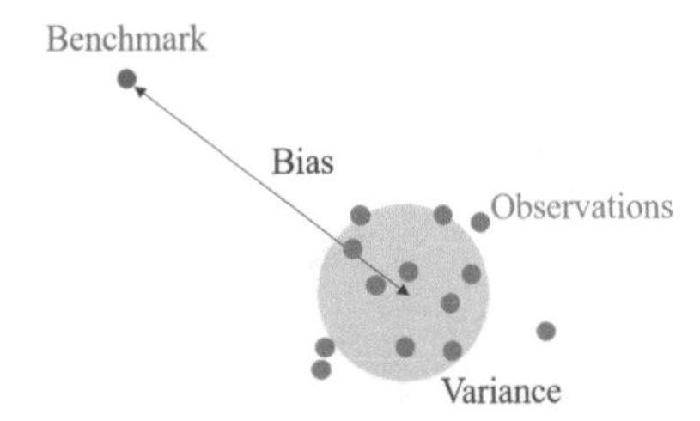

Fig. 32.2 Statistical properties of deviation

about outliers, i.e., those exhibiting large deviations from the general trend. In this case, the deviation is not an obstacle of science, but an exciting point of entry to new findings of heterogeneity and uniqueness.

32.3 Statistical Properties of Deviation

If we observe a deviation only once or a couple of times, we are not confident to make an affirmative claim, because we are unsure if it is large enough and/or statistically significant to be treated with extra attention. If we do observe the deviation of the same kind many times (Fig. 32.2), we may be able to acquire sufficient evidence to tell the nature of the deviation. In particular, we can quantify its uncertainty with *variance*, and determine if it indicates *bias* by measuring its direction and magnitude.

32.4 Strategies for Dealing with Deviation

The analysis of different types of deviations helps us choose the proper strategy to deal with them in research. Here we describe four strategies, which form a spectrum from most avoiding to most welcoming:

Elimination or mitigation. This strategy might be the most intuitive one. Depending on how well the nature of deviation is understood and how much information is available, the impacts of deviation will be eliminated or mitigated.

Acceptation and evaluation. We no longer perceive the deviation as a nuisance that must be eradicated. Instead, we accept it as an inherent property of the real world or the model we adopt. Then, we would like to evaluate and understand it, through, e.g., quantifying its magnitude and uncertainty.

Investigation and explanation. With an understanding that an exception (outlier) typically indicates something that is unknown, surprising, or even exciting, we want to look into it and give it an explanation.

Exploration and utilization. Finally, we entirely reverse the role of deviation. Instead of seeing it as an enemy, we see it as an ally that can help us discover new knowledge.

To achieve this harmony, we need to shift our mind and think outside the box. In the next section, we will show application examples to demonstrate how deviation can help urban scientists in an innovative and beneficial way.

32.5 When Deviation Fuels Urban Science

32.5.1 Deviation as Outliers: Exploring Uniqueness of Places

In a study conducted by Zhang et al. (2019a), street view images were used to estimate the temporal patterns of human mobility on streets in Beijing, China via machine learning techniques. They employed the hourly volume of taxi trips as a proxy for the human mobility pattern and assumed that the streetscape depicted in street-level imagery reflects urban functions. A deep convolutional neural network (DCNN) was trained to predict the hourly volume of taxi trips based on street view images. Figure 32.3a shows a plot of the relationship between the actual number of taxi trips and the corresponding mean absolute error (MAE) of the predictions from the DCNN model. Generally, the MAE increases steadily with the number of taxi trips (Pearson's $r = 0.91$); however, there are two outliers popping up, shown as the red dots.

The authors of the study chose to not ignore the two outliers, but instead to give them a further investigation. They found that for the two outliers, Liuli Bridge (Fig. 32.3b) and Beijing South Station (Fig. 32.3c), each represents a special local traffic situation. At Liuli Bridge, the taxi flow was underestimated (1577 actual vs. 937 predicted), possibly due to the additional taxi traffic brought about by the coach station that was not reflected in the street view images. At Beijing South Station, the taxi flow was overestimated (2167 actual vs. 2736 predicted), because a considerable amount of potential taxi customers got diverted to subway, as this is where two subway lines intersect. Albeit seemingly frustrating at the beginning from the perspective of modeling, these deviations from the general trend eventually led to findings about the place uniqueness and offered insights about Beijing's traffic pattern at a local scale.

32.5.2 Deviation as Misclassification: Measuring Visual Similarity Between Cities

Zhang et al. (2019b) proposed a framework to measure the visual similarity among (and the distinctiveness of) 18 selected cities based on a confusion matrix (Fig. 32.4). By feeding millions of geo-tagged photos obtained from social media, a DCNN was trained to recognize the city from which the photos were taken. To evaluate the

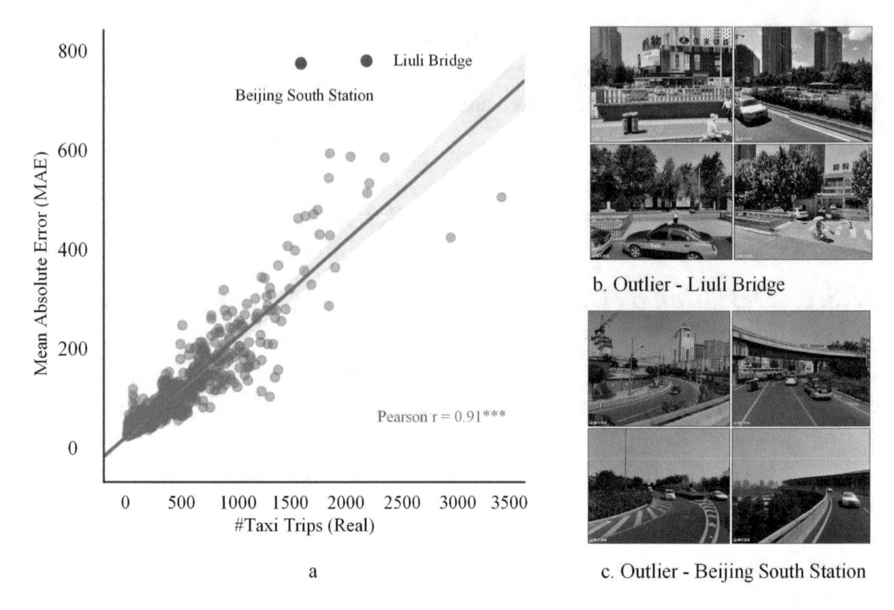

Fig. 32.3 Estimating the temporal variation of taxi flow on a street based on street view images using a DCNN. **a** MAE increases steadily with the number of taxi trips with two outliers. **b** Image samples of Liuli Bridge. **c** Image samples of Beijing South Station. *Source* Adapted from Zhang et al. (2019a)

performance of the model, a confusion matrix was generated to record the identification results. In Fig. 32.4a, each cell contains the value of ratio N_{ij}/N_i, where N_i is the number of all photos actually of city *i*, and N_{ij} is the number of photos in N_i that were identified as city *j* by the model.

Commonly, one would simply use this matrix to assess the performance of the proposed model. Zhang et al. (2019b), however, utilized the deviation information in the matrix to evaluate visual features of the cities. If a diagonal value of the confusion matrix considerably deviates from its theoretical extremum, 100%, it means the city is not easy to be identified visually—the diagonal value actually can be used to quantify the visual distinctiveness of that city. On the other hand, if an off-diagonal value considerably deviates from its theoretical extremum, 0%, it indicates a good chance that a city is to be misclassified to another. If two off-diagonal values at the symmetrical positions are both high, these two cities are easily to be mutually misclassified, i.e., they have a high visual similarity (Fig. 32.4b).

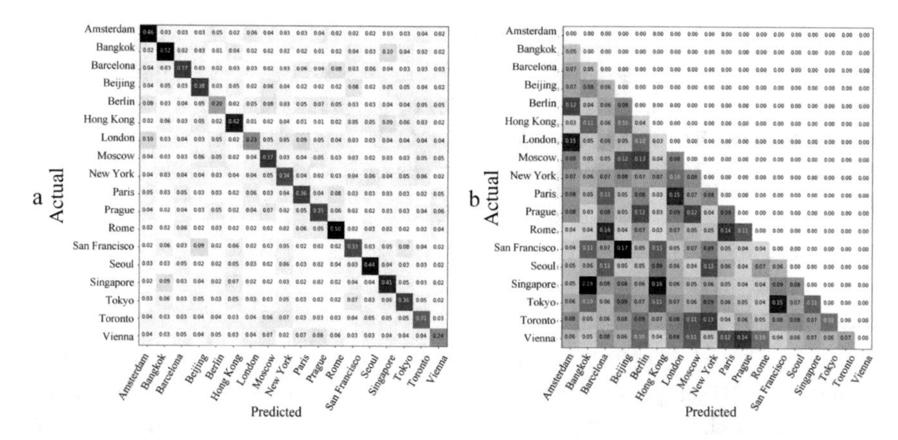

Fig. 32.4 Measuring similarity and distinctiveness of cities using a confusion matrix. **a** Confusion matrix. **b** Similarity matrix. *Source* Adapted from Zhang et al. (2019b)

32.5.3 *Deviation as a Weak Replication: Revealing the Heterogeneity of the World*

Figure 32.5 shows a study of predicting house prices based on Google Street View imagery using a DCNN model. When both training set and testing set were from the same city, either Boston or Detroit, the models were able to successfully associate house price with visual appearance (Fig. 32.5a, b). However, when the model was trained with the data from Detroit and then applied to Boston, it greatly overestimated house prices in Boston, and what was even worse, it reversed the actual spatial pattern of Boston's house prices between downtown and suburbs.

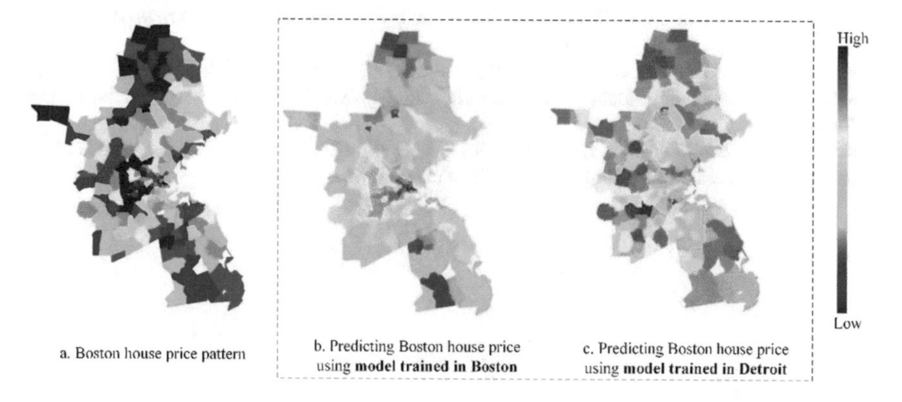

Fig. 32.5 Revealing the heterogeneity of the real estate market by weak replication. **a** Actual spatial distribution of Boston house prices. **b** Predicted Boston house prices from the model trained with the data from **Boston**. **c** Predicted Boston house prices from the model trained with the data from **Detroit**. *Image credit* Kang, Yuhao (University of Wisconsin-Madison)

Obviously, the model trained with the data from Detroit does not work well for Boston. However, such a deviation unravels the distinctive tastes in the real estate market of the two cities: the visual quality of a neighborhood is less priced-in in Boston than in Detroit, while a "downtown-look" is a price booster in Boston but a price killer in Detroit. In this way, the deviation becomes a valuable information source.

32.6 Conclusion

There are different types of deviations that can demonstrate different characteristics. On the one hand, they may cost additional wisdom and endeavors to meet the challenge they impose to research. On the other hand, they can be valuable information sources for researchers to uncover more about the phenomenon of interest. Within a context of urban science, this perspective is opening up new possibilities in changing the way we perceive and understand cities and in revealing the heterogeneity and general laws of the world. We hope this chapter can serve as an initiation of further discussion and exploration on this topic.

References

Goodchild, M. F., & Li, W. (2021). Replication across space and time must be weak in the social and environmental sciences. *Proceedings of the National Academy of Sciences, 118*(35), e2015759118.

Hansen, B. E. (2021). *Econometrics*. Department of Economics, University of Wisconsin-Madison.

Yuan, Y., Lu, Y., Chow, T. E., Ye, C., Alyaqout, A., & Liu, Y. (2020). The missing parts from social media-enabled smart cities: Who, where, when, and what? *Annals of the American Association of Geographers, 110*(2), 462–475.

Zhang, F., Wu, L., Zhu, D., & Liu, Y. (2019a). Social sensing from street-level imagery: A case study in learning spatio-temporal urban mobility patterns. *ISPRS Journal of Photogrammetry and Remote Sensing, 153*, 48–58.

Zhang, F., Zhou, B., Ratti, C., & Liu, Y. (2019b). Discovering place-informative scenes and objects using social media photos. *Royal Society Open Science, 6*(3), 181375.

Chapter 33
Variants of Location-Allocation Problems for Public Service Planning

Yunfeng Kong

Abstract This chapter presents some variants of the location-allocation problems (LAPs) with additional criteria for service planning such as partial coverage of service demand, contiguous service areas, and equal service areas. The variants arise in applications such as the selection of facility sites for the "15-minute city", the delineation of public service areas, and the provision of some emergency services in the COVID-19 pandemic. The criteria are formulated as linear inequalities and thus can be added to the classical LAP models. It is challenging to solve those variants, since LAPs are known to be nondeterministic polynomial time hard (NP-hard), and the new criteria may impose further obstacles to the analytical solution. At the end of the chapter, I discuss possible methods to solve the variants.

Keywords Location-allocation problem · Partial coverage · Contiguous service area · Mathematical model · Solution method

33.1 Introduction

The location-allocation problems (LAPs) have been extensively investigated since 1960s. All such problems aim to optimally locate a set of facilities and assign all the demand to facilities. They have been widely used in both public and private facility planning, such as schools, healthcare centers, warehouses, and logistic centers. For problem definition, mathematical formulation, algorithm design, and applications of the LAPs, please refer to Eiselt and Marianov (2011) and Laporte et al. (2015).

The real-world site selection often requires additional criteria to achieve an effective and efficient service system. This chapter presents three new criteria for public service planning: partial coverage of service demand, contiguous service areas, and equal service areas. The partial coverage means that the model allows a part of the demand from a demand point to be not covered by the facilities within the specified impedance threshold. It is a way to balance the service cost and spatial access. Such

---

Y. Kong (✉)
College of Geography and Environmental Science, Henan University, Kaifeng 475000, Henan, China
e-mail: yfkong@henu.edu.cn

B. Li et al. (eds.), *New Thinking in GIScience*,
https://doi.org/10.1007/978-981-19-3816-0_33

a balance is a major consideration in service planning, e.g., in realizing the so-called "15-minute city", which aims to achieve the goal that most human needs and many desires can be met by facilities located within a travel distance of 15 min (Ministry of Housing and Urban–Rural Development of the People's Republic of China, 2018). The contiguous service area means that the service area of a facility should be formed by spatially contiguous areal units. It is one of the essential issues in service planning (Kong 2021b), and also a necessity to consider service management and related policy making (Daskin, 2011). The equality of service areas means that the demands in the service areas are equal, and thus the work tasks can be equally assigned to the service suppliers. In case of the lockdown in some cities in China over COVID-19 outbreak, a set of facilities with equal and contiguous service areas can be used to deploy emergency services such as the medicine distribution, the supply of daily necessities, and the large-scale nucleic acid detection of COVID-19 (Kong, 2021a). In the following sections, LAPs with these new criteria are mathematically formulated as mixed integer linear programming (MILP) models. The possible solution methods for the new problems are discussed. Finally, two concluding remarks are highlighted.

33.2 Variants of Location-Allocation Problems

33.2.1 Classical Location-Allocation Problems

Let I be a set of candidate locations for opening facilities, and J be a set of customers. Each facility at location i has a fixed cost f_i and a maximum service capacity s_i. Each customer j has a demand d_j. The distance from customer j to the facility located at i is known, denoted as d_{ij}, and the cost for satisfying the demand of customer j from the facility at i is given, denoted as c_{ij}. The classical location-allocation problems can be formulated by defining deferent decision variables, objective functions, and constraints as follows (Laporte et al., 2015).

$$f_1 = \sum_{i \in I} \sum_{j \in J} c_{ij} x_{ij} \tag{33.1}$$

$$f_2 = \sum_{i \in I} f_i y_i + \sum_{i \in I} \sum_{j \in J} c_{ij} x_{ij} \tag{33.2}$$

$$f_3 = \sum_{i \in I} f_i y_i \tag{33.3}$$

$$\sum_{i \in I} x_{ij} = 1, \quad \forall j \in J \tag{33.4}$$

$$\sum_{i \in I, d_{ij} \leq r} y_i \geq 1, \quad \forall j \in J \tag{33.5}$$

$$\sum_{i \in I} y_i = P \tag{33.6}$$

$$x_{ij} \leq y_i, \quad \forall i \in I, \ j \in J \tag{33.7}$$

$$\sum_{j \in J} d_j x_{ij} \leq s_i y_i, \quad \forall i \in I \tag{33.8}$$

$$x_{ij} = [0, 1], \quad \forall i \in I, \ j \in J \tag{33.9}$$

$$x_{ij} = \{0, 1\}, \quad \forall i \in I, \ j \in J \tag{33.10}$$

$$y_i = \{0, 1\}, \quad \forall i \in I \tag{33.11}$$

The p-median problem (PMP) minimizes the objective function (33.1) subject to constraints (33.4), (33.6), (33.7), (33.10) and (33.11). The capacitated PMP (CPMP) minimizes the objective function (33.1) subject to constraints (33.4), (33.6), (33.8), (33.10) and (33.11).

The uncapacitated facility location problem (UFLP) targets the minimization of objective function (33.2) subject to constraints (33.4), (33.7), (33.9) and (33.11). The capacitated version (CFLP) minimizes the objective function (33.2) subject to constraints (33.4), (33.8), (33.9) and (33.11). In single-source CFLP (SSCFLP), each customer can only be served by one facility. It targets the minimization of objective function (33.2) subject to constraints (33.4), (33.8), (33.10) and (33.11). Since the number of facilities, the facility locations, and the demand allocations must be considered simultaneously, the SSCFLP might be the hardest problem of LAPs.

Given a maximum service radius r, the location set covering problem (LSCP) aims to search for a minimum number of locations for opening facilities. It minimizes the objective function (33.3) subject to constraints (33.5) and (33.11).

33.2.2 Partial Coverage Location-Allocation Problems

Spatial accessibility is one of the most important indicators to evaluate service quality and equality. In urban planning, the service radius has been widely used to guide the service design. Deferent service radiuses are recommended by planning authorities for different public services. Given a maximum service radius r, the constraints (33.4) can be replaced with constraints (33.12) in PMP, CPMP, UFLP, CFLP and SSCFLP. In constraints (33.12), each demand can only be covered by a facility with the service radius r, which will improve the spatial access to services. The LAPs with a maximum service radius are denoted as *r*PMP, *r*CPMP, *r*UFLP, *r*CFLP and *r*SSCFLP.

$$\sum_{i \in I, d_{ij} \leq r} x_{ij} = 1, \quad \forall j \in J \tag{33.12}$$

The location problems with a service radius can effectively improve spatial access to services. However, under this condition the number of facilities required to cover all demand will increase dramatically. For example, in a city with a population of about 0.71 million and a size of 132 km^2, a conventional SSCFLP would locate 8 community healthcare centers to cover all service demand, with an average travel distance of 1.47 km. However, using the *r*SSCFLP with a maximum service radius of 2.0 km, 24 centers are required.

It is possible to balance the service quality and service-supply efficiency using partial coverage of demand. Partial coverage for LSCP, PMP, UFLP has been investigated with case studies (Cordeau et al., 2019; Daskin & Owen, 1999; Nozick, 2001; Vasko, 2003). Methods such as Lagrangian heuristic and Benders decomposition were used to solve the partial coverage location problems. Experiments show that the numbers of required facilities in the partial coverage solutions change dramatically with different service radiuses and percentages of demand coverage.

A constraint on partial coverage (33.13) can be added in the classical PMP, CPMP, UFLP, CFLP, and SSCFLP. The parameter μ denotes a percentage of service demand that must be covered within the service radius *r*. Thus, the partial coverage LAPs are denoted as μPMP, μCPMP, μUFLP, μCFLP and μSSCFLP, respectively.

$$\sum_{i \in I} \sum_{j \in J, d_{ij} \leq r} d_j x_{ij} \geq \mu \sum_{j \in J} d_j \tag{33.13}$$

Partial coverage LSCP (μLSCP) can be formulated as the minimization of objective function (33.3) subject to (33.11) and (33.14)–(33.16). The decision variable z_j indicates whether costumer *j* is covered by at least one facility or not.

$$\sum_{i \in I, d_{ij} \leq r} y_i \geq z_j, \quad \forall j \in J \tag{33.14}$$

$$\sum_{j \in J} d_j z_j \geq \mu \sum_{j \in J} d_j \tag{33.15}$$

$$z_j = \{0, 1\}, \quad \forall j \in J \tag{33.16}$$

Note that in problems such as μCPMP, μCFLP, and μSSCFLP, all demand must be satisfied by conditions (33.4). In these problems, it is a necessity to cover all demand for public services. Meanwhile, a small part of demand is allowed to be covered outside the service radius, which will decrease the number of facilities required and thus improve the service-supply efficiency. As a result, these variants have application potentials in service planning toward the "15-minute city".

33.2.3 *Location-Allocation Problems with Contiguous Service Areas*

The delineation of the service area for a facility is a must in the assessment and/or planning of some public services. A service area is considered *contiguous* if one can travel between any two points in the area without crossing its boundaries. Such a continuity is a convenience to the management of schools (Caro et al., 2004), healthcare centers (Emiliano et al., 2017), disaster shelters (Hu et al., 2014), and many other facilities. LAPs with contiguous service areas aim to simultaneously locate the facilities, allocate demand units to the facilities, and delineate contiguous service areas for the facilities.

Let J be a set of spatial units in a geographical area. Let I, a subset of J ($I \subseteq J$), be candidate units for locating facilities. The variables f_i, s_i, d_j, d_{ij}, and c_{ij} have the same meanings as defined earlier. The formulas (33.1)–(33.16) can also be applied to LAPs associated with geographical areas. At the same time, it is possible to extend the problems by adding contiguity constraints on facility service areas. A flow-based model was proposed for the p-Regions problem (Duque et al., 2011) and has been adaptively formulated for service area problem (Kong, 2021b) and districting problem (Kong, 2021b; Kong et al., 2019). Let a_{jk} indicate whether unit j and k share a border, and N_j be a set of units that are adjacent to unit j ($N_j = \{k | a_{jk} = 1\}$). Let f_{ijk} be decision variables that indicate the flow volume from unit j to unit k in service area i, and the flow model for SSCFLP can be formulated as follows:

$$f_{ijk} \leq |J| * x_{ij}, \quad \forall i \in I,\ j \in J,\ k \in N_j \tag{33.17}$$

$$f_{ijk} \leq |J| * x_{ik}, \quad \forall i \in I,\ j \in J,\ k \in N_j \tag{33.18}$$

$$\sum_{k \in N_j} f_{ijk} - \sum_{k \in N_j} f_{ikj} \geq x_{ij}, \quad \forall i \in I,\ j \in J \backslash i \tag{33.19}$$

$$f_{ijk} \geq 0, \quad \forall i \in I,\ j \in J,\ k \in N_j \tag{33.20}$$

In the flow-based model, the service area contiguity is ensured by establishing a flow route from each spatial unit to its facility unit within its service area. Constraints (33.17) and (33.18) ensure that a flow may only be passed through neighboring units in the same district. If unit i does not serve as the center unit, constraints (33.19) state that one unit of flow must be created from this unit, and finally flows to its service-supply unit.

The flow model (33.17)–(33.20) can be added to CPMP, SSCFLP, rCPMP, rSSCFLP, μCPMP, and μSSCFLP. These problems with contiguous service areas are denoted as cCPMP, cSSCFLP, crCPMP, crSSCFLP, $c\mu$CPMP, and $c\mu$SSCFLP, respectively.

33.2.4 The Equal Districting Problem

The equal districting problem (EDP) arises in applications such as political redistricting, police patrol area delineation, sales territory design and some service systems design (Kong, 2021a). The EDP is the problem of grouping geographic areas into P districts that have equal quantities of voters, work tasks, service demands, or other indicators. The key criteria in the EDP include equality, contiguity, and compactness (Kalcsics, 2015).

The EDP can be formulated as an equal-capacitated PMP with contiguous service areas. Location-allocation models were proposed by Hess et al. (1965), Hojati (1996), and George et al. (1997) for political districting. This approach was further investigated by Kong et al. (2019) and Kong (2021a). The facility locations can be considered as district centers. As a result, it is convenient to measure district contiguity and compactness using the district centers (Kalcsics, 2015).

Let J be a set of spatial units in a geographical area, and d_j is an attribute of unit j. Let I be a set of candidate units for locating facilities. If we assume that all the spatial units can be candidate units, then it is obvious that $I = J$. Let $\overline{Q} = \sum_{i \in J} d_j / P$ be the facility capacity, and then the capacity constraints (33.8) can be replaced by (33.21). Note that P is the number of districts. Constraints (33.21) confirm that the demands allocated to facilities are almost equal with a predefined error ε, e.g. $\varepsilon = 5\%$. The equal-capacitated PMP minimizes the objective function (33.1) subject to (33.4), (33.6), (33.8), (33.10), (33.11), and (33.21). The equal-capacitated PMP with contiguous service areas minimizes the objective function (33.1) subject to (33.4), (33.6), (33.8), (33.10), (33.11), and (33.17)–(33.21).

$$(1 - \varepsilon)\overline{Q} y_i \leq \sum_{j \in J} d_j x_{ij} \leq (1 + \varepsilon)\overline{Q} y_i, \quad \forall i \in I \tag{33.21}$$

Using the location-allocation method, the criteria for EDP, such as the district equality, compactness, and contiguity, are satisfied by the constraints (33.21), the objective function (33.1), and the flow model (33.17)–(33.20), respectively. In addition, the facility locations serve as the centers of districts.

33.2.5 Problem Properties

The solutions of a partial coverage LAP may change significantly from case to case due to different settings for parameters r and μ (Daskin & Owen, 1999; Nozick, 2001). Given a p-median problem instance, its optimal objectives satisfy the inequalities (33.22), (33.23), and (33.24). The CPMP, SSCFLP, and their variants share the same properties.

$$f_{PMP} = f_{rPMP(r=\infty)} \leq f_{\mu PMP(r,\mu)} \leq f_{\mu PMP(r,\mu=100\%)} = f_{rPMP(r)} \tag{33.22}$$

$$f_{\mu PMP(r_1,\mu)} \leq f_{\mu PMP(r_2,\mu)}, \quad \forall r_1 > r_2 \tag{33.23}$$

$$f_{\mu PMP(r,\mu_1)} \leq f_{\mu PMP(r,\mu_2)}, \quad \forall \mu_1 < \mu_2 \tag{33.24}$$

$$f_{CPMP} = f_{rCPMP(r=\infty)} \leq f_{\mu CPMP(r,\mu)} \leq f_{\mu CPMP(r,\mu=100\%)} = f_{rCPMP(r)} \tag{33.25}$$

$$f_{\mu CPMP(r_1,\mu)} \leq f_{\mu CPMP(r_2,\mu)}, \quad \forall r_1 > r_2 \tag{33.26}$$

$$f_{\mu CPMP(r,\mu_1)} \leq f_{\mu CPMP(r,\mu_2)}, \quad \forall \mu_1 < \mu_2 \tag{33.27}$$

$$f_{SSCFLP} = f_{rSSCFLP(r=\infty)} \leq f_{\mu SSCFLP(r,\mu)} \leq f_{\mu SSCFLP(r,\mu=100\%)} = f_{rSSCFLP(r)} \tag{33.28}$$

$$f_{\mu SSCFLP(r_1,\mu)} \leq f_{\mu SSCFLP(r_2,\mu)}, \quad \forall r_1 > r_2 \tag{33.29}$$

$$f_{\mu SSCFLP(r,\mu_1)} \leq f_{\mu SSCFLP(r,\mu_2)}, \quad \forall \mu_1 < \mu_2 \tag{33.30}$$

For the μLSCP, μPMP, μCPMP and μSSCFLP, the service quality and service efficiency are controlled by the two parameters: r and μ. The PMP, CPMP and SSCFLP ($r = \infty$ and $\mu = 0\%$) emphasize the solution efficiency. The service quality in terms of spatial access is considered in LSCP, rPMP, rCPMP, and rSSCFLP ($\mu = 100\%$). However, if a small r value is given, an instance of the problem may be infeasible, or its objective may be very high. In real-world service planning, a satisfactory solution may be obtained by tuning r and μ.

An LAP with contiguous service areas may have its objective raised, and also have the facility locations changed.

33.3 Possible Approaches to Solving the Models

It is challenging to solve the LAPs, since they are known to be nondeterministic polynomial time hard (NP-hard). In the past 60 years, countless algorithms have been proposed for the LAPs (Turkoglu & Genevois, 2020). The solution methods can be classified into four categories: exact methods, Lagrangian relaxation-based heuristics (LHs), local search based or evolutionary based heuristics/metaheuristics, and hybrid algorithms. Most existing algorithms for LAPs can be adapted to solve the new problems discussed in this chapter. Nevertheless, for the hardest problems such as μSSCFLP, it is still challenging to solve large instances.

The general set covering problem (SCP) and the specific LSCP have been extensively investigated since 1960s (Farahani et al., 2012). The general-purpose MIP solvers like CPLEX are competitive for solving small and some large LSCP instances (Caprara et al., 2000). The author's experiments show that many real-world LSCP/μLSCP instances can be effectively solved by Gurobi Optimizer 9 on a desktop computer. For very large LSCP/μLSCP instances, it is necessary to design an LH, metaheuristic, or hybrid algorithm.

The fast implementation (Resende & Werneck, 2007) of the interchange method (Teitz & Bart, 1968) might be the best choice for solving PMP. The metaheuristics based on variable depth neighborhood structure or interchange method are more effective in a reasonable computational time. The sampling technique is helpful to solve very large PMP instances (Mu & Tong, 2020). These existing methods might be useful for rPMP and μPMP, but their performance remains uninvestigated.

Various Lagrangian relaxation-based heuristics (LHs) for PMP, CPMP, UFLP, CFLP, and SSCFLP have been proposed since 1970s. LHs are usually simple, and also have an advantage of providing a lower bound to evaluate the incumbent solution. As a result, LHs are widely used to generate initial solutions for metaheuristics. It is worth to explore the Lagrangian relaxation techniques for μLSCP μCPMP μCFLP, and μSSCFLP.

Exact methods such as branch-and-bound, branch-and-cut, column generation and Benders decomposition have been widely used to solve various LAPs. Along with the rapid progress in mixed integer linear programming, it is an easy way to solve LAP instances by commercial or open-source mixed-integer programming (MIP) solvers. Existing experiments show that many location problem instances could be exactly solved in a reasonable time by CPLEX Optimizer or Gurobi Optimizer.

Table 33.1 shows some problem solutions obtained by Gurobi 9.1.2 on a desktop computer with Intel Core I7-6700 CPU 3.40-GHz, and 8-GB RAM. The instances were created by the author based on rural and urban geographic data. The solution quality in terms of MIPGap and computational time show that the optimal or near-optimal solutions could be found by commercial MIP solvers. Experiments also show that the performance of MIP solver depends on the problem type, instance size, and problem parameters. The instance complexity also depends on supply–demand ratio and the cost structure. Since the techniques from metaheuristics have

Table 33.1 Selected problem instances and their solutions found by Gurobi optimizer

Problem	Instance size ($\|I\| * \|J\|$)	Problem parameters	MIPGap	Time (s)
LSCP	2214 * 2214	$r = 1.0$ km	4.70%	7200.00
μLSCP	2214 * 2214	$r = 1.0$ km, $\mu = 90\%$	Optimal	146.11
PMP	1276 * 1276	$P = 50$	Optimal	377.10
rPMP	1276 * 1276	$r = 3.0$ km, $P = 50$	Optimal	2036.67
μPMP	1276 * 1276	$r = 3.0$ km, $\mu = 80\%$, $P = 50$	Optimal	4068.82
cPMP	324 * 324	$P = 20$	Optimal	3390.67
CPMP	146 * 2999	$P = 25$	Optimal	4973.70
rCPMP	146 * 2999	$r = 2.0$ km, $P = 25$	Optimal	1039.75
μCPMP	146 * 2999	$r = 1.0$ km, $\mu = 80\%$, $P = 25$	Optimal	6695.39
cCPMP	33 * 1276	$P = 20$	Optimal	2564.45
$c\mu$CPMP	297 * 297	$r = 4.0$ km, $\mu = 80\%$, $P = 20$	Optimal	6648.62
SSCFLP	33 * 1276	–	0.01%	7200.00
cSSCFLP	33 * 1276	–	0.16%	7200.00
rSSCFLP	198 * 2999	$r = 1.5$ km	Optimal	1605.55
μSSCFLP	146 * 2999	$r = 1.0$ km, $\mu = 80\%$	1.84%	7200.00
μcSSCFLP	297 * 297	$r = 4.0$ km, $\mu = 80\%$	14.71%	7200.00
EDP	324 * 324	$\varepsilon = 5\%$, $P = 15$	Optimal	4973.77

been incorporated in the MIP solvers, the state-of-the-art solvers might be a good choice for solving many real-world problems.

A large LAP instance is hard to solve; however, it could be solved by repeatedly searching for small parts of the instance. Accordingly, the matheuristic, which explores large neighborhoods by a MIP solver, may be a promising method for high-complexity problems such as CPMP, SSCFLP and their variants.

33.4 Concluding Remarks

Novel variants of classical LAPs are valuable for modern public service planning. First, the service cost and spatial access of service can be balanced by the two parameters in partial coverage location problems. Second, the location-allocation method for equal districting is effective to design some service systems.

Most variants of the classical LAPs are difficult to solve. The current competitive solution methods for LAPs should be reinvestigated and adapted to solve the variants. The matheuristic might be an efficient way to solve the new problems. In addition, application-oriented GIS tools should be developed for planning practitioners.

Acknowledgements Research partially supported by the National Natural Science Foundation of China (No. 41871307).

References

Caprara, A., Toth, P., & Fischetti, M. (2000). Algorithms for the set covering problem. *Annals of Operations Research, 98*, 353–371.
Caro, F., Shirabe, T., Guignard, M., et al. (2004). School redistricting: Embedding GIS tools with integer programming. *Journal of the Operational Research Society, 55*(8), 836–849.
Cordeau, J. F., Furini, F., & Ljubi, I. (2019). Benders decomposition for very large scale partial set covering and maximal covering location problems. *European Journal of Operational Research, 275*(3), 882–896.
Daskin, M. S. (2011). *Service science*. Wiley.
Daskin, M. S., & Owen, S. H. (1999). Two new location covering problems: The partial P-center problem and the partial set covering problem. *Geographical Analysis, 31*(3), 217–235.
Duque, J. C., Church, R. L., & Middleton, R. S. (2011). The p-regions problem. *Geographical Analysis, 43*(1), 104–126.
Eiselt, H. A., & Marianov, V. (Eds.). (2011). *Foundations of Location Analysis*. Springer Science.
Emiliano, W. M., Telhada, J., & Carvalho, M. D. (2017). Home health care logistics planning: A review and framework. *Procedia Manufacturing, 13*, 948–955.
Farahani, R. Z., Asgari, N., Heidari, N., et al. (2012). Covering problems in facility location: A review. *Computers and Industrial Engineering, 62*, 368–407.
George, J. A., Lamar, B. W., & Wallace, C. A. (1997). Political district determination using large-scale network optimization. *Socio-Economic Planning Sciences, 31*(1), 11–28.
Hess, S. W., Weaver, J. B., Siegfeldt, H. J., et al. (1965). Nonpartisan political redistricting by computer. *Operations Research, 13*(6), 998–1006.
Hojati, M. (1996). Optimal political districting. *Computers and Operations Research, 23*(12), 1147–1161.
Hu, F., Yang, S., & Xu, W. (2014). A non-dominated sorting genetic algorithm for the location and districting planning of earthquake shelters. *International Journal of Geographical Information Science, 28*(7), 1482–1501.
Kalcsics, J. (2015). Districting problems. In G. Laporte, S. Nickel, F. S. da Gama (Eds.), *Location Science* (pp. 595–622). Springer.
Kong, Y. (2021a). A hybrid algorithm for the equal districting problem. In: G. Pan, et al. (Eds.), *Spatial data and intelligence. SpatialDI 2021. Lecture Notes in Computer Science* (Vol. 12753). Springer.
Kong, Y. (2021b). An iterative local search based hybrid algorithm for the service area problem. *Computational Urban Science, 1*, 19.
Kong, Y., Zhu, Y., & Wang, Y. (2019). A center-based modeling approach to solve the districting problem. *International Journal of Geographical Information Science, 33*(2), 368–384.
Laporte, G., Nickel, S., & da Gama, F. S. (Eds.). (2015). *Location science*. Springer.
Ministry of Housing and Urban–Rural Development of the People's Republic of China. (2018). *Urban residential area planning and design standards* (GB50180-2018).
Mu, W., & Tong, D. (2020). On solving large p-median problems. *Environment and Planning B: Urban Analytics and City Science, 47*(6), 981–996.
Nozick, L. K. (2001). The fixed charge facility location problem with coverage restrictions. *Transportation Research Part E: Logistics and Transportation Review, 37*(4), 281–296.
Resende, M., & Werneck, R. F. (2007). A fast swap-based local search procedure for location problems. *Annals of Operations Research, 150*, 205–230.
Teitz, M. B., & Bart, P. (1968). Heuristic methods for estimating the generalized vertex median of a weighted graph. *Operations Research, 16*, 955–961.
Turkoglu, D. C., & Genevois, M. E. (2020). A comparative survey of service facility location problems. *Annals of Operations Research, 292*, 399–468.
Vasko, F. J. (2003). A large-scale application of the partial coverage uncapacitated facility location problem. *Journal of the Operational Research Society, 54*, 11–20.

Chapter 34
Smart, Sustainable, and Resilient Transportation System

Zhong-Ren Peng, Wei Zhai, and Kaifa Lu

Abstract The transportation system, particularly the surface transportation system, has been evolving, albeit slowly. But that evolution has been exacerbated recently toward a smarter, more sustainable, and more resilient system. Connected and automated technologies enable vehicles smarter; electrical vehicles, and shared mobility, particularly shared micromobility and microtransit system make the transportation systems more sustainable; adaptive and resilient infrastructure planning and design makes the transportation infrastructure system more resilient. These changes represent the future of transportation system, a smarter, more sustainable, and more resilient system, with mobility on demand.

Keywords Transportation system · Smart · Sustainability · Resilience · Equity

34.1 Introduction

Transportation system plays a vital role to meet the travel needs of society and is an indispensable component in the urban fabric. Generally, the evolution of transportation system could be a good manifestation of urban development at different stages. Modern cities are embracing a smarter future but also facing more serious external challenges, i.e., climatic threats, equity issues and sustainability concerns. This naturally puts forward new requirements of future mobility options towards a smarter, more sustainable, and more resilient transportation system to combat these challenges. Meanwhile, technological development and innovations provide more

---

Z.-R. Peng (✉) · K. Lu
International Center for Adaptation Planning and Design (iAdapt), College of Design, Construction and Planning, University of Florida, Gainesville, FL 32611-5706, USA
e-mail: zpeng@ufl.edu

K. Lu
e-mail: kaifa.lu@ufl.edu

W. Zhai
School of Architecture and Planning, University of Texas at San Antonio, San Antonio, TX 78207, USA
e-mail: wei.zhai@utsa.edu

B. Li et al. (eds.), *New Thinking in GIScience*,
https://doi.org/10.1007/978-981-19-3816-0_34

potential solutions, such as the emergence of automated vehicles, electric vehicles, and shared micro-mobility that could offer full mobility of demand. The rest of this chapter aims to present some visions of future transportation systems and innovative ideas on how to promote a smarter, sustainable, and resilient transportation system to adapt to future technology innovations, urban development, and climate change.

34.2 Smarter and More Intelligent

Intelligent Transportation System (ITS) has experienced rapid development over the last decades that can help improve mobility, enhance safety, and promote sustainability (Lin et al., 2017). As a typical product of ITS, the connected and automated vehicle (CAV) system is a transformative technology that consists of interconnected and automated vehicles through vehicle to X (V2X) technologies.

34.2.1 Automated Vehicles

The rapid evolution of autonomous technology in the field of automotive and computer vision has made it possible to use automated vehicles for passenger and freight transportation. As a result, automated vehicles have attracted worldwide attention for their huge potentials in reducing traffic congestion, enhancing road safety, and promoting equity for people who are not able to drive. Autonomy characteristics of automated vehicles are generally categorized into six levels (Iclodean et al., 2020), no automation (L0), driver assistance (L1), partial automation (L2), conditional automation (L3), high automation (L4) and full automation (L5). Current automated vehicles serving public roads have reached the Level 3 and 4 of automation, including Apollo Baidu in China, EasyMile EZ10 and Navya Arma in France, and Olli Local Motors in USA (Iclodean et al., 2020). However, the vast majority of these automated vehicles emerge from startups, and there is still a long way to go towards full automation and implementation. The issues are not just limited to autonomy technologies. There are challenges associated with ethnic and legislation issues as well. Nevertheless, full autonomous vehicles will be the future even though there is a bumpy road ahead.

34.2.2 Connected Vehicles and V2X Technologies

V2X technologies refer to a number of different communication technologies serving for vehicles to communicate with other on-road vehicles (connected vehicles) (V2V) and roadside infrastructures (V2I), pedestrians (V2P), etc. V2I captures vehicle-generated traffic data, advisories from infrastructures to the vehicle that inform the driver of safety, mobility, or environment-related conditions. Furthermore, future deployment of V2I infrastructures would be more likely located alongside or integrated with existing ITS equipment (U.S. DOT, 2021). V2V enables the connected vehicles to wirelessly exchange information about speed, location, and heading of each vehicle. The advancement of V2V technologies allows truly connected vehicles, such as receiving omni-directional messages up to 10 times per second, creating a 360° "awareness" of other vehicles in proximity, and detecting dangers within a range of more than 300 m (U.S. DOT, 2021). V2P encompasses the interactions on a broad set of road users including pedestrians, cyclists, children in strollers, people using wheelchairs, passengers onboarding and offboarding buses, etc. (U.S. DOT, 2021). V2P will transmit these interactions to the connected vehicles in real time for rapid response and safe driving. These technology innovations will make vehicles connected and smarter, and they will significantly reduce or even eventually eliminate vehicle crashes.

34.3 More Sustainable

In addition to increased mobility, future development of transportation system should consider longer-term impacts such as climate change to reduce vehicle emissions, decarbonize mobility and promote sustainability while maintaining a normal level of services for citizens. Generally, sustainable transportation can make a positive contribution to the environmental, social, and economic sustainability of the cities and communities they serve. Many efforts have been taken to emphasize the necessity of making transport more sustainable, ranging from electric vehicles, shared mobility (i.e., micromobility and microtransit), sustainable transportation, and land use planning.

34.3.1 Electric Vehicles

Electric vehicles (EVs) are gaining increasing interests and will become the majority of vehicles on the road in the near future as a more sustainable mode of transportation with less greenhouse gas emissions than traditional vehicles. As revealed by Farid et al. (2021), EVs consume less energy per unit distance, emit less carbon dioxide into the atmosphere and could even shift the emissions to generate new power for EV themselves. There are lots of mature EV companies that produce a significant number of EVs, such as Tesla and Chevrolet in USA, Toyota in Japan, NIO, XPeng Motors and BYD in China (Shao et al., 2021). However, the wide adoption of EVs highly relies on the development level of the supporting EV infrastructures, i.e., the spatial distribution of charging stations in cities. Therefore, how to accelerate the deployment of EV infrastructures is becoming the biggest barrier in promoting the usage of EVs. This requires joint efforts of different stakeholders including EV companies, federal, state, and local governments, as well as residents.

34.3.2 Microtransit

Microtransit is a form of demand-responsive transportation mode, especially serving for the first- or last-mile transit demands. Microtransit service offers a highly flexible route and schedule of minibus vehicles (e.g., autonomous vehicles, connectors, circulators, and dial-a-ride, etc.) usually shared with other passengers (Abduljabbar et al., 2021). The emergence of microtransit greatly extends the efficiency and accessibility of the existing transit system by satisfying people's travel demand for the first or last mile that traditional transit cannot cover. Furthermore, an efficient microtransit service may substitute many traditional public transit services, particularly in areas with an imbalance of supply and demand (Driverseat, 2021). According to Hazan et al. (2019), microtransit contributes to a 15%–30% decrease in traffic and carbon emissions. Another important goal of microtransit is to further expand the existing transit network's geographic and demographic reach towards more equity, usually serving populations in low-density areas, in the socioeconomically disadvantageous segments, and the elderlies who lack other reliable transportation options. Microtransit service required changes and support of current transportation policies, but once approved, it is much easier to implement without requiring extra transportation facilities. Frost and Sullivan's research (Driverseat, 2021) shows that the microtransit market is growing enormously, estimated from $2.8 billion in 2017 to $551.61 billion in 2030. That is almost 200% growth in the market.

34.3.3 Micromobility

Micromobility is another new sustainable transportation mode to satisfy people's travel demands on the first and last mile (Abduljabbar et al., 2021). Micromobility usually refers to a wide range of small and lightweight vehicles operating at speeds typically below 15 mph (~24 km/h) and driven by users personally, which is in contrast with microtransit that relies on drivers of vehicles or minivans to transport passengers, at least before the autonomous vehicles are fully adopted. Micromobility devices include bicycles, e-bikes, electric scooters, electric skateboards, shared bicycles, and electric pedal-assisted bicycles, etc., which are widely available in many cities across the globe. Micromobility not only enhances accessibility to the public transportation system and leads to possible modal shifts away from private vehicles, but also changes urban mobility patterns and travel behaviors. The wide emergence of micromobility accelerates the transition from the existing transportation system into a more sustainable one due to fewer carbon emissions brought by micromobility devices (Oeschger et al., 2020). Since the micromobility devices are all road vehicles, existing road infrastructures can still be used without further investment except extra addition of some basic cycle lanes, ramps, and docking stations. It is worth mentioning that the eruption of pandemic further stimulates the micromobility market due to its fewer points of contact and ease of maintaining physical distancing, which is considered as less risky than other mobility options (McKinsey & Company, 2020). This will also be an important trend of shared mobility in the post-pandemic period.

34.4 More Resilient

34.4.1 Definition of Resilient Transportation

The current transportation system is plagued by disruptions, both natural and man-made, including extreme weather events, major accidents, and equipment or infrastructure failures, which significantly impact the normal operation of the system and cause tremendous infrastructure damages and economic costs. Fortunately, efforts are being made to enhance the resilience of the transportation system in the full life cycle of transportation planning, design, construction, and operations. A more resilient transportation system represents the ability of a transportation system to move people around in the face of one or more major obstacles with minimal disturbance, or the system can fully and quickly recover from disturbance with minimal function reduction and economic costs.

Vulnerable transportation assets. Roads, bridges, tunnels, rails, airports, and other transportation facilities, in inland locations as well as in coastal communities, can be vulnerable to climate-related events. For example, storm-related flooding—exacerbated by rising sea levels in coastal cities—can close tunnels, subway stations, low-lying roads, and marine cargo facilities, either temporarily or permanently. Flooding from increasingly frequent heavy downpours can disrupt traffic, damage culverts, and reduce the service life of stormwater infrastructure. High temperatures can accelerate the deterioration of pavement on roads and runways, and cause failures of railroad and subway tracks.

Planning for transportation resilient systems. As transportation infrastructure is usually intended to last 50 years or longer, today's transportation planners may be able to save time, money, and traffic-induced headaches in the long run by anticipating and planning for future conditions. Unless decision makers address the resilience of transportation infrastructure and operations in a changing climate, new conditions are likely to mean higher costs, greater disruption, and more damage to transportation infrastructure in urban communities. While existing transportation infrastructure was designed to handle a broad range of conditions based on historic climate, the frequency and intensity of some extreme weather events are increasing. Transportation planners are likely to face difficult choices about how and where to invest resources to bolster or replace existing infrastructure. Strategies that transportation departments might use in adapting to climate change include: (1) Integrate climate change considerations into asset management. (2) Strengthen or abandon infrastructure that is vulnerable to flooding. (3) Raise standards for the resilience of new infrastructure. (4) Add redundant infrastructure to increase system resiliency. (5) Promote zoning, insurance, and disaster recovery policies that discourage development in vulnerable areas.

34.4.2 Resilient Transportation System

The transportation system is traditionally designed to function in regular weather conditions, while extreme weather threats were only considered to a limited extent. As a result, extreme weather events such as snowstorms, hurricanes, and tornadoes, can degrade transportation system performance significantly. Considering that the transportation system is one of the most important critical infrastructure systems, its failure could lead to cascading consequences in economic, social, and financial systems. Many of these events are unforeseeable, making it much difficult for practitioners to protect the transportation infrastructure. Extreme weather is projected to grow more frequently as climate change continues, and its consequences on transportation will become more severe (Koetsee and Rietveld, 2009). Deliberate attacks such as sabotage, terrorism, and acts of war are also examples of external stressors. Unfortunately, terrorist attacks took place on practically every mode of transportation.

As a result, policymakers and researchers are paying more attention to the effects of disasters on transportation infrastructures. Disaster responses, such as evacuation operations during disasters, are all dependent on the transportation system's effectiveness. The inherent strength of the transportation system to recover from detrimental consequences is referred to as resilience. Some emerging studies have developed resilience quantification frameworks and strategic mitigation methodologies to quantify a transportation system's resilience to meet this need. These efforts also attempt to improve system resilience and shorten the time it takes to recover from a disaster (Faturechi & Miller-Hooks, 2015).

In traditional transportation systems, the deployment of intelligent transportation system elements such as connected and autonomous vehicles (CAVs) and adaptive safety solutions might affect the resilience phenomena and system performance metrics. CAV technologies have advanced significantly in recent years, with deployment predicted within the next decade. Many studies have quantified the robustness of the conventional transportation system (i.e., a system that does not use CAVs) (Chen & Miller-Hooks, 2012; Faturechi & Miller-Hooks, 2014). In addition, some recent studies have looked at how CAVs operate in natural catastrophe situations and in mixed-traffic environments (Zhu & Ukkusuri, 2016).

34.4.3 Interconnections Between Automated Vehicles and Resilience

As introduced above, smart transportation systems, such as connected and automated vehicles, will transform urban development and lead to innovative governance of cities. But what happens when things become completely chaotic, like during or after an earthquake or other natural disaster when traffic lights are out, other drivers are panicked, debris is everywhere, and people are running around hysterically. It requires future researchers to take steps to prepare for that kind of scenario, by simulating what human drivers will do in those situations. In practice, the automated fleet has already encountered bits and pieces of what the cars might experience during a natural disaster, even though the problems would be amplified in an actual emergency. For example, cars frequently encounter broken traffic lights. In construction zones and during accidents, there are changes in the road conditions that may not show up on maps. The safety of automated vehicles was called into deeper question in 2020 after an autonomous vehicle from Uber, which had a human operator in the driver seat, killed a pedestrian in Tempe, Arizona. The victim was walking with her bike at night outside of a crosswalk (Kohli and Chadha, 2019). In a natural disaster, conditions would become even more unpredictable. It requires automated vehicles to reason through new situations in real-time.

Automated vehicles will be used in areas that may be prone to earthquakes, floods, and blizzards in the future. The cars' ability to assess such threats and adapt to changes in the environment will be critical in keeping passengers safe during an emergency. Automakers should also be working on safety guidelines that define how automated vehicles react in the face of natural disasters and mayhem. Is the car going to slow down? Or will it seek alternate paths to avoid further disruptions? It requires researchers to dig into these uncertainties in their research.

34.5 Look into the Future

Future smart and sustainable transportation system. To combat the serious challenges brought by population growth and ongoing urbanization, the surface transportation system is evolving into a smarter and more sustainable one. This could be jointly reflected by future development of road traffic and rail traffic: (1) Highly autonomous vehicles will be more available on the roads with the support of improved vehicle automation technologies and information exchange between vehicle and infrastructure. (2) More shared transportation systems based on Mobility on Demand (MOD) will be promptly developed (U.S. DOT, 2021), which is also usually accompanied by fewer vehicle ownerships. These trends of road traffic suggest that the prevalence of autonomous vehicles, microtransit and micromobility options will reduce vehicle ownerships so that people can call them at any time when needed. Now some cities, like Tampa's Downtowner system (Smith et al., 2018), have started to promote and implement the Mobility on Demand services. (3) An increase of high-speed rail roads and trains will be another prevailing trend in the future surface transportation system due to the demands for increased mobility and accessibility between cities, the desire for less carbon emissions per capita.

Future resilient transportation system. First, vehicles will become smarter. It can automatically sense, identify hazardous environment, and automatically control the vehicle movement to adapt for the changing environments. When faced with natural disasters, we believe future automated vehicles will take defensive and precautionary behaviors. Instead of attempting to promptly calculate a safe route through icy roads or floods, the smart vehicle may first identify the potential risk and redirect itself or pause to find the best solutions to this emergency. Second, there will be sensors on the road and through roadside devices, to collect real time information about flooding, ice, extreme heat, extreme precipitation, and to create automatic alerts which will be sent to vehicles through V2X technologies. An automated fleet will constantly learn about its surroundings, keep an eye out for developments that could make driving extremely dangerous. Furthermore, real-time notifications through connected networks can be used to send information to surrounding automobiles, which is very beneficial for avoiding unexpected road closures, municipal events, crashes, and other hazards. External gear in automated cars will be capable of closely monitoring changes in the car's surroundings. Every unit, for example, would include many LIDAR sensors, which could improve 3D visibility in both low-light (dark) and extremely bright settings. Radar may also be used by automated cars to detect objects in foggy or snowy situations. The environment and road are captured clearly by cameras mounted all around the car. Virtual testing software could be another viable option for preparing automated vehicles for different disasters. Third, the adaptive planning, design, construction, and operation of transportation infrastructure will make the transportation system more resilience.

Future 3-D transportation system. In addition to surface transportation systems discussed above, future transportation system should also involve both aerial and underground transportation system, e.g., airbus and large scale underground tunnel: (1) Airbus delivers innovative solutions to pioneer transportation mode in the aerospace, spanning aircraft and unmanned helicopter, which can make a positive contribution to more convenient and sustainable multimodal mobility system, though it is currently largely a conceptual product (Petrescu et al., 2017). The Lilium Lake Nona Vertiport, which provide an experimental hub for an all-electric, vertical take-off and landing aircraft in Florida, offers a glimpse of this exciting future (TAVISTOCK, 2021). Generally, urban air mobility network leverages the sky to provide more flexible travel services for city dwellers from any possible origins to destinations over larger geographical areas than the existing public transportation system. (2) Another future transportation system would be the adoption of the underground tunnel with the vision of improving mobility and reducing surface traffic congestion. Boston's Big Dig provides an early example that brought a significant drop (62%) in the total vehicle hours from 1995 to 2003 and is now providing around $168 million per year in time and cost savings to travelers (MDOT, 2021). With significant cost reduction in tunnel building, underground tunnel system may be another important development trend in the future 3-D transportation system.

34.6 Conclusions

This chapter presents some thoughts about how future transportation system evolves into smarter, more sustainable, and more resilient one. Accordingly, we attempt to put forward several potential requirements about the supporting infrastructures against different modes of transportation to better adapt to future urban development, particularly under the adverse impacts of climate change. Connected and automated vehicles, electric vehicles, microtransit vehicles and micromobility devices, as well as airbus and underground transportation system, all have exhibited huge potentials in pursuing sustainable and resilient urban transportation system and provide Mobility of Demand services. The most important thing at present is how to develop and achieve the technologies needed for future transportation system and how to increase people's awareness and adoption of these smart and sustainable future modes of transportation. Undoubtedly, there is still a long way to go, for example, how to upgrade feasible automation technologies, how to design and plan sustainable and resilient supporting infrastructures, how to address equity and policy issues are some of the challenges in urgent need for solutions to enable the development of a smarter, more sustainable, and more resilient future transportation system.

References

Abduljabbar, A. L., Liyanage, S., & Dia, H. (2021). The role of micro-mobility in shaping sustainable cities: A systematic literature review. *Transportation Research Part D: Transport and Environment, 92*, 102734.

Chen, L., & Miller-Hooks, E. (2012). Resilience: An indicator of recovery capability in intermodal freight transport. *Transportation Science, 46*(1), 109–123.

Driverseat. (2021). *The future of microtransit*. https://driverseatinc.com/2021/05/31/the-future-of-microtransit/.

Farid, A. M., Viswanath, A., Al-Junaibi, R., Allan, D., & Van der Wardt, T. J. T. (2021). Electric vehicle integration into road transportation, intelligent transportation, and electric power systems: An Abu Dhabi case study. *Smart Cities, 4*, 1039–1057.

Faturechi, R., & Miller-Hooks, E. (2014). Travel time resilience of roadway networks under disaster. *Transportation Research Part B: Methodological, 70*, 47–64.

Faturechi, R., & Miller-Hooks, E. (2015). Measuring the performance of transportation infrastructure systems in disasters: A comprehensive review. *Journal of Infrastructure Systems, 21*(1), 04014025.

Hazan, J., Lang, N., Wegscheider, A. K., & Fassenot, B. (2019). On-demand transit can unlock urban mobility. *Boston Consulting Group, Automotive Industry, Cities of the Future, Transportation and Logistics*. https://www.bcg.com/publications/2019/on-demand-transit-can-unlock-urban-mobility

Iclodean, C., Cordos, N., & Varga, B. O. (2020). Autonomous shuttle bus for public transportation: A review. *Energies, 13*, 2917.

Koetse, M. J., & Rietveld, P. (2009). The impact of climate change and weather on transport: An overview of empirical findings. *Transportation Research Part D: Transport and Environment, 14*(3), 205–221.

Kohli, P., & Chadha, A. (2019). Enabling pedestrian safety using computer vision techniques: A case study of the 2018 uber inc. self-driving car crash. In *Future of Information and Communication Conference* (pp. 261–279). Springer.

Lin, Y., Wang, P., & Ma, M. (2017). Intelligent transportation system (ITS): Concept, challenge and opportunity. In *IEEE 3rd International Conference on Big Data Security on Cloud, IEEE International Conference on High Performance and Smart Computing, and IEEE International Conference on Intelligent Data and Security* (pp. 167–172).

Massachusetts Department of Transportation (MDOT). (2021). *The Big Dig: Project background*. https://www.mass.gov/info-details/the-big-dig-project-background.

McKinsey & Company. (2020). *The future of micromobility: Ridership and revenue after a crisis*. https://www.mckinsey.com/industries/automotive-and-assembly/our-insights/the-future-of-micromobility-ridership-and-revenue-after-a-crisis.

Oeschger, G., Carroll, P., & Caulfield, B. (2020). Micromobility and public transport integration: The current state of knowledge. *Transportation Research Part D: Transport and Environment, 89*, 102628.

Petrescu, R. V., Aversa, R., Akash, B., Corchado, J., Apicella, A., & Petrescu, F. I. (2017). Home at airbus. *Journal of Aircraft and Spacecraft Technology, 1*(2), 97–118.

Shao, X., Wang, Q., & Yang, H. (2021). Business analysis and future development of an electric vehicle company—Tesla. In *Proceedings of the 2021 International Conference on Public Relations and Social Sciences* (Vol. 586, pp. 395–402).

Smith, M. C., Bauer, J., Edelman, M., & Matout, N. (2018). Transportation systems management and operations in smart connected communities. *Repository & Open Science Access Portal*. FHWA-HOP-19-004. https://rosap.ntl.bts.gov/view/dot/43561

TAVISTOCK. (2021). *Lilium Lake Nona Vertiport*. https://tavistockdevelopment.com/project/lilium-lake-nona-vertiport/

United States Department of Transportation (U.S. DOT). (2021). *Intelligent Transportation Systems Joint Program Office*. https://www.its.dot.gov/v2i/, https://www.its.dot.gov/factsheets/mobilityondemand.htm

Zhu, F., & Ukkusuri, S. V. (2016). An optimal estimation approach for the calibration of the car-following behavior of connected vehicles in a mixed traffic environment. *IEEE Transactions on Intelligent Transportation Systems, 18*(2), 282–291.

Chapter 35
The "Here and Now" of HD Mapping for Connected Autonomous Driving

Liqiu Meng

Abstract This chapter is dedicated to the concept, characteristics, components, and structures of high-definition (HD) maps and their state of development for autonomous driving. The self-driving vehicle is essentially a rolling supercomputer, controlled more by its software than its hardware. HD maps assume a decisive control role in guiding such a vehicle safely and efficiently through a dynamic environment. Compared to standard maps, HD maps are fundamentally different in terms of generation procedure, map content, map scale and target users. HD mapping is analytically composed of three elements—the "here" mapping, the "now" mapping and the integrated "here" and "now" mapping. The main tasks associated with each element are demonstrated with best practice examples. Key research challenges include extraction of meaningful driving scenarios, edge-case modeling in the absence of training data, predicting contextual human behavior, and safety-first decision making in moral dilemmas.

Keywords HD mapping · Causal relationship · Edge-case modeling · Embodied cognition · Safety-first decision making

35.1 Autonomous Driving as a Connected Supercomputer

Research of autonomous driving began in the 1980s and led to the design of first driverless cars such as Navlab from Carnegie Mellon University (1986), and VaMP from Munich University of the Federal Armed Forces in cooperation with Mercedes-Benz in the EUREKA-PROMETHEUS project (1987–1995). These pioneering prototypes relied on computer vision to travel long distances with little human intervention. However, the performance and computational power required in real-world scenarios make them too expensive and cumbersome for commercial production, apart from the fact that traffic regulations were not ready for self-driving vehicles back then.

L. Meng (✉)
Cartography and Visual Analytics, Technical University of Munich, 80333 Munich, Germany
e-mail: liqiu.meng@tum.de

B. Li et al. (eds.), *New Thinking in GIScience*,
https://doi.org/10.1007/978-981-19-3816-0_35

Thanks to rapid improvements in computational efficiency, steady advances in sensor technology and machine learning algorithms, and growing data in the last decade, research in this field has experienced a new surge and resulted in numerous powerful models of self-driving cars as demonstrated by Google, Tesla, and Uber and so on. The new approach embraces disruptive innovations and is dominated by Silicon Valley automakers.

A self-driving vehicle is essentially a supercomputer on wheels embedded in a dynamic living environment and confined to the road infrastructure. Its physical architecture and wiring harness bring together thousands of components related to sensors and actuators, braking systems or engine control. In addition to built-in components that work together to move the wheels, and ergonomically designed components for bodily comfort, a self-driving vehicle is typically equipped with sensors all around the vehicle, e.g. cameras, Radar, LiDAR, Ultrasonic, Global Navigation Satellite System (GNSS) and Inertial Measurement Unit (IMU), to record the motion and provide a 360° view of the environment.

More radical changes in autonomous driving have occurred on the software side. Advanced Driver Assistance Systems (ADAS) offer a range of standard services from navigation guide, congestion information, lane departure warning, distance control to parking aids. Together with value-added services including infotainment, telematics, and infrastructure networking, they provide users with a high level of satisfaction. New features of self-driving vehicles related to safety, security, and global climate change goals push further software proliferation for internal and external communications (Charette, 2021).

Depending on the sensor setup, a self-driving vehicle generates data streams of more than a gigabyte per minute for the supercomputer to learn while driving. The learning is guided by high-definition maps or HD maps. With some analog to high-definition television (HDTV) representing a set of television standards with higher vertical, horizontal, or temporal resolution over Standard Definition Television (SDTV), HD maps refer to maps with much higher resolutions than standard maps, in-car navigation maps or smartphone maps.

35.2 Characteristics of HD Maps

HD maps have revolutionized standard maps in multiple ways. First of all, HD maps sense and make sense at the same time and in real time. Second, their contents are sharply focused on "here and now" and connected with a living environment. Third, they are ground truth models at a scale of nearly 1:1. Finally, they are made by machines for machines. The HD mapping procedure can be analytically decomposed into three elements—the "here" mapping, the "now" mapping and the integrated "here and now" mapping.

35.2.1 The "Here" Mapping

The "here" mapping deals with the positioning of the vehicle and the interpretation of its local environment in centimeter-level resolution. Main tasks are spatial data fusion, annotation, and AI-supported understanding. Spatial data fusion aims to construct 3D local environments. The data sources to be fused include street view images, point clouds, signals from GNSS and/or IMU, remote sensing images, and street maps with details about road surface, lane placement, road boundaries, and roadside features. Annotation is a time-intensive preparation. A large number of ground features such as navigation marks, traffic signs, barriers, construction sites etc., and mobile features including vehicles, humans and other fine-grained moving objects that may occur in driving environments should be manually labelled. AI-supported understanding relies on deep-learning algorithms, annotated training data, and expert knowledge to identify all safety-relevant features, their locations and topological relationships. Figures 35.1 and 35.2 show some examples of "here" maps.

The data sources for the "here" mapping are unevenly distributed. Street maps at coarse scales are being repeatedly collected and cross-verified by government agencies, private vendors, and volunteers. With increasing resolutions, road and navigation information becomes more valuable, but also more scarce and harder to access for technical and data policy reasons. Close collaboration between automakers, research institutions and lawmakers in different world regions is needed to promote the creation and shared accessibility of centimeter-level global "here" maps.

Fig. 35.1 A "here" map of varying granularity from Woven Planet Level 5. https://medium.com/wovenplanetlevel5/https-medium-com-lyftlevel5-rethinking-maps-for-self-driving-a147c24758d6, date of access: 2022.06.09

Fig. 35.2 A "here" map with labelled features by Mobileye

35.2.2 The "Now" Mapping

The "now" mapping aims to capture the dynamic characteristics of moving vehicles at a temporal resolution between 10 ms and 100 ms, involving tasks of temporal data modeling and understanding. First of all, temporal data from connected sensors in and around the vehicle and in the road infrastructure should be fused to derive accurate time stamps for all possible vehicle operations situated in driving scenarios. Other digital documents containing time-dependent statistical data about traffic volume, population flow, weather condition, air quality etc. can be deployed to supplement the temporal modeling. Moving conditions for the next seconds or minutes may be estimated by prefetching data from standard maps, digital landscape models, weather forecast models and other connected vehicles. Second, each vehicle operation such as lane switching, braking, accelerating, highway entry and exit, cut-in etc. should be explicitly modelled as an event indicating changes in position and topological relationships with other road features. Finally, AI-supported methods are applied to detect events and their meanings.

A showcase of modeling and understanding the "cut-in" event is illustrated in Figs. 35.3 and 35.4. Figure 35.3a demonstrates how a vehicle CV deviates from its longitudinal direction at time T_0 with an angle Ω and starts to cut in at the velocity V_i between the ego vehicle EV and the leading vehicle LV. It leaves footprints at three consecutive time points. The test data containing cut-ins is captured in real traffic by running a vehicle from LiangDao GmbH at different times along different streets as shown in Fig. 3b. The possible causes for cut-ins are categorized and represented as a knowledge graph in Fig. 35.4. By detecting various cut-in events in test data, a model is built to predict where and when a cut-in is most likely to occur and what causes with which relative weights may trigger the event.

The data sources for the "now" mapping are sparse and dispersed. The finer the temporal resolution, the more valuable the data source, which is also technically more difficult to obtain and maintain. Nonetheless, it is possible to create test scenarios with limited scopes and to share the experiences and lessons learnt.

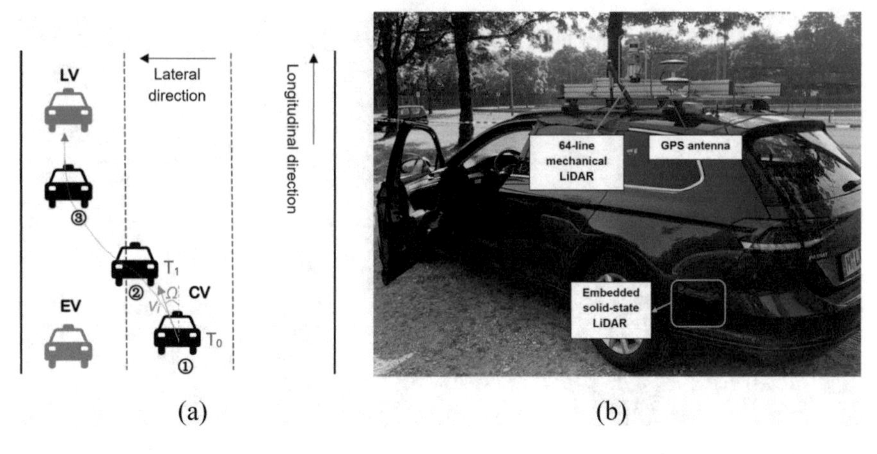

Fig. 35.3 **a** a typical scenario with cut-in operation; **b** the vehicle used by LiangDao GmbH to capture scenario data (Deng, 2021)

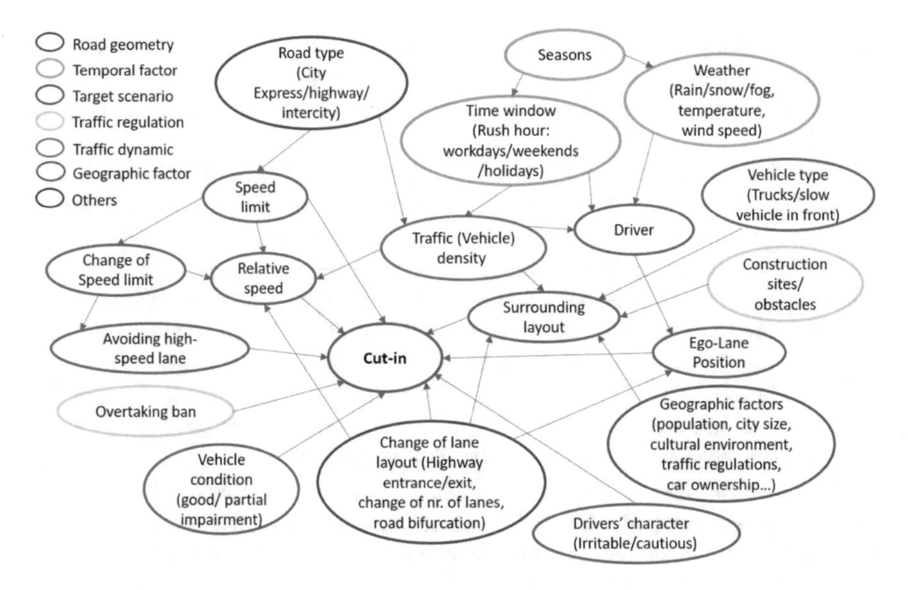

Fig. 35.4 A knowledge graph of causes for the cut-in operation (Deng, 2021)

35.2.3 The Integrated "Here and Now" Mapping

The integrated "here and now" mapping aims to represent driving scenarios at the highest spatio-temporal resolution and link them to traffic rules. Figure 35.5 demonstrates Tesla's "Fully Self-Driving Beta" in a safety–critical operation guided by the HD map. The synchronized "here" and "now" mapping can be achieved by accumulating test data in real traffic with controlled settings corresponding to Level 4,

Fig. 35.5 Tesla's "Full Self-Driving Beta" guided by the HD map during a left turn in real traffic (https://youtu.be/uClWlVCwHsI from 2021.04.11)

i.e. the second highest level on the automation scale from Level 0–5 defined by the Society of Automotive Engineers (SAE). A large number of automakers and specialized technology companies are committed to data collection, following different strategies.

BMW launched in partnership with DXC a platform 2019 to collect data by its test and customer fleet. More than 500 million km in test areas in Munich and Shanghai were completed in 2021. Volkswagen couple with Microsoft is massively expanding its fleet's mileage and environmental data, taking advantage of more than 10 million new vehicles per year (Tiedemann & Nagel, 2021). TomTom's fleet across Europe, the US, Japan, and South Korea, generated more than 400,000 km of HD map coverage in 2019, when TomTom started to use Volvo's test vehicles with multiple sensors to update and extend HD maps (Kayla, 2019). Similarly, CARMERA was teamed up with Toyota to first map downtown Tokyo and then expand to other complex urban environments, such as New York City (Billington, 2019).

With the aim to reach a seamless global coverage of HD maps, HERE created the OneMap Alliance (Ellis, 2019). Crowdsourced data from cell phones and production vehicles from various automotive companies are combined with high-end sensory data from industrial surveying, leading to improved spatiotemporal accuracy of HD maps. Atlatec creates HD maps by driving its vehicle on every lane of a road multiple times and then extracting consistent global positions with an absolute accuracy of 3 cm for 95% of the roads covered. Meanwhile, 3D models are constructed from driving environments for simulation purposes, focusing on the relative positional accuracy of the vehicle (Dahlström, 2020).

Mobileye developed a low-cost crowdsourcing approach called Road Experience Management (REM). Its self-driving cars equipped with cameras run in different environments, different countries and different continents and send data to the cloud in small packets—less than 10 kilobytes per kilometer—corresponding to a small

Fig. 35.6 Mobileye's HD-mapping with hands-free driving vehicle in Munich: in the urban area at 41 km/h (left) and on a highway at 124 km/h (right)

footprint around the vehicle. As shown in Fig. 35.6, this approach allows a continuous update of its RoadBook—a scalable database of HD maps with high local accuracy and rich semantic layer of driving culture and traffic rules.

Processing real-time data streams for HD mapping requires enormous computing power. No wonder, Tesla presented on its "AI Day 2021" a Dojo supercomputer with a performance of around 1.1 exaflops (Romero, 2021). Research is also being conducted on data compression, filtering, and classification. Based on training data from test drives in public areas and from closed-vehicle-in-the-loop simulations, the project "KIsSME" at Karlsruhe Institute of Technology aims to create AI-models and selectors that may help reduce data amount and expand the scenario catalog. The Bertrandt Group is developing a labeling tool "Bertrandt Data Labeler", which allows marking the most relevant objects and identifying meaningful driving scenarios as training data for AI-supported HD mapping (Tiedemann & Nagel, 2021).

The HD mapping procedure usually consists of the following steps:

- preprocessing of 3D point cloud and camera images to create a precise trajectory of the vehicle, an aligned 3D point cloud, and a coarse 3D model, e.g., using simultaneous localization and mapping (SLAM) algorithms;
- segmentation of preprocessed data to derive the ground surface, and distinctive 2D and 3D geometries in the driving environment;
- recognition of meaningful 2D and 3D ground features, e.g., using trained AI models;
- semantic enrichment of recognized features with traffic rules for safe mobility; and
- integration of real-time information, human behavior models and domain knowledge.

HD maps can be divided into tiles of a uniform size to fit the driving environment (Kayla, 2019). The features of each tile can be structured in layers (Chellapilla, 2018), depending on how often they are updated, how important they are for safe driving, whether they need to be shared with other vehicles, etc. Many HD mapping platforms conduct edge computing locally while data is being collected. This helps enhance driving robustness in all situations at Level 3, even without a network connection (Tiedemann & Nagel, 2021).

35.3 Research Challenges of HD Mapping

Statistics on the number of hands-free mileage and the number of disengagements per driving kilometer of a vehicle can be used to evaluate the self-driving performance and to encourage the transparent release of official "test mileage and disengagement" reports, which may improve public trust in autonomous driving. Nevertheless, current HD mapping platforms suffer from a common bottleneck of insufficient data for safety–critical scenarios despite billions of kilometers driven. The number of traffic rules is quite limited, but their contextual combinations may become innumerable. Each specific scenario faced by a self-driving vehicle is literally a socio-technical system in which other vehicles, cyclists, pedestrians, and obstacles exhibit varying degrees of agility and accountability.

35.3.1 Edge-Case Modeling

In the recent decade, AI technologies have made breath-taking advantages in image understanding, language translation, speech generation for chat bots, game playing, protein folding and more. Self-driving cars are getting smarter too, but at a much slower development pace in the final stage when the responsibility shifts entirely from the human driver to the machine (Piper, 2020). Unlike technologies developed in closed labs before being released to the world, self-driving vehicles are experimented on public roads. Each collision disturbs the public perception. Although crashes with serious injuries are rare and the causes for each fatal accident are quickly identified (Walker, 2020), there is no practical way to determine a statistically convincing crash rate. Testing edge cases involving accidents and extreme behavior of traffic participants on public streets is both dangerous and unethical.

The lack of training data for machine learning can be mitigated by simulations based on small samples from real traffic in combination with traffic rules. Simulations can also serve as training tools for the public and legislators to gain insight into self-driving technologies and their rules for different stakeholders and different purposes.

The Association for Standardization of Automation and Measuring Systems (ASAM) issued a number of Open Standards in 2018 as shown in Fig. 35.7. Being

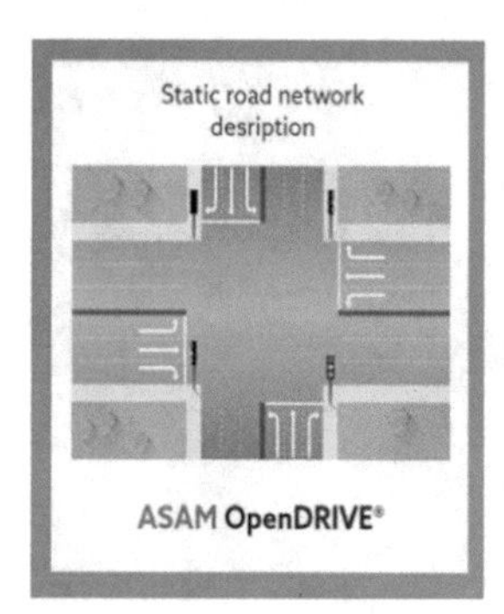

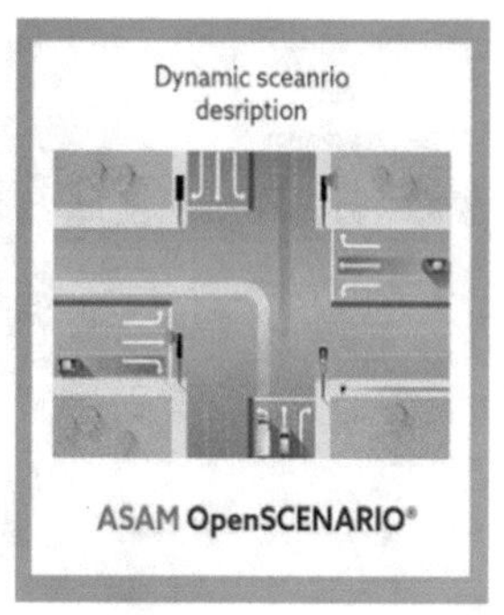

Fig. 35.7 OpenX Standards by ASAM for simulation

based on other public standards such as UML, XML and CORBA, ASAM standards are platform neutral and can be used to create virtual driving environments and validate automated driving functions. The free sample datasets are now increasingly provided in ASAM format for simulation purposes (Dahlström, 2020).

Microsoft introduced CausalCity—a simulation environment for complex driving scenarios, as shown in Fig. 35.8 (McDuff et al., 2021). Volunteers can use the provided Python code and baseline code to generate complex scenarios and explore the causal relationships.

Woven Planet Level 5 developed an agent-based simulation platform to predict the behavior of traffic agents, i.e. other vehicles around a self-driving vehicle. An open dataset containing motion data of traffic agents, the movement logs and annotations from cars, cyclists, pedestrians, and other traffic agents is used to train the algorithm. Figure 35.9 illustrates the detected traffic agents and their predicted behavior at the next moment.

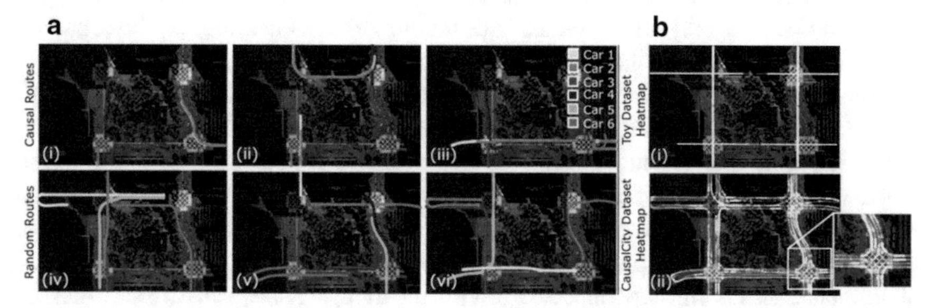

Fig. 35.8 CausalCity for the simulation of complex driving scenarios. https://www.microsoft.com/en-us/research/blog/causalcity-introducing-a-high-fidelity-simulation-with-agency-for-advancing-causal-reasoning-in-machine-learning, date of access: 2022.06.09

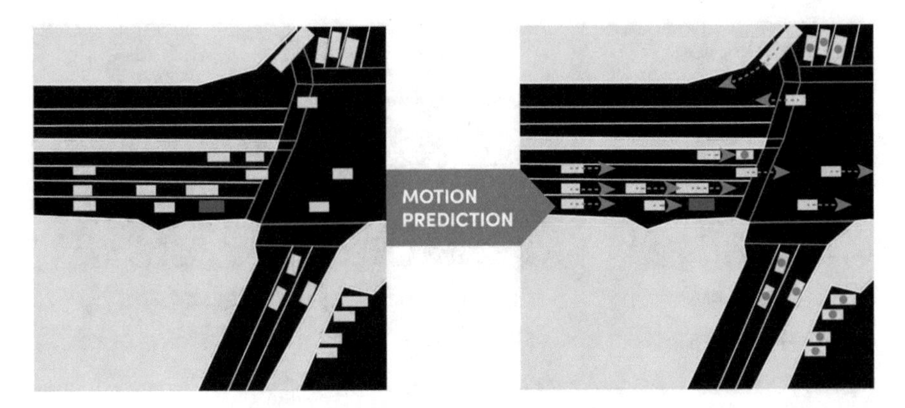

Fig. 35.9 The dynamic environment of a self-driving car marked in purple: detected traffic agents as small rectangles (left), predicted motion marked by arrows (right) (https://level-5.global/data/prediction), date of access: 2022.06.09

35.3.2 Decision Making for Safety-First Driving

By training on real and simulated driving scenarios, machine-learning algorithms will be able to discover a growing number of meaningful events including their occurrence patterns, thus enhancing the intelligence of HD maps. Nevertheless, machines are not yet good at imitating humans to make sound decisions in emergency. Humans are capable of avoiding fatal accidents even without thinking and learning experience. In the absence of information, humans rely on instinct and subconscious intuition to survive in seemingly desperate situations. This innate human decision-making mechanism cannot yet be fully replaced by HD maps generated from learning-based algorithms. Although self-driving vehicles guided by HD maps have increasingly taken on human traits, can reliably operate in normal traffic and avoid most types of collisions, they still miss humans' gut feeling. This remaining gap has sparked heated debates related to the trustworthiness of AI.

Human driving in real traffic requires embodied cognition of the environment with mind–body coordination. It involves emotions, desires, sense of self and its relation with other participants. Human driver "is inescapably affected by the immediate who, what, where, when, and perhaps why" (Tversky, 2012). Driving past a place can be a reminder of the representative mindsets and landmarks. The verbal comments, gestures, or even facial expressions of the participants in- and outside the vehicle may all influence the driving experience. Likewise, the living environment of a self-driving vehicle is not just about the changing geometries, but also about communications with other road features and connected vehicles in and beyond the field of view. The affordances of the tangible body of the vehicle and its sensors change with the real traffic.

Current simulation platforms focus more on modeling relationships between vehicles than on understanding contextual human behavior. Due to inadequate information about this latter part, the extent to which self-driving vehicles can learn to behave like humans, including obeying and violating traffic rules, remains unknown. It becomes difficult when self-driving vehicles must weigh options involving risks to human lives. A number of empirical studies attempt to address this problem with emphasis on safety-first decision making.

Human opinions on how machines should make decisions in moral dilemmas outside the legal realm, as well as discussions of potential consequence scenarios have been crowdsourced in a project called "The Moral Machine"[1]. Aptiv issued Rulebooks (Censi et al., 2019) as a pre-ordered set of rules to guide the behavior of self-driving cars. These rules are derived from behavior learning including rare edge cases. Rules that ensure human safety are the highest priority, while those related to comfort and progresses come last. Such a hierarchy may also support the development of (inter)national regulations and serve the informed public discourse. NVIDIA released in 2019 the Safety Force Field (SFF) as an augmented element for collision-avoidance validation and verification on its self-driving platform. Using sensor data and simulations of highway and urban driving scenarios, SFF focuses on braking and steering constraints and determines actions of eliminating collisions to keep vehicles and other road participants in safety. Mobileye introduced a model of Responsibility-Sensitive-Safety (RSS). Usually, the self-driving vehicle obeys traffic rules and maintains a balance between safety and utility. But the RSS can also guide the vehicle to protect itself from the unpredictable human behavior, and avoid accidents by violating one or more traffic rules, provided this does not cause another collision.

Stilgoe (2021) compared two approaches: safety-in-numbers and safety-by-design. The approach of safety-in-numbers regards safety as a technical goal with a measurable property, e.g. mileage statistics without serious injury. The approach of safety-by-design treats the autonomous driving as a safety–critical system with the vehicle as one of many related components. It provides a starting point to design. It may "geo-fence" vehicles to prevent them from straying into overly unpredictable spaces in real traffic, or it may change system settings to ensure safe driving, for example, by restricting the actions of other road participants or upgrading road infrastructure. Given the two approaches, HD maps are anticipated to guide the safety-first decision making as a journey between the starting point and the goal.

35.4 Concluding Remarks

Connected autonomous driving is a cutting-edge AI technology that takes public areas as its testing ground and therefore must meet extremely high safety requirements in order to be accepted by the public. While the shift from Level 0, 1, 2, 3 to 4 on the SAE scale of automation has been incrementally realized, the final shift from Level 4–5 is a disruptive one, with the entire responsibility falling on the self-driving vehicle. A safe, secure, liable, environmentally friendly and enjoyable autonomous

[1] www.moralmachine.net, date of access: 2022.06.09.

mobility within a legal and ethical framework will claim years of globally coordinated research and development by legislators, automakers, HD-mapmakers, providers of computing platforms and the public. With increasingly gained knowledge from edge-case modeling, contextual human behavior, and safety-first decision making, and its integration into HD maps, humans will remain in the loop, but more as interactive passengers and road participants than supervisors.

References

Billington, J. (2019, March 1). Toyota's self-driving spin-off trials new HD mapping technology. *Autonomous Vehicle International. ADAS.*

Censi, A., Slutsky, K., Wongpiromsarn, T., Yershov, D., Pendleton, S., Fu, J., & Frazzoli, E. (2019). Liability, ethics, and culture-aware behavior specification using Rulebooks. https://arxiv.org/pdf/1902.09355.pdf

Charette, R. N. (2021). How software is eating the car. *IEEE Spectrum, 2021*(6), 7.

Chellapilla, K. (2018, October 15). Rethinking maps for self-driving. *Woven Planet Level, 5.*

Dahlström, T. (2020). How accurate are HD maps for autonomous driving and ADAS simulation? *Atlatec Blog, 2020*(10), 22.

Deng, Y. H. (2021). *A novel prediction model: towards an efficient route-planning for acquisition of representative scenarios.* Master's thesis, Technical University of Munich, 106 p

Ellis, C. (2019, April 12). Mapping the world: solving one of the biggest challenges for autonomous cars. *Techradar.pro*

Kayla, M. (2019). What are HD maps, and how will they get us closer to autonomous cars? *IoT times, 2019*(9), 16.

McDuff, D., Song, Y., Vemprala, S., Vineet, V., Ma, S., & Kapoor, A. (2021). CausalCity: Introducing a high-fidelity simulation with agency for advancing causal reasoning in machine learning. *Microsoft Research Blog, 2021*(06), 29.

Piper, K. (2020, February 28). It's 2020. Where are our self-driving cars? *Vox Future Perfect Newsletter*

Romero, A. (2021, September 5). Tesla AI day 2021 review—Part 1: The promise of full self-driving cars. *Towards data science.*

Stilgoe, J. (2021). How can we know a self-driving car is safe? *Ethics and Information Technology, 23*, 635–647. https://doi.org/10.1007/s10676-021-09602-1

Tiedemann, Y., & Nagel, P. (2021, November 11). Autonomes Fahren - Wie Entwickler den Datenbergen zu Leibe rücken, *automotiveIT*.

Tversky, B. (2012). Spatial cognition embodies and situated. In: *The Cambridge Handbook of Situated Cognition.* Cambridge University Press, Chapter 12, pp. 201–216

Walker, A. (2020). Are self-driving cars safe for our cities? *Transportation, 2020*(1), 8.

Chapter 36
Modelling Teleconnections in Land Use Change

Yimin Chen and Xia Li

Abstract Land teleconnections refer to the supply–demand relationship of land between distant countries/regions and its socio-environmental impacts. Modeling land teleconnections is critical for understanding the environmental and social consequences arising from land use and consumption. In this chapter, we first explain a widely used analytical tool for the quantification of land teleconnections, and briefly describe several data sources of global/national trade. We then discuss three potential research themes of land teleconnections for future work, such as identifying the functional characteristics of land use, evaluating the land-related impacts of changing consumption patterns, and associating land teleconnections with sustainability.

Keywords Land teleconnections · MRIO model · Land use change · Sustainability

36.1 Introduction

Despite the scarcity of land resources, modern economic developments can match land supply and demand through interregional exchanges and trade. Land resources in one region can be consumed by another region (Meyfroidt & Lambin, 2009). As a result, the consumption pattern or policy change in one region can exert significant impacts on the land use in other (distant) regions (Meyfroidt et al., 2013). This phenomenon has been called "teleconnection", a term originally used to refer to climate phenomena correlated over large geographic distances (Seto et al., 2012).

Modeling land teleconnections is critical for understanding the environmental and social consequences arising from land use and consumption. In an increasingly globalized world, land use change is driven not only by local factors but also by demands for goods and services in remote areas. Some regions export goods and

Y. Chen
School of Geography and Planning, Sun Yat-sen University, Guangzhou 510275, Guangdong, China
e-mail: chenym49@mail.sysu.edu.cn

X. Li (✉)
School of Geographic Sciences, East China Normal University, Shanghai 200241, China
e-mail: lixia@geo.ecnu.edu.cn

B. Li et al. (eds.), *New Thinking in GIScience*,
https://doi.org/10.1007/978-981-19-3816-0_36

services if a comparative advantage exists in local production, while other regions import commodities that would otherwise incur a greater opportunity costs if they are produced in the regions of consumption (Yu et al., 2013). Although such trading allows efficient use of land, a potential consequence is the inequity in the interregional trade. For instance, wealthy countries usually have large amounts of land-related goods and services to import from other countries (Weinzettel et al., 2013). However, the production of such goods and services often aggravates the environmental problems in less-developed regions, such as deforestation (DeFries et al., 2010), air pollution (Lin et al., 2014), water shortage (White et al., 2018), and food insecurity (Marselis et al., 2017). Such inequity is a major concern in achieving the 17 Sustainable Development Goals (SDGs) proposed by the United Nations (2015).

In this chapter, we first explain a widely used analytical tool for the quantification of land teleconnections, and briefly describe several data sources of global/national trade. We then discuss three potential research themes of land teleconnections for future work, such as identifying the functional characteristics of land use, evaluating the land-related impacts of changing consumption patterns, and associating land teleconnections with sustainability.

36.2 Quantification of Land Teleconnections

A useful analytical tool to quantify land teleconnections is the multiregional input–output (MRIO) model (Miller & Blair, 2009). The core of the MRIO model is the interregional trade matrices that depict the from-to flows of goods and services among different economic sectors and among the interlinked geographic regions. With these matrices, modelers can track the land resources that are embodied in inter-regional trades. In this section, we first explain the basics of the MRIO model, and then demonstrate how to extend the MRIO model to quantify the embodied land resources in inter-regional trade.

Assuming a system that consists of m interlinked regions with n sectors, the total output of sector i in region r is:

$$x_i^r = \sum_{s=1}^{m} \sum_{j=1}^{n} z_{ij}^{rs} + \sum_{s=1}^{m} y_i^{rs} + e_i^r \tag{36.1}$$

where x_i^r is the total output of sector i in region r in a monetary unit. From the perspective of production and consumption, x_i^r is used in three ways. A part of x_i^r is used as the direct input to support the production of goods and services in other sectors, represented by $\sum_{s=1}^{m} \sum_{j=1}^{n} z_{ij}^{rs}$. Here z_{ij}^{rs} represents the direct input in sector j in region s derived from sector i in region r. Another part of x_i^r is directly consumed, represented by $\sum_{s=1}^{m} y_i^{rs}$. Here y_i^{rs} represents the final consumption of sector i in region s supplied by region r. Finally, e_i^r represents the flow toward foreign countries, i.e., the amount of the output of sector i in region r exporting to other countries in the world.

To facilitate the calculation, a direct input coefficient is further defined to represent the amount of input from sector i in region r required to produce one monetary unit output of sector j in region s:

$$a_{ij}^{rs} = \frac{z_{ij}^{rs}}{x_j^s} \tag{36.2}$$

Equation (36.1) can be rewritten as:

$$x_i^r = \sum_{s=1}^{m} \sum_{j=1}^{n} a_{ij}^{rs} x_j^s + \sum_{s=1}^{m} y_i^{rs} + e_i^r \tag{36.3}$$

The above equation can be transformed to a matrix to describe the entire input–output system. With $X^* = [x_i^r]$ and $A^* = \left[a_{ij}^{rs}\right]$, $\sum_{s=1}^{m} \sum_{j=1}^{n} a_{ij}^{rs} x_j^s$ becomes $A^* X^*$. Furthermore, if $Y^* = [y_i^{rs}, e_i^r]$, Eq. (36.3) can be rewritten as:

$$X^* = A^* X^* + Y^* \tag{36.4}$$

Here X^*, A^*, and Y^* represent the matrices of the output, direct input coefficients, and final demand, respectively. Assuming that A^* is constant, Eq. (36.4) can be changed to:

$$X^* = \left(I - A^*\right)^{-1} Y^* \tag{36.5}$$

where I is an identity matrix, and $(I - A^*)^{-1}$ is known as the Leontief inverse matrix (Loentief, 1936), which reveals the amount of direct and indirect inputs needed to satisfy one unit of final demand in monetary values. Therefore, Eq. (36.5) can be used to calculate the changes in sectoral outputs driven by changes in the final demand.

To explore the embodied land used for the production of goods and services, Eq. (36.5) can be further extended with a land use coefficient matrix:

$$L_k^* = l_k^* \left(I - A^*\right)^{-1} Y^* \tag{36.6}$$

where L_k^* is the matrix representing the total area of the land use k for each region, and l_k^* represents the amount of direct land use k to produce one monetary unit of sectoral output. In the matrix L_k^*, the elements L_k^{rs} represent the required total area of land use k in region r to satisfy the consumption in region s. The total area of land use k in region r can be further disaggregated to:

$$L_k^{r\cdot} = L_k^{rr} + \sum_{r \neq s} L_k^{rs} + L_k^E \tag{36.7}$$

where L_k^{rr} represents the local consumption of land use k in region r, while $\sum_{r \neq s} L_k^{rs}$ represents the total area (i.e., outflow) of land use k in region r used to satisfy the consumption in other regions. L_k^E is the amount of land use k in region r exported to foreign countries/regions. Similarly, the total area of land use k to satisfy the consumption in region r can be expressed as

$$L_k^{\cdot r} = L_k^{rr} + \sum_{r \neq s} L_k^{sr} \tag{36.8}$$

where $\sum_{r \neq s} L_k^{sr}$ represents the inflow of land use k from other regions used for satisfying the consumption in region r.

The main data for land-based MRIO analysis include global/national MRIO tables and land use maps. Data sources of global MRIO tables include, for examples, the Global Trade Analysis Project (GTAP) database and the Globally Environmentally Extended Multiregional Input–output Tables (EXIOBASE). The most recent GTAP database (version 10) contains data of global bilateral trades between 121 countries and 65 sectors for the reference years of 2004, 2007, 2011, and 2014 (Aguiar et al., 2019). The EXIOBASE database provides two forms of data, namely the monetary form and the hybrid form (using mass or energy units) (Merciai & Schmidt, 2018). An important feature of the EXIOBASE database is the wide environmental extensions, including 417 emission categories. The most recent reference year in the EXIOBASE database is 2011. For China, data that can be used to construct the MRIO model are available for the years 2002, 2007, 2010, 2012, 2015, and 2017 (Liu et al., 2014; Zheng et al., 2021).

36.3 Potential Research Themes in Study of Land Teleconnection

36.3.1 Identifying the Functional Characteristics of Land Use

An important issue in land-based MRIO analysis is that the current classification schemes of land use are not fully compatible with the classification of economic sectors in the interregional trade data required by MRIO. Large-scale land use maps derived from remotely sensed images often use the classification schemes defined by, for examples, IGBP, FAO, USGS and other organizations/institutions. These classification schemes focus mainly on the physical characteristics of land units. These land use maps also hold a rural–urban dichotomy, lumping urban areas into a single class (e.g., 'impervious surface', 'artificial surface', or 'urban'), ignoring the complicated functional characteristics of the land in urban areas (Seto et al., 2012). The MRIO matrices, however, are about goods and services flows from one economic sector to another (e.g., from the textile sector to the clothing sector). Land use maps with only a single 'urban' class do not tell where activities of different economic sectors (e.g., textile, clothing, machinery manufacturing, etc.) take place, and thus cannot support land-based MRIO modeling. To solve this problem, researchers have tried to disaggregate the single-class urban land area using proxy information, such as employment data (Chen et al., 2019) or referring to the urban land survey data for a representative area (Yu et al., 2013). However, such disaggregation may contain considerable uncertainty and bias.

Recent advance in urban remote sensing may help bridge the gap between the lack of details of (urban) land use maps and the need of functional characterization of MRIO. Unlike traditional land cover mapping that depends mainly on remotely sensed images, many recent studies of urban land use mapping have incorporated information from the so-called social sensing data (e.g., Points-of-Interests or POI). Based on these two data sources, machine learning methods can be developed to infer the actual use of urban land. For instance, Gong et al. (2020) and Chen et al. (2021) integrated multiple sources of social sensing data, such as social media data, mobile device data, and volunteered geographic information, with several types of remotely sensed images to infer urban land use types, and have provided an open access China urban land use map. Nevertheless, further research is still required to have the land use classification to be fully compatible with the economic sectors.

36.3.2 Evaluating the Impact of Changing Consumption Patterns on Land Use

Several studies have applied the MRIO analysis to the quantification of land embodied in inter-regional trades. For instances, Yu et al. (2013) revealed that developed countries have large shares of their total land use for consumption purposes displaced to other countries, especially the land for non-agricultural products. By contrasts, developing countries appropriate less global land used for the exports of non-agricultural products, but in turn use much of their land to satisfy the demand of agricultural products in foreign countries. Marselis et al (2017) found that despite the low land availability in regions suffering undernourishment, they export large amounts of embodied agricultural land to regions not suffering undernourishment. Our previous work (Chen et al., 2019) summarized the pattern of embodied land in China's domestic interregional trades. We found that the west-to-east and north-to-south "flows" of embodied agricultural land are prevalent, and the regions of Northeast and Northwest China are the primary exporters of agricultural land. These analyses, which focus on the intensity and direction of 'land flows', are fundamental to comprehensively evaluate the impacts of changing consumption patterns.

Among the land-related impacts, those that are caused by food consumption have received wide attentions. A considerable proportion of the output of the worldwide agricultural land is exchanged through international trade, forming a global food system. Therefore, dietary changes in food-importing countries will inevitably exert environmental impacts in food-exporting countries. Through MRIO analysis one can explore how to improve the environment in the food-exporting countries by encouraging more environment-friendly diets in the food-importing countries. In a recent research, Osei-Owusu et al. (2022) applied the MRIO analysis to evaluate the impacts of different food consumption scenarios and suggested a dietary change toward less meat and dairy diets to achieve a significant reduction in agricultural land footprint and greenhouse gas emissions. With the ever-growing population and rapid urbanization in global south, the trade-off between the food consumption and the environment degradation becomes a rising concern in many countries, for which the scenario-based MRIO analysis can be helpful in seeking solutions.

36.3.3 Associating Land Teleconnections with Sustainability

The sustainability requires a balance among the environment, society, and economy. In an increasingly globalized world, achieving the Sustainable Development Goals (SDGs, reference) should explicitly take into account the processes that link distant regions rather than seeing individual regions separately. In particular, asymmetric transfers of resources (including virtual "land exportation") in international trade often increase environmental burden in poor countries and hamper their socio-environmental sustainability. This has been explained by the theory of unequal ecological exchange and confirmed by empirical evidences (Dorninger et al., 2021). An important research theme regarding the unequal ecological exchange is the impact of international trade on sustainable development. Xu et al. (2020) found that, according to the data of 1995–2009, although international trade had positively affected global sustainable development, gains in sustainability took place mainly in developed countries rather than developing countries. Without international trade, however, developed countries would reach lower levels of sustainability compared with developing countries. These findings imply the greater responsibility of developed countries in the global sustainability transformation, and the change toward more sustainable lifestyles is a feasible approach. In this sense, research regarding teleconnections can help identify environmentally sound and socially accepted lifestyles to reduce land and other material footprints.

Another research topic is to develop policy formulation and assessment tools for exploring the strategy to achieve the goal of sustainable development. Wang et al. (2020) have proposed a model by integrating linear programming and MRIO analysis. This model can generate the pathways of industrial restructuring that aims at achieving multiple, conflicting goals in the sustainable development, such as maximizing employment, minimizing resources use (e.g., energy, water, land, etc.), and minimizing emissions/pollutions. These goals cover the dimensions of environment, society, and economy, and the proposed model is to find the optimal solution to balance these goals. The model has a great potential in examining sustainability policy interactions between different countries and different dimensions (e.g., environment, society, and economy). Particularly, for China's recent commitment on realizing the carbon–neutral in 2060, the model can be used to take land teleconnections into account in the policy implementation.

36.4 Concluding Remarks

In this chapter, we explain the concept of land teleconnections, which refers to the supply–demand relationship of land between distant countries/regions and its socio-environmental impacts. One of the useful tools for studying land teleconnections is the land-based MRIO model. Analysis using land-MRIO model allows tracing the land-related impacts along the supply chain that often involves multiple sectors and multiple distant countries/regions. With this method, several important research themes of land teleconnections can be explored. We identify three of them, including identifying the functional characteristics of land use, evaluating the land-related impacts of changing consumption patterns, and associating land teleconnections with sustainability. They cover several fields such as scientific data production, food consumption, scenario analysis, land-driven carbon-neutrality, and policy evaluation, which are fundamental to the progress toward global sustainable development.

References

Aguiar, A., Chepeliev, M., Corong, E., McDougall, R., & Mensbrugghe, D. (2019). The GTAP data base: Version 10. *Journal of Global Economic Analysis, 4*(1), 1–27.

Chen, B., Xu, B., & Gong, P. (2021). Mapping essential urban land use categories (EULUC) using geospatial big data: Progress, challenges, and opportunities. *Big Earth Data, 5*(3), 410–441.

Chen, Y., Li, X., Liu, X., Zhang, Y., & Huang, M. (2019). Quantifying the teleconnections between local consumption and domestic land uses in China. *Landscape and Urban Planning, 187*, 60–69.

DeFries, R. S., Rudel, T., Uriarte, M., & Hansen, M. (2010). Deforestation driven by urban population growth and agricultural trade in the twenty-first century. *Nature Geoscience, 3*(3), 178.

Dorninger, C., Hornborg, A., Abson, D. J., Von Wehrden, H., Schaffartzik, A., Giljum, S., Engler, J.-O., Feller, R. L., Hubacek, K., & Wieland, H. (2021). Global patterns of ecologically unequal exchange: Implications for sustainability in the 21st century. *Ecological Economics, 179*, 106824.

Gong, P., Chen, B., Li, X., Liu, H., Wang, J., Bai, Y., Chen, J., Chen, X., Fang, L., & Feng, S. (2020). Mapping essential urban land use categories in China (EULUC-China): Preliminary results for 2018. *Science Bulletin, 65*(3), 182–187.

Lin, J., Pan, D., Davis, S. J., Zhang, Q., He, K., Wang, C., Streets, D. G., Wuebbles, D. J., & Guan, D. (2014). China's international trade and air pollution in the United States. *Proceedings of the National Academy of Sciences, 111*(5), 1736–1741.

Liu, W., Tang, Z., & Chen, J. (2014). *Theory and practice for building multi-regional input–output table for 30 provinces in China in 2010*. China Statistics Press.

Liu, X., Yu, L., Cai, W., Ding, Q., Hu, W., Peng, D., Li, W., Zhou, Z., Huang, X., & Yu, C. (2021). The land footprint of the global food trade: Perspectives from a case study of soybeans. *Land Use Policy, 111*, 105764.

Loentief, W. (1936). Quantitative input and output relations in the economic system of United States. *Review of Economics and Statistics, 18*(3), 105–125.

Marselis, S. M., Feng, K., Liu, Y., Teodoro, J. D., & Hubacek, K. (2017). Agricultural land displacement and undernourishment. *Journal of Cleaner Production, 161*, 619–628.

Merciai, S., & Schmidt, J. (2018). Methodology for the construction of global multi-regional hybrid supply and use tables for the EXIOBASE v3 database. *Journal of Industrial Ecology, 22*(3), 516–531.

Meyfroidt, P., & Lambin, E. F. (2009). Forest transition in Vietnam and displacement of deforestation abroad. *Proceedings of the National Academy of Sciences, 106*(38), 16139–16144.

Meyfroidt, P., Lambin, E. F., Erb, K.-H., & Hertel, T. W. (2013). Globalization of land use: Distant drivers of land change and geographic displacement of land use. *Current Opinion in Environmental Sustainability, 5*(5), 438–444.

Miller, R. E., & Blair, P. D. (2009). *Input–output analysis: Foundations and extensions*. Cambridge University Press.

Osei-Owusu, A. K., Towa, E., & Thomsen, M. (2022). Exploring the pathways towards the mitigation of the environmental impacts of food consumption. *Science of the Total Environment, 806*, 150528.

Seto, K. C., Reenberg, A., Boone, C. G., Fragkias, M., Haase, D., Langanke, T., Marcotullio, P., Munroe, D. K., Olah, B., & Simon, D. (2012). Urban land teleconnections and sustainability. *Proceedings of the National Academy of Sciences, 109*(20), 7687–7692.

United Nations. (2015). *Transforming our world: the 2030 Agenda for Sustainable Development*. United Nations: New York, NY, USA.

Wang, J., Wang, K., & Wei, Y.-M. (2020). How to balance China's sustainable development goals through industrial restructuring: A multi-regional input–output optimization of the employment–energy–water–emissions nexus. *Environmental Research Letters, 15*(3), 034018.

Weinzettel, J., Hertwich, E. G., Peters, G. P., Steen-Olsen, K., & Galli, A. (2013). Affluence drives the global displacement of land use. *Global Environmental Change, 23*(2), 433–438.

White, D. J., Hubacek, K., Feng, K., Sun, L., & Meng, B. (2018). The water-energy-food nexus in East Asia: A tele-connected value chain analysis using inter-regional input–output analysis. *Applied Energy, 210*, 550–567.

Xu, Z., Li, Y., Chau, S. N., Dietz, T., Li, C., Wan, L., Zhang, J., Zhang, L., Li, Y., & Chung, M. G. (2020). Impacts of international trade on global sustainable development. *Nature Sustainability, 3*(11), 964–971.

Yu, Y., Feng, K., & Hubacek, K. (2013). Tele-connecting local consumption to global land use. *Global Environmental Change, 23*(5), 1178–1186.

Zheng, H., Bai, Y., Wei, W., Meng, J., Zhang, Z., Song, M., & Guan, D. (2021). Chinese provincial multi-regional input–output database for 2012, 2015, and 2017. *Scientific Data, 8*(1), 1–13.

Chapter 37
Progresses and Challenges of Crime Geography and Crime Analysis

Lin Liu

Abstract Crime Geography and spatial analysis of crime has gained great momentum lately, coupled with the advancement of geographic information science (GIScience) and big data in human mobility. According to (Liu in Oxford Bibliographies in Geogr 2021), crime geography and crime analysis normally cover spatio-temporal crime pattern **d**etection, crime **e**xplanation, crime **p**rediction, crime **p**revention and crime intervention **a**ssessment. The acronym of DEPPA captures these five elements. Pattern detection uncovers spatio-temporal patterns of crime distribution, such as crime hotspots. Crime explanation aims to discern major contributing factors based on multivariate regression modeling and machine learning. Crime prediction forecasts future crime patterns using machine learning and other predictive methods. Crime prevention devises targeted intervention strategies such as hot spot policing, based on historical and future crime patterns. Assessment examines the effectiveness of crime prevention, to find out if crime is reduced in the targeted area and whether the nearby areas are affected by the intervention. This chapter summarizes some of the latest progresses and challenges of crime geography and crime analysis along the issues of the unit of analysis and spatial scale, comparison analysis, new data and new variables, crime prevention and assessment, and the spatio-temporal mismatch problem.

Keywords Crime analysis · Big data · GIS · Crime prevention · Ambient population

37.1 Unit of Analysis and Spatial Scale

The spatial units of analysis vary by country. Census block, census block group, census tract, and neighborhood areas are the typical spatial units of analysis in America. European countries tend to use units similar to those of America. Communities (Juwei, or Shequ), Jiedao or Paichusuo are the typical units in China. A Jiedao

L. Liu (✉)
Department of Geography and GIScience, University of Cincinnati, Cincinnati, OH 45221-0131, USA
e-mail: Lin.Liu@uc.edu

B. Li et al. (eds.), *New Thinking in GIScience*,
https://doi.org/10.1007/978-981-19-3816-0_37

typically has a Paichusuo, the smallest official police unit, and covers multiple communities. It should be noted that a unit in China has much higher population than its counterpart in America. An alternative to the census units is the use of grids for crime analysis. The dimension of these grids typically ranges from 100 to 1000 m or even larger.

Several recent trends have emerged. One is the shift to finer units such as individual street address location, taking advantage of street view images (Zhou et al., 2021). These studies are closely related to the literature on Crime Prevention Through Environmental Design (CPTED). Another trend is the expansion of the unit of analysis to an entire city, to examine crime distribution at the provincial or national levels. The challenge of such large-scale studies is the lack of theoretical support, as most of the environmental criminology theories are aimed at the city level. Still another is the multiple level modeling that considers variables from more than one level of spatial units. For example, a model based on census block group combines variables at the block group level and the larger neighborhood level.

There have been limited studies related to the modifiable area unit problem in crime geography and crime analysis. For example, Steenbeek and Weisburd (2016) examined the variability of crime across different spatial units in The Hague, 2001–2009. So far, no major breakthroughs have been reported. Preliminary results have shown that perception variables derived from street view images, such as safety and liveliness, are only applicable to small units such as block or block group but not larger units such as tracts or neighborhoods, because within unit variation may surpass between unit variation for larger units.

In addition to spatial units, temporal units also play an important role in crime geography and crime analysis. Similar to the aggregation of crime events to areal units, crime events are typically aggregated to longer time intervals. Most studies are based on the count of the crime events in a whole year or even multiple years. Some use seasonal data, to reveal seasonable changes in crime. More and more recent studies distinguish between day and night, weekday and weekend.

A fundamental guideline for choosing the appropriate unit of analysis in space and time is to ensure that the between unit variation is larger than the within unit variation. This guideline is applicable to both the crime data itself and the explanatory data.

37.2 Comparison Analysis

Most crime studies are conducted within a single city, by focusing on one or two crime types. Lately much more attention has been directed to the comparison of multiple crime types in a single city, comparison of crime in time, and comparison across multiple cities (Weisburd, 2015). On the comparison of spatial patterns of crime, it is possible that one place is a hotspot for one crime type but a cold spot for another. Multiple crime types may coexist in the same areas. These spatial patterns may change over time. The other is the comparison of contributing factors. Different crime types may have different contributing factors. For example, ambient population drives theft, but not assault. Likewise, contributing factors may vary across cities. For example, the impact of high schools and bars on crime is very different between Chinese cities and American cities. High schools in China tend to be under strict surveillance and thus do not generate much crime. Bars in china are typically located in affluent neighborhoods and they do not attract as much crime as the bars in American cities. These comparisons are typically based on multi-variate regression analysis, although the latest machine learning tools may also have some explanatory power.

Comparison analyses have yielded some interesting results. Existing theories seem to be applicable to all nations, although some explanatory variables may have to be replaced or adjusted. For example, the variable of racial heterogeneity is widely used in America, but it is not applicable to China, as the Chinese population is dominated by the Hans. As an alternative, the makeup of migrant workers is often used in China. Unfortunately, there exists few evidence that the comparison analysis has made fundamental theoretical contributions.

37.3 New Data and New Variables

Many new data sources have become available during the past ten years. The most notable ones include trajectory data, mobility data from smart phones, geotagged social media data, surveillance cameras, street view images, and nightlight satellite images. These data are often considered part of the big data. Trajectory data can be generated from GPS tracking and communications between cell phones to nearby signal towers (Song et al., 2018). The former would have a higher locational precision than the latter. Many smart phone APPs, especially those on Android phones, track the locations of the phones. Specialty companies would collect and compile mobility data. Examples of such include Safegraph in America and Unicom Smart Footprint data in China. Social media data may have geotags. For example, about 3% to 5% of tweets are geotagged with precise coordinates, a place, or a bounding box. Surveillance cameras can be used to count people or even track individual suspects. Street view images can reveal detailed physical features for environmental auditing and people counting. Night light images can show lights at night, revealing economic activities and ambient populations (Liu et al., 2021).

These new data have been used to represent ambient population, which include both residents and visitors. Ambient population has been proven to be a superior representation of potential victims of theft, in comparison with census population. Some of the data above can generate people's movement trajectories, representing the activities of the individuals or groups. This has led to a major advancement in operationalizing the routine activity theory, a well-known environmental criminology theory.

While the benefits of these new data and new variables are obvious, the potential downsides need to draw our attention as well. One is individual's privacy. Individual's trajectory should not be revealed. Most of the recent publications are based on aggregated group data, to protect individual's privacy. The other pertains to confidentiality. A confidentiality agreement is typically signed with the data provider and must be honored in research practice.

37.4 Crime Prevention and Assessment

Results of crime analysis can potentially serve as a guidance for developing targeted crime prevention strategies. For example, spatio-temporal patterns of crime can reveal hotspots and hot time periods, which can be used to guide hot spot policing, by directing more police resources to the hotspots and hot time periods. Results of multivariate regression models tell the relative importance of individual factors to crime. More attention could be given to the most important factors. Ideally intervention should be directed to the factors that cause crime. However, conventional regression analysis cannot truly reveal causal effects. Causal analyses in crime are rare and conducted in virtual environments, although difference in differences and propensity score matching have been applied to study the effect of a natural intervention, such as the opening of new subway line and bus stops (Liu et al., 2020). More attention and effort are needed for causal crime analysis in the future. Scholars should be cognizant that not all factors can be intervened. To make crime analysis more applicable in practice, collaboration with the practitioners is needed at onset of the research design to identify intervenable factors.

Targeted intervention usually reduces crime in the target area. It may also reduce the crime in the neighboring areas. Conversely, it may push crime to the neighboring areas. The former is termed diffusion of benefits, and the latter displacement of crime. In addition to spatial displacement, crime can displace in crime type, victims, and approach to crime, etc. All these need to be examined wholistically in assessing the overall effectiveness of the crime prevention strategies.

37.5 The Spatio-temporal Mismatch Problem

Most crime events have precise location and time, but explanatory data are typically available in relatively large spatial units and not updated often. This leads to an inherent mismatch between the high spatio-temporal resolutions of crime variables and the low resolutions of explanatory variables. Traditional approaches would typically reduce the higher resolutions of crime data to match the lower resolution of data on social and built environments. For example, individual crime events could be aggregated annually to the census tract, to match those of the census variables.

Fortunately, the advancement of big data has brought opportunities of developing solutions to the spatio-temporal mismatch problem. The aforementioned big data are capable of generating high resolution data in both space and time, which can help explain crime in greater granularities. Therefore, we do not have to reduce the resolutions of the crime data anymore. Recent studies have developed crime models for three-hour time periods of a day (Song et al., 2018), and crime models based on 150 m x 150 m grids and address locations (Zhou et al., 2021). These high-resolution crime models have generated new findings and could potentially add to the existing theories.

37.6 Conclusion

This chapter serves as a summary on progresses and challenges of crime geography and crime analysis. An appropriate unit of analysis in crime analysis should ensure that the between unit variation is larger than the within unit variation. While the spatio-temporal mismatch problem is inevitable, recent advancement in big data has been able to generate variables in fine spatio-temporal resolutions that better match those of the crime data. To make crime analysis more applicable, scholars need to pay more attention to the causal and intervenable factors that drive crime. While the chapter is aimed at crime geography and crime analysis, it can also shed lights on related fields such as public health that also analyzes point-based event data.

References

Liu, L. (2021). GIS and crime analysis. *Oxford Bibliographies in Geography*. https://doi.org/10.1093/OBO/9780199874002-0233

Liu, L., Zhou, H. L., & Lan, M. X. (2021). Agglomerative effects of crime attractors and generators on street robbery? An assessment by Luojia 1–01 satellite nightlight. *Annals of the American Association of Geographers*. https://doi.org/10.1080/24694452.2021.1933888

Liu, L., Lan, M. X., Eck, J. E., & Kang, E. L. (2020). Assessing the effects of bus stop relocation on street robbery. *Computers, Environment and Urban Systems, 80* (101455). https://doi.org/10.1016/j.compenvurbsys.2019.101455

Song, G. W., Liu, L., Bernasco, W., Xiao, L. Z., Zhou, S. H., & Liao, W. W. (2018). Testing indicators of risk populations for theft from the person across space and time: The significance of mobility and outdoor activity. *Annals of the American Association of Geographers., 108*(5), 1–19.

Steenbeek, W., & Weisburd, D. (2016). Where the action is in crime? An examination of variability of crime across different spatial units in The Hague, 2001–2009. *Journal of Quantitative Criminology, 32*(3), 449–469.

Weisburd, D. (2015). The law of crime concentration and the criminology of place. *Criminology, 53*(2), 133–157.

Zhou, H. L., Liu, L., Lan, M. X., Zhu, W. L., Zhu, Song, G. W., Jing, F. R., Zhong, Y. R., Su, Z. H., & Gu, X. (2021). Using Google street view imagery to capture micro built environment characteristics in drug places, compared with street robbery. *Computers, Environment and Urban Systems, 88* (101631). https://doi.org/10.1016/j.compenvurbsys.2021.101631

Chapter 38
GIS Empowered Urban Crime Research

Yijing Li and Robert Haining

Abstract The chapter looks at the contribution Geographical Information Systems (GIS) have made to research into the spatial and temporal patterns of urban crime and criminality, indicating areas of possible growth in the future and sketching the policy relevance of these developments for practical policing and research. It discusses four main GIS empowered crime research elements using a "3W1H" framework to help organize ideas: "Who (2P: Police and Public)", "What (2C: Crime and Context)", "Why (2W: When and Where)" and "How (2D: Data and Design)". We highlight how interactive data visualisation leads to better public communication, essential for successful policing, as well as contributing to research agendas.

Keywords GIS · Urban crime · Spatial and temporal patterns · Crime mapping · Data-driven

38.1 Introduction

Study into urban crime can be dated back to at least the nineteenth Century but this field gathered particular momentum in the 1920s and 1930s when links between neighbourhood social conditions and crime and criminality were systematically investigated by what is referred to as the "Chicago School" (see for example Shaw & McKay, 1931). This early work was undertaken within the discipline of sociology but in the years since, the study of crime and criminality has attracted interest from many other disciplines including geography, psychology, public health, behavioural science, and economics whilst also establishing itself as a field of study (criminology) in its own right (Becker, 1968; George, 1978; Hakim and Rengert, 1981;

Y. Li (✉)
Lecturer in Urban Informatics, Department of Informatics, King's College London, London WC2B 4BG, UK
e-mail: yijing.li@kcl.ac.uk

R. Haining
Department of Geography, Emeritus Professor in Human Geography, University of Cambridge, Cambridge CB2 3EN, UK
e-mail: bob.haining@geog.cam.ac.uk

B. Li et al. (eds.), *New Thinking in GIScience*,
https://doi.org/10.1007/978-981-19-3816-0_38

Sorenson and Pilgrim, 2000). For extensive and recent reviews of the field see the various Oxford Handbooks of Criminology series (2012, 2017, 2018). The technical challenges associated with crime research has also attracted interest from computer scientists, with GIS assuming an important place in the handling and displaying of geographically referenced crime data, data that are routinely collected and digitized by local police forces (Li, 2015). Ever since the widespread utilisation of GIS packages and relevant computational techniques in the 1990s (Wortley & Mazerolle, 2008), researchers have been able to rapidly map and query crime data, illuminating geographical (spatial and territorial) and spatial–temporal patterns and trends often in near real-time limited only by the speed with which crime-site data can be input into computer systems (Spencer and Ratcliffe, 2005). The "marriage" of cutting-edge machine learning and data visualisation techniques in recent decades has added further impetus to the field.

Such advances in urban crime data analysis (both the interpretation and presentation of crime data) have enabled wider and better communication between police forces and the general public, drawn the public's attention to local safety issues, disseminated knowledge about local police services and provided up-to-date information on progress, facilitated the promotion of crime prevention strategies and improved the efficiency of routine policing practices. In light of these, this chapter will illustrate "GIS-empowered" urban crime research using a **3W1H** donut-gram (Fig. 38.1) to indicate the broad classes of questions that are of interest:

1. **W**hat: what is the potential for urban crime research and what are the future directions for the subject? This strand can be streamed into:

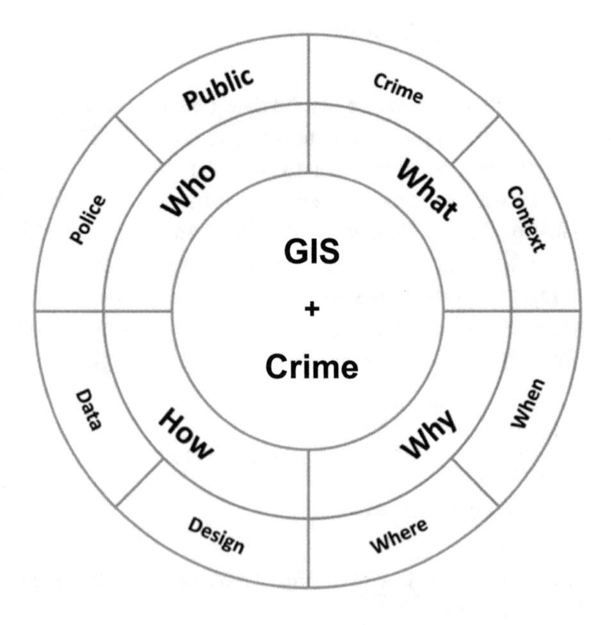

Fig. 38.1 Urban Crime Research: the 3W1H Donut-gram

- **C**rime: evolution of long-standing crime research agendas into, for example, offender behaviors and victimization in geographical space and time and changes due to structural shifts associated with post-pandemic economies, such as changes to working practices (more home working and less commuting), and linked to
- **C**ontext: contextual (e.g., environmental) influences on crime and criminality in urban areas. Changes in urban structure and how this will impact on crime and criminality.

2. **W**hy: Why is it important to empower urban crime research with GIS techniques? In particular it contributes to work into spatio-temporal crime patterns research:
   - **W**hen: the crime-peak hour(s) of the day, days of the week, seasonal and temporal trends in crime incidents in urban areas.
   - **W**here: the detection of hot spots and "crime streets" (micro-spaces associated with specific street segments that have very high levels of crime and criminality) in urban spaces for the purpose of directly targeting police resources.
3. **W**ho: Who would such research benefit? Beneficiaries will include not only policy practitioners like police agencies, but also ordinary citizens.
   - **P**olice: evidence-based policing strategies as well as space and time specific crime-combating measures.
   - **P**ublic: transparent information communication and interaction.
4. **H**ow: How GIS will better empower urban crime research in the coming decades?
   - **D**ata: emerging big data and multiple new sources of data (e.g., from social media, smart phones, remotely sensed images, etc.)
   - **D**esign: innovative methodologies, augmenting traditional spatial statistical techniques such as regression and cluster detection methods, for instance machine learning and artificial intelligence.

38.2 Urban Crime Literature

We start by considering the first "W" (What) of the urban crime research donut-gram, and its two streams crime and context, and the second "W" (Why) and its two streams identifying crime patterns over place and time. There is a well-developed literature investigating the consequences of urban development on crime and criminality, ever since Durkheim (1897) coined the term "anomie" to describe social alienation, and the Chicago School's research into social disorganisation and its influence on urban criminality (Messner and Rosenfeld, 1997; Chamlin and Cochran, 1995; Savolainen, 2000; Bernburg, 2002; Kim and Pridemore, 2005).

38.2.1 Crime Embedded in the Urban Context

In the course of exploring urban crime influences and changes in those influences over time, the majority of criminological or sociological theories have focused on changes in the amount and types of crime and the importance of such variables (varying over time and between places) as poverty, changes in economic inequality, criminal opportunities, cultural conflicts, weakened social control and social disorganization (Durkheim, 1897, Cloward and Ohlin, 1960, Kim and Pridemore, 2005). Theories that include institutional anomie theory (Messner and Rosenfeld, 1997) and social capital theory (Coleman, 1988; Sampson et al., 1999), place emphasis on the mediating role of social cohesion and the strength of cultural values that do not equate "success" with "money". They are consistent with viewpoints from Chicago School theorists in emphasising neighborhood structure and its links to levels of crime. For example, Shaw and McKay (1931) were most concerned with the deleterious effects of racial and ethnic heterogeneity, residential mobility, and low socioeconomic status on an area's ability to prevent crime. Together with other factors like family disruption (Sampson and Groves, 1989), relative poverty (Messner, 1982), and racial segregation (Krivo and Peterson, 1996), these theories work together to provide a basis for examining changing crime patterns in urban areas.

Since the 1970s, urban crime research has evolved becoming no longer the preserve of criminologists, sociologists, and police practitioners, but has also received interdisciplinary input from geographers and computer scientists who have paid particular attention to crime influences emanating from the urban contextual environment. For instance, Storch (1979) reviewed Gurr's (1977) work on evaluating urban crimes in London, Stockholm, and Sydney since 1930 using statistical records to derive associations between increases in crime and civil order and trying to uncover the influences of institutional and political factors. Research had shown that crime intensities, changes, complexities, and dynamics have been significantly affected by the particular urban contexts, as do the corresponding crime-countering measures and policies, which require account to be taken of local contexts and specific local crime trajectories. A question of particular interest currently is how social and economic changes, set in motion by the covid pandemic will play out in terms of urban crime patterns (Ceccato et al. 2021).

38.2.2 Crime Patterns

Urban crime is not a collection of random incidents in space and time, and for this reason has attracted increasing numbers of researchers and practitioners to explore the observed patterns in time and space, with the aims of gaining a better understanding of the dynamics of crimes, and to prepare more accurate crime forecasts in order to devise preventive strategies and measures.

Time-series analysis and various related methods had been employed by statisticians, sociologists and mathematicians amongst others to describe the crime trends in target urban areas, by year, by season, by week as well as daily, hourly, or even every 15 min (Lauritsen and White, 2014; Andresen and Malleson, 2015; Felson and Poulsen, 2003; OJJDP, 2016; Williams and Coupe, 2017), for not only descriptive and exploratory purposes, but also looking into the temporal predictions to support policing and patrolling assignments.

GIS and spatial analysis techniques are proving to be essential for studying criminal activity, especially in detecting crime hot spots (Chainey, 2020) or hot street (micro) segments (Tom-Jack et al., 2019) in different urban contexts. Recently, Murray et al. (2001) highlighted the novel capability of GIS and spatial analysis approaches for examining crime in urban regions using a case study in Brisbane, Australia; Ratcliffe (2010) noted that mainstream research in spatial criminology lies in the study of spatial and temporal crime patterning, and prediction; Chainey (2020: p43) mapped the high crime concentration micro-places in New York City (USA), Montevideo (Uruguay) and Rio de Janeiro (Brazil). Intensive case studies have enriched our understanding of urban crime and its context (**W**hat), the identification of crime patterns in place and time (**W**hy), and the development of urban crime research, drawing on more diversified data sources and evolved GIS methodologies (**H**ow), in the expectation of empowering crime research towards better police service practices to facilitate crime reduction and prevention (**W**ho).

38.3 GIS Empowered Crime Research and Practice

Ever-evolving GIS methodologies have empowered the realisation ("**H**ow") of crime research and practice. The use of GIS in crime research has strong roots in practical policing challenges in the 1990s most notably when the New York City Police Department (Harris, 1999) started to replace traditional but cumbersome pinpoint maps with a computerised crime mapping system. However, the intellectual root for using spatial technology goes further back. For example, in the 1960s, Jacobs (1961) "eyes on the street" theory drew attention to the importance of the physical environment on some forms of criminal behaviour. City planners, by paying attention to small scale urban layout could facilitate informal surveillance which would in turn discourage the motivated offender. Urban renewal in the 1950s and 1960s had not, in Jacob's view, paid sufficient attention to how urban layout might have an impact on a range of crimes from burglary to street crime. Good environmental (structural) urban design will enhance the public's community guardianship role and play an important part in responding to rising crime levels. GIS with its ability to manage spatial (i.e., land use, census, and transportation) databases clearly had a role to play in developing "safer urban spaces" as well as play an important role in crime pattern detection (section 38.2.2) and crime mapping in space and time. There is potential to link these two roles as part of a spatial decision support system to monitor the consequences of urban re-design on crime patterns.

38.3.1 Crime Mapping for Police Services

Digital crime mapping, for example the widely acknowledged COMPSTAT model (McDonald, 2002), was initially employed to support police decision making, manage patrol operations (Chainey and Ratcliff, 2013), and make significant progress in improving and making more efficient police services in the face of increasing 'demand' and spiralling costs (Craglia et al., 2000). COMPSTAT was credited with having played an important role in reducing crime levels in New York City, evolving into a prospective crime mapping tool (Hart et al., 2020) to support predictive policing often based on the use of spatial analytical techniques, for example Kernel Density Estimation (KDE) (Bailey and Gatrell, 1995) and Risk Terrain Modelling (RTM) (Caplan and Kennedy, 2010).

Computer-assisted crime mapping has improved the efficiency of police services significantly. For instance, Reaves (2010) compared the percent of local police departments using computers by 2007 and found that police departments in urban areas especially serving populations of over 250,000, had utilised GIS 100% for crime analysis and crime mapping. Besides, the use of GIS combined with spatial analysis tools has contributed to the development of various crime mapping applications which assisted with patrol dispatching, community policing and resource planning.

38.3.2 Crime Mapping for Public Engagement

In the twenty-first Century, driven by advances in computer science and data (geo)visualisation techniques, crime mapping has become more supportive for public sectors enabling the provision of information quickly to citizens (e.g., crime dashboard in the city of London) supporting fulfilling the goal of transparent communication as a key element of open government promises. For example, Chainey and Tompson (2012) affirmed the policy impacts from UK police forces' adoption of an online crime mapping tool in 2008, with the fruits of it improving engagement with and empowerment and promotion of public service transparency and accountability. However, such interactive communication normally places high requirements on the citizens' ability to interpret the data. This in turn necessitates careful cartographic visualisation of the information to be communicated to reduce the risk of any confusion arising because of the lack of direct human interaction. It proposed the integration of crime mapping systems with social media to facilitate instant and prompt communication of local crime issues for the benefit of citizens. However as has been remarked, involving social media is not without its challenges and potential for misinformation.

38.4 What Next?

The digital era is ushering in a period where very large and complex data sets are available, providing new opportunities and motivation to develop new and innovative ways for conducting urban crime research (Hart et al., 2020) and supporting policing services utilizing emerging data & state-of-the-art methods (**H**ow).

38.4.1 Emerging Data

The linked development of both emerging new sources of fine-grained urban crime and other relevant data and data mining techniques have together contributed to enhanced understanding of urban crime (Zhao and Tang, 2018). Here are some examples:

1. **I**nternet-of-Things (IoT) data: social media for example geo-located Twitter sentiment data and Foursquare data have been utilised together with other types of data including weather data for example, to make crime predictions (Wang and Li, 2021; Wang et al., 2012; Chen et al., 2015). The mobile data being used to simulate the mobile population has become more widely deployed (Bogomolov et al., 2014; Rosés et al., 2021). There are also many open-source databases that have been made available for researchers to explore spatial–temporal crime patterns from a comparative perspective, e.g., the Crime Open Database (CODE) recording 10 largest US cities' crimes over 11 years by type (Ashby, 2019), and the multiple sources of UK crime datasets (Tompson et al., 2015).
2. **T**racking data: GPS data has been used to simulate populations at risk from crime (Kikuchi et al., 2012) in micro-places, or record police patrols and micro-place-based intervention effects (Hutt et al., 2021), with the latter accompanied by GPS and Body Worn Recorder video data (BWV). Other locational tracing data such as that provided by taxi data (Vomfell et al., 2018), cell phone tracking (Song et al., 2019) and Google location data (Valentino-DeVries, 2019) are potentially valuable data sources in this context.
3. **I**mage data: recently remotely sensed data have attracted interest as inputs into crime and policing research as a way of avoiding the high costs of manual data collection, because of its increasing abundance and accessibility at high resolution. Najjar et al. (2018) investigated the use of deep learning to predict crime rates from raw satellite imagery for the purpose of promoting urban safety. Wu et al. (2018) trained both street view and satellite images with crime data in San Francisco, to predict the relative crime risks at different locations. Patino et al. (2014) used urban fabric descriptors computed from very high spatial resolution imagery to assess whether neighbourhood design and condition has a quantifiable imprint, as suggested it should by "broken windows" theory. They analysed the relationships between land cover, structure, texture descriptors and intra-urban homicide rates in Medellin, Colombia. Other work, such as that by Wolfe and Mennis (2012), Woodworth et al. (2014) and Liu et al. (2020), have used satellite imagery to estimate burglary density, assess the effect of

vegetation cover and nightlight respectively, on urban crime; Ceccato (2021) and colleagues are exploring the use of built form and other indexes derived from remotely sensed imagery together with artificial intelligence to predict crime patterns in Stockholm (an ongoing project funded by FORMAS).

4. Video footage data: Ashby (2017) used data from the British railway network to show that CCTV is a powerful investigative tool for many types of crime, and this finding was further endorsed by Lindegaard and Bernasco's (2018) work on suggesting the value of camera recordings as part of the investigative and criminological tool kit. Thomas et al. (2021) reviewed 162 CCTV schemes on crime prevention cases systematically across 15 countries over the past five decades illustrating the global expansion and internationalization of these technologies.

However, such emerging data have been exposed to challenges in terms of their accessibility, accountability, comparability, reliability, generalisability, interpretability, and representativeness. Such data need linking to advanced computational methods if they are to be "distilled" into useful information and knowledge and then communicated to stakeholders (including the public where relevant) in easy-to-understand language.

38.4.2 State-Of-The-Art Methods

Whilst GIS is an important part of the enabling technology for "spatial thinking", which is an essential underpinning to urban crime research and police response given the territorial as well as temporal nature of policing, it is important to recognize that GIS can support work with spatially referenced data in at least three distinct ways (Burrough & McDonnell, 1998): (1) as a powerful set of digital "tools": "*…for collecting, storing, retrieving at will, transforming and displaying spatial data from the real world for a particular set of purposes*" (p. 11); (2) as a database management system: "*a computer based set of procedures used to store and manipulate geographically referenced data*". (p. 11). Arguably these are the two main uses of GIS in crime research and policing practice up to the current time where sometimes advanced spatial statistical methods and geo-visualization have been integrated into the GIS for such activities as crime hotspot detection and geographic profiling. When a deeper understanding of crime patterns is needed, researchers have often made use of advanced statistical techniques such as spatial regression models (Anselin, 2009), geographically weighted regression (Cahill and Mulligan, 2007) and Spatio-temporal Bayesian modeling (Hu et al., 2018, Haining and Li 2020; Law et al. 2020).

A third and relatively under-developed (to date) use of GIS is as (3) a spatial decision support system (SDSS) which involves the integration of spatially referenced data in a problem-solving environment. This may involve inputting crime data, a range of urban data including socio-economic, physical infrastructure and police activity data, in order to evaluate different policing strategies in terms of efficiencies and outcomes [e.g., crime reduction currently of concern to society; elimination of hotspots; evaluation of a crime reduction programme (Li et al. 2013)].

In face of the ever-increasing urban data volumes and diverse sources, state-of-the-art interdisciplinary methodologies have been increasingly adopted by urban crime researchers.

Machine learning and AI techniques ("deep learning") have in recent decades been widely utilised to (1) detect crime hotspots (Kounadi et al., 2020; Nair and Gopi, 2020) together with spatial analytical techniques such as, adaptive kernel density estimation methods; (2) predict crime incident locations using for example Convolutional Neural Network (CNN) to train millions of imagery datasets (Najjar et al., 2018; Wu et al., 2018); and (3) optimise police patrol routing (PPR) using a (hybrid) Genetic Algorithm (GA), linear programming, local search and routing policies (Dewinter et al., 2020). Cichosz (2020) has summarised how algorithms like correlation analysis, random forest, linear regression, negative binomial regression, logistic regression, naive Bayes classifiers, SVM, neural network, decision trees, *k*-NN, polynomial regression, autoregression, clustered continuous conditional random field, and gradient boosting have been adapted by different cities in their policing efforts.

However, returning to the "**3W1H**" framework with which we began, such emerging data and state-of-the-art algorithms (including data visualization techniques and interactive dashboard platforms) which will certainly feature in next generation urban crime research and associated policy initiatives, will not be, in themselves, sufficient to serve future needs and expectations. Rather a key feature of that future must involve comprehensive, open, and transparent public participation and communication between stakeholders. This will in turn require a more comprehensive and systematic integration of GIS into urban crime research and urban policing. In summary, urban crime research in the coming decades will engage with wider aspects of the social and economic ecosystem. The aim must be to respond to the public's concerns and progress a whole society urban safety agenda that recognizes the needs of different groups defined by, for example, their ethnicity or their gender. This can only be taken forward through joint efforts involving academics, police agencies, private sector entrepreneurs and governance policymakers. It will be essential to communicate with all stakeholders, listening and responding to their fears and concerns in flexible ways and in language that they can understand.

References

Andresen, M. A., & Malleson, N. (2015). Intra-week spatial-temporal patterns of crime. *Crime Science, 4*, 12. https://doi.org/10.1186/s40163-015-0024-7

Ashby, M. P. J. (2017). The value of CCTV surveillance cameras as an investigative tool: An empirical analysis. *European Journal on Criminal Policy and Research, 23*, 441–459. https://doi.org/10.1007/s10610-017-9341-6

Ashby, M. P. J. (2019). Studying crime and place with the crime open database. *Research Data Journal for the Humanities and Social Sciences, 4*(1), 65–80. https://doi.org/10.1163/24523666-00401007

Bailey, T., & Gatrell, A. (1995). *Interactive spatial data analysis*. Longman Scientific and Technical.

Becker, G. S. (1968). Crime and punishment: An economic approach. *Journal of Political Economy, 76*(2), 169–217.

Bernburg, J. G. (2002). Anomie, social change and crime. A theoretical examination of Institutional-Anomie theory. *The British Journal of Criminology, 42*, 729–742.

Bogomolov, A., Lepri, B., Staiano, J., Oliver, N., Pianesi, F., & Pentland, A. (2014). Once upon a crime: Towards crime prediction from demographics and mobile data. In: *Proceedings of the 16th ACM international conference on multimodal interaction (ICMI)*, pp. 427–434. https://doi.org/10.1145/2663204.2663254

Bruinsma, G. J. N., & Johnson, S. D. (2018). *The Oxford handbook of environmental criminology*. Oxford University Press.

Burrough, P. A., & McDonnell, R. A. (1998). *Principles of geographical information systems* (pp. 333). Oxford University Press, Nova York.

Caplan, J. M., & Kennedy, L. W. (2010). *Risk terrain modeling manual: Theoretical framework and technical steps of spatial risk assessment*. Rutgers Cener on Public Security.

Cahill, M., & Mulligan, G. (2007). Using geographically weighted regression to explore local crime patterns. *Social Science Computer Review, 25*(2), 174–193. https://doi.org/10.1177/0894439307298925

Ceccato, V. (with A. Nascetti, and R.Haining). (2021). Developing the use of remote sensing data in safety planning. Project funded by FORMAS January 2021–December 2023.

Ceccato, V., Kahn, T., Herrmann, C., & Ostlund. A. (2021). Pandemic restrictions and spatiotemporal crime patterns in New York, Sao Paulo and Stockholm. *Journal of Contemporary Criminal Justice*, Sept. https://doi.org/10.1177/10439862211038471

Chainey, S., & Tompson, L. (2012). Engagement, empowerment and transparency: Publishing crime statistics using online crime mapping. *Policing: A Journal of Policy and Practice, 6*(3), 228–239. https://doi.org/10.1093/police/pas006

Chainey, S., & Ratcliffe, J. (2013). *GIS and crime mapping*. Wiley.

Chainey, S. (2020). *Understanding crime: Analyzing the geography of crime*. ESRI Press.

Chamlin, M. B., & Cochran, J. K. (1995). Assessing Messner and Rosenfeld's Institutional Anomie Theory: A Partial Test. *Criminology, 33*(3), 311–330.

Chen, X., Cho, Y., & Jang, S. Y. (2015). Crime prediction using Twitter sentiment and weather. *2015 Systems and Information Engineering Design Symposium, 2015*, 63–68. https://doi.org/10.1109/SIEDS.2015.7117012

Cichosz, P. (2020). Urban crime risk prediction using point of interest data. *ISPRS International Journal of Geo-Information, 9*, 459. https://doi.org/10.3390/ijgi9070459

Cloward, R. A., & Ohlin, L. E. (1960). *Delinquency and opportunity: A theory of delinquent gangs*. New York: Free Press.

Coleman, J. (1988). Social capital in the creation of human capital. *American Journal of Sociology, 94*, S95–S120.

Craglia, M., Haining, R., & Wiles, P. (2000). A comparative evaluation of approaches to urban crime pattern analysis. *Urban Studies, 37*(4), 711–729.

Dewinter, M., Vandeviver, C., Vander Beken, T., & Witlox, F. (2020). Analysing the police patrol routing problem: A review. *ISPRS International Journal of Geo-Information, 9*, 157–173. https://doi.org/10.3390/ijgi9030157

Durkheim, E. (1897). *Suicide: A study in sociology*. Translated by Spaulding, J. A., & Simpson, C. London: Routledge and Kegan Paul.

Felson, M., & Poulsen, E. (2003). Simple indicators of crime by time of day. *International Journal of Forecasting, 19*(4), 595–601. https://doi.org/10.1016/S0169-2070(03)00093-1

George, D. E. (1978). The geography of crime and violence: A spatial and ecological perspective. In: *Association of American Geographers: Resource papers for college geography* (Vol. 78, Issue 1).

Haining, R. P., & Li, G. (2020). *Modelling spatial and spatial-temporal data: A Bayesian approach* (p. 608). CRC Press.

Hakim, S., & Rengert, G. F. (1981). *Crime spillover*. Sage Publications.

Harris, K. (1999). *Mapping crime: Principles and practice*. NIJ.

Hart, T. C., Lersch, K. M., & Chataway, M. (2020). *Space, time, and crime*. Carolina Academic Press.

Hu, T., Zhu, X., Duan, L., & Guo, W. (2018). Urban crime prediction based on spatio-temporal Bayesian model. *PLoS ONE, 13*(10), e0206215. https://doi.org/10.1371/journal.pone.0206215
Hutt, O. K., Bowers, K., & Johnson, S. D. (2021). The effect of GPS refresh rate on measuring police patrol in micro-places. *Crime Science, 10*(3). https://doi.org/10.1186/s40163-021-00140-1
Jacobs, J. (1961). *The death and life of great American cities*. Vintage Books.
Law, J., Quick, M., & Jadavji, A. (2020). A Bayesian spatial shared component model for identifying crime-general and crime-specific hotspots. *Annals of GIS, 26*(1), 65–79.
Kikuchi, G., Amemiya, M., & Shimada, T. (2012). An analysis of crime hot spots using GPS tracking data of children and agent-based simulation modeling. *Annals of GIS, 18*(3), 207–223. https://doi.org/10.1080/19475683.2012.691902
Kim, S., & Pridemore, W. A. (2005). Poverty, socioeconomic change, institutional anomie, and homicide. *Social Science Quarterly, 86*, 1377–1398.
Kounadi, O., Ristea, A., Araujo, A. et al. (2020). A systematic review on spatial crime forecasting. *Crime Science, 9*(7). https://doi.org/10.1186/s40163-020-00116-7
Krivo, L. J., & Peterson, R. D. (1996). Extremely disadvantaged neighborhoods and urban crime. *Social Forces, 75*(2), 619–650.
Lauritsen, J. L., & White, N. (2014). Seasonal patterns in criminal victimization trends. *U.S. Department of Justice Office of Justice Programs Bureau of Justice Statistics*: NCJ245959. https://bjs.ojp.gov/content/pub/pdf/spcvt.pdf
Li, G., Haining, R. P., Richardson, S., & Best, N. (2013). Evaluating the no-cold calling zones in Peterborough England: Application of a novel statistical method for evaluating neighbourhood policing methods. *Environment and Planning A, 45*(8), 2012–2026.
Li, Y. (2015). *Geography of crime in China since the economic reform of 1978*. Cambridge Scholars Publishing.
Liebling, A., Maruna, S., & McAra, L. (2017). *The Oxford handbook of criminology*. Oxford University Press.
Lindegaard, M. R., & Bernasco, W. (2018). Lessons learned from crime caught on camera. *Journal of Research in Crime and Delinquency, 55*(1), 155–186. https://doi.org/10.1177/0022427817727830
Liu, L., Zhou, H., Lan, M., & Wang, Z. (2020). Linking Luojia 1–01 nightlight imagery to urban crime. *Applied Geography, 125*, 102267. https://doi.org/10.1016/j.apgeog.2020.102267
Maguire, M., Morgan, R., & Reiner, R. (2012). *The Oxford handbook of criminology*. Oxford University Press.
McDonald, P. (2002). *Managing police operations: Implementing the New York crime control model-CompStat*. Belmont, VA. Wadsworth.
Messner, S., & Rosenfeld, R. (1997). Political restraint of the market and levels of criminal homicide: A cross national application of institutional-anomie theory. *Social Forces, 75*, 1393–1416.
Messner, S. F. (1982). Poverty, inequality, and the urban homicide rate: Some unexpected findings. *Criminology, 20*(1), 103–114.
Murray, A. T., Mcguffog, I., Western, J. S., & Mullins, P. (2001). Exploratory spatial data analysis techniques for examining urban crime. *British Journal of Criminology* 309–329.
Nair S. N., & Gopi, E. S. (2020). Deep learning techniques for crime hotspot detection. In: Kulkarni, A. & Satapathy, S. (Eds.), *Optimization in machine learning and applications. Algorithms for intelligent systems*. Springer. https://doi.org/10.1007/978-981-15-0994-0_2
Najjar, A., Kaneko, S., & Miyanaga, Y. (2018). Crime mapping from satellite imagery via deep learning. *ArXiv*, abs/1812.06764.
OJJDP Statistical Briefing Book. 2016. Available at: https://www.ojjdp.gov/ojstatbb/offenders/qa03401.asp?qaDate=2016. Released on October 22, 2018.
Patino, J. E., Duque, J. C., Pardo-Pascual, J. E., & Ruiz, L. A. (2014). Using remote sensing to assess the relationship between crime and the urban layout. *Applied Geography, 55*, 48–60. https://doi.org/10.1016/j.apgeog.2014.08.016
Ratcliffe, J. (2010). Crime mapping: Spatial and temporal challenges. In: Piquero, A. R., Weisburd, D. (Eds.), *Handbook of quantitative criminology*. Springer, pp. 5–24
Reaves, B. A. (2010). *Local police departments*, 2007. (NCJ 231174). BJS.

Ros'es, R., Kadar, C., & Malleson, N. (2021). A data-driven agent-based simulation to predict crime patterns in an urban environment. *Computers, Environment and Urban Systems, 89*, 101660.
Sampson, R. J., & Groves, W. B. (1989). Community structure and crime: Testing social-disorganization theory. *American Journal of Sociology, 94*(4), 774–802.
Sampson, R. J., Morenoff, J., & Earls, F. (1999). Beyond social capital: Spatial dynamics of collective efficacy for children. *American Sociological Review, 64*(5), 633–660.
Savolainen, J. (2000). Inequality, welfare state, and homicide: Further support for the institutional anomie theory. *Criminology, 38*, 1021–1042.
Shaw, C., & McKay, H. D. (1931). Social factors in juvenile delinquency. *Report on the Causes of Crime*. National Commission on Law Observance and Enforcement. Government Printing Office.
Song, G., Bernasco, W., Liu, L., Xiao, L., Zhou, S., & Liao, W. (2019). Crime feeds on legal activities: Daily mobility flows help to explain thieves' target location choices. *Journal of Quantitative Criminology, 35*, 831–854. https://doi.org/10.1007/s10940-019-09406-z
Sorensen, J. R., & Pilgrim, R. L. (2000). An actuarial risk assessment of violence posed by capital murder defendants. *Journal of Criminal Law & Criminology, 90*, 1251–1270.
Spencer, C., & Ratcliffe, J. (2005). Mapping and analysing change over time. *GIS and Crime Mapping* 223–256. https://doi.org/10.1002/9781118685181.ch8
Storch, R. D. (1979). Review: The study of urban crime. In: Gurr, T. R. (1977). (Reviewed Work), *Rogues, rebels, and reformers. A political history of urban crime and conflict*. *Social history* (Vol. 4, Issue 1), pp. 117–122. https://www.jstor.org/stable/4284864
Thomas, A., Piza, E., Welsh, B., & Farrington, D. (2021). The internationalization of CCTV surveillance: Effects on crime and implications for emerging technologies. *International Journal of Comparative and Applied Criminal Justice*. https://doi.org/10.1080/01924036.2021.1879885
Tom-Jack, Q., Bernstein, J., & Loyola, L. (2019). The role of geoprocessing in mapping crime using hot streets. *ISPRS International Journal of Geo-Information, 8*, 540. https://doi.org/10.3390/ijgi8120540
Tompson, L., Johnson, S., Ashby, M., Perkins, C., & Edwards, P. (2015). UK open source crime data: Accuracy and possibilities for research. *Cartography and Geographic Information Science, 42*(2), 97–111. https://doi.org/10.1080/15230406.2014.972456
Valentino-DeVries, J. (2019). Tracking phones, google is a dragnet for the police. *The New York Times*. April 13, 2019. https://www.nytimes.com/interactive/2019/04/13/us/google-location-tracking-police.html
Vomfell, L., Härdle, W. K., & Lessmann, S. (2018). Improving crime count forecasts using Twitter and taxi data. *Decision Support Systems, 113*, 73–85. https://doi.org/10.1016/j.dss.2018.07.003
Wang, X., Gerber, M. S., & Brown, D. E. (2012). Automatic crime prediction using events extracted from Twitter posts. In: Yang S. J., Greenberg A. M., & Endsley M. (Eds.), *Social computing, behavioral—Cultural modeling and prediction*. SBP 2012. Lecture Notes in Computer Science, Vol. 7227. Springer. https://doi.org/10.1007/978-3-642-29047-3_28.
Wang, Z., & Li, Y. (2021). Could social medias reflect acquisitive crime patterns in London? *Journal of Safety Science and Resilience*. https://doi.org/10.1016/j.jnlssr.2021.08.007
Williams, S., & Coupe, T. (2017). Frequency versus length of hot spots patrols: A randomised controlled trial. *Cambridge Journal of Evidence Based Policing, 1*, 5–21. https://doi.org/10.1007/s41887-017-0003-1
Wolfe, M. K., & Mennis, J. (2012). Does vegetation encourage or suppress urban crime? Evidence from Philadelphia, PA. *Landscape and Urban Planning, 108*(2–4), 112–122. https://doi.org/10.1016/j.landurbplan.2012.08.006
Woodworth, J. T., Mohler, G. O., Bertozzi, A. L., & Brantingham, P. J. (2014). Non-local crime density estimation incorporating housing information. *Philosophical Transactions of the Royal Society A: Mathematical, Physical and Engineering Sciences, 372*(2028). https://doi.org/10.1098/rsta.2013.0403
Wortley, R., & Mazerolle, L. (2008). *Environmental criminology and crime analysis: Situating the theory, analytic approach and application*. Portland, OR: Willan Publishing.
Wu, J., Hui, J., & Xian, R. T. W. (2018). *Utilization of street view and satellite imagery data for crime prediction*. http://cs230.stanford.edu/projects_winter_2020/reports/32644967.pdf
Zhao, X., & Tang, J. (2018). Crime in urban areas: A data mining perspective. *ACM SIGKDD Explorations Newsletter, 20* (1). https://doi.org/10.1145/3229329.3229331

Chapter 39
GIS in Building Public Health Infrastructure

Ge Lin

Abstract This chapter recounts personal anecdotes of augmenting public health infrastructure using GIS. I start with what motivated me to work primarily on real world problems in public health data integration and disease surveillance. In the process of enhancing spatial data capacity and GIS solutions, a key to success is being flexible while working with public health programs for their other needs. Enabling visualization of multiple disease maps for multiple programs is a good starting point from maps to mechanisms. We must continuously build other community health datasets to snowball programs' inquiries. Expanding GIS capabilities in disease cluster detection is technologically easy, but organizationally challenging, which requires domain knowledge and design thinking. In the Big Data era, GIS bridges precision neighborhood health with precision medicine to improve population health at both individual and neighborhood levels.

Keywords Data infrastructure · Disease detection · Geocoding · Neighborhood · Precision · Spatial cluster

39.1 Introduction

Medicine treats individual patients, whereas public health treats people and their communities as 'patients'. A physician diagnoses a patient's illness by observing, questioning, and lab testing. A set of data points from these inquires generates a differential diagnosis. Sometimes, many patients with similar disease symptoms keep coming from a few neighborhoods, but most physicians can hardly relate their residential locations. Noticing this deficiency, Dr. Stephen Spann, the former Chair of the Department of Family and Community Medicine at Baylor College of Medicine asked me to make some census data maps for resident doctors and tutor them on how to download and interpret the 2000 US Census tract data. That was 2004 when I first anecdotally experienced neighborhood health while interacting with resident

G. Lin (✉)
Society Hub, Hong Kong University of Science and Technology, Clear Water Bay, Kowloon, Hong Kong, China
e-mail: gelin@ust.hk

B. Li et al. (eds.), *New Thinking in GIScience*,
https://doi.org/10.1007/978-981-19-3816-0_39

doctors. Dr. Spann told me that he normally would have an image of his patient and the family, but he wished to have a mirror image of the patient community during diagnosis and treatment.

Public health data at the community level is a mirror of community health. GIS not only bridges data to the mirror, but also feeds images to it. Just like a person routinely looks at a mirror before going to work, people should periodically check their neighborhood health mirror that may consist of (1) aggregated behavior risk and disease prevalence and incidence, (2) access to health and community health assets, (3) community or neighborhood socioeconomic status (SES). Even though I applied GIS in all these areas in the late 1990s, I rarely engaged in frontline practice during the first 10 years on the geography faculty.

Nobel prize winners are top scientists, the Nobel committee, however, judges them not only on the scientific merit of their discoveries, but also on the benefit they confer to humankind. One day in 1992, I showed my advisor that I had extended his Buffon's needle paper (Rogerson, 1990) from one-dimension migration distance to two-dimension migration area (he did one integral in calculus, and I did two). He glanced at my derivations and said: there are many societal problems. You would get more mileage if you spent your energy on solving real world problems. That comment sowed the seed of benefiting society in my heart. Now, 30 years from that conversation, I still draw satisfaction from working on real world problems. In the following, I share a success story about integrating spatial data with public health data. I then provide case lessons from implementing disease cluster detections. Finally, I offer some thoughts about how to put GIS into public health infrastructure for precision neighborhood health.

39.2 Integrating Geocoding and GIS into Public Health Data Infrastructure

State public health program data are mostly based on personal records, some from program participants and some from registries, such as birth, death, cancer, trauma. To place individuals into their communities, we need to geocode their addresses into geographic point locations (latitude and longitude). When geocoded location information is linked to census tract data for at-risk populations, SES, and other neighborhood variables, program specialists can assess place-based social determinants of health–potentially preventable differences in health by systematically placing socially disadvantaged groups at a further disadvantage on health.

Geocoding in the 2000s, however, was not as streamlined as it is today. When we geocoded multiple program data with millions of records, we opted to link all program data and generate a large address dataset with deduplicated address records. In this process, we also set up a federated data warehouse consisting of program data marts: each program owns its data and programs communicate the same patient records via a master patient index. To safeguard person-level data, we only let one person access all programs' identifiable information. This operation was sustained for many years (until the COVID-19 pause).

The linked data marts greatly eased inter-program data queries. In the old days, a program coordinator would write down the name of the participant of interest and go to another program to find out more. The data warehouse made this process seamless. Two programs can simultaneously look into many participants and many health conditions. In a conventional public health program inquiry, a map of obesity may be displayed as a result of an 'obesity geocoding project'. Our enterprise approach is to geocode all program data. Occasionally, multiple programs come to meet us requesting census tract data maps for many chronic conditions. Each map is consistently classified according to an age-adjusted rate. If a program is interested in a hot spot, it can ask relevant programs to see if the hot spot is collocated with other disease conditions. It is often that one program's influence on a policy intervention is small, but multiple programs' influences are much stronger. In addition, when programs and community stakeholders see these maps, they often deepen their assessments from maps to mechanisms. They ask for more maps, such as SES and health asset deficits (e.g. grocery store ratings, food deserts. street safety, primary care clinics). While providing poverty or other SES indicators are instantaneous, most community health asset data need to be added separately.

It may appear that we did this for efficiency, but our goal from the very beginning was to enhance programs' missions: improving the health of people and their communities. That is why we were flexible. What started with a multi-program geocoding project, ended up in expanding data linkage among programs. In fact, for most GISers, database manipulation and data linkage are routine. This skill allows us to link non-geographic data with a gentle learning curve. In addition, information technology changes relatively fast for efficiency or cost-saving, whereas programs' missions normally do not change much. If we are flexible to help programs achieve their missions, programs would work iteratively with us to deepen the inquiry.

39.3 Disease Cluster Detection in Disease Surveillance

In this section, I present three cases progressively to show that we could expand spatial disease detection to aspatial to meet program needs.

Case 1, Cluster detection needs a good framing question. After I moved to the School of Public Health in Nebraska in 2008, I was excited about opportunities to apply spatial statistical methods in disease cluster detections. Over the years, I carried out about 8–10 cancer cluster detection tasks for the state cancer registry. I detected at least a half dozen site-specific cancer clusters (breast, colorectal, and thyroid cancers, etc.). When brief detection reports were turned to the cancer surveillance program, none of the detected clusters resulted in further epidemiological investigations. Sometimes, a detected cluster area was too big. Sometimes an age-adjusted risk might not be sufficient. For example, when detecting a disease cluster from a municipal waste landfill, those in proximity to it tend to be in low SES too. Perhaps most importantly, we—GISers or biostatisticians–are normally handed with data for a narrowly defined analytical question; after the analysis, we hand the result back. Although each transaction has a meeting, the lack of domain knowledge intimidates us to speak. A statistically significant test result without a good framing question would not make us feel rewarding. Domain knowledge helps us to reframe questions either for further inquiries or for methodological developments.

Case 2, Proactive in occupational health risk detection. Then it came to COVID-19 daily surveillance back in March and April of 2020 when I took the task of providing early warnings for COVID-19 outbreaks in counties adjacent to Nebraska. Around April 9, 2020, I found that Greeley, Morgan County in Colorado had an outbreak attributable to a meat processing plant. I sent an alert to the state occupational health coordinator, who was curious about how I found out. Anyhow, everyone was too busy to act on the tip. Two weeks later, a state wide COVID-19 outbreak occurred among meatpacking plants, and I was asked to assist outbreak investigations over the weekend. This experience taught me that detections without instigation criteria already in place would hardly draw attention. Certainly, it was not the first time I assessed occupational risk in Nebraska, but it was the closest to having some impact on COVID-19 prevention among meat processing workers. Previously, I was peripherally involved in an elevated cancer investigation among individuals working for the same unit of an institution. It was found that with just one additional case, the risk would be statistically significant. Since the inquiry was sensitive to the media, the investigation was limited to a very small circle and was tabled for not significant. However, it is risky to continue business as usual for that work unit, and the institution should proactively address potential exposures, rather than waiting for the bomb to drop.

Case 3, GIS-based disease detection needs to extend to network-based data. A final case is what I witnessed closely in Clark County Nevada (Las Vegas) about an outbreak of acute non-viral hepatitis of unknown etiology in November and December of 2020. Five previously healthy children aged 7 months–5 years were hospitalized and then transferred to Utah for acute liver failure liver transplantation. Acute liver failure among young children is rare in the United States, and acute non-viral hepatitis of unknown etiology is extremely rare. Hence two such cases in the time span would be unusual, let alone five. Could the current space–time surveillance system alert health authorities earlier? The answer is no, as cases spread all over the metro area. Even if it did signal clustering along the time dimension, it would not be able to timely pin down the source. It turned out that a Utah hospital surgeon alerted Clark County at transfer number 3. A follow-up investigation found that a brand of alkaline bottled water was linked to this outbreak investigation 4 months later (Ruff et al., 2021). If the water supply information were in the system, and the geographic surveillance could be extended to the supply network, the system could potentially spot and source track the outbreak earlier. This case is not unique, as supply chain-based toxic exposures and food poisoning are mostly source-tracked after the events. As more and more goods are delivered at homes, flow-based network cluster detection will likely become more prominent.

The above cases suggest that we need to implement scientific knowledge in public health surveillance by educating ourselves (GISers) and decision makers, so that disease cluster detection is a collective decision process rather than a relay process. John Snow, the father of epidemiology, believed that cholera was waterborne even before the 1854 Cholera Outbreak in London. He used a dot map to show that the water pump on Broad Street was the source of infection. With his early belief, reaching perhaps a half number of dots on the map would be convincing. However, shutting off the pump and convincing the local health authority was not timely. It required the help of a local priest, and Snow's cholera etiology was recognized years later. Ignaz Semmelweis, the father of hand hygiene, found that post-delivery mortality was substantially higher from physician-deliveries than midwife-deliveries. He attributed this to a lack of handwashing among physicians. However, his recommended handwashing was branded crazy, only to be recognized 20 years later. On the flip side, when we (GISers) find something unusual in surveillance, we are often not sure of its importance due to a lack of disease-specific knowledge. If we systematically put specific disease risks into the surveillance system rather than waiting for convinced decision makers, some disease outbreaks could be ameliorated. Scaling up technology is relatively easy, but implementing science or putting alert triggering criteria (for investigation) requires consensus among public health practitioners (domain experts), IT experts (GISers in this context), decision-makers, and often stakeholders. This calls for design thinking, and the bar is very high, as we are dealing with automatic instigations against false alarms.

While stressing the importance of extending our skill to non-geography problems, we should also be open-minded about stretching existing GIS functionalities to uncomfortable zones of non-geographic disease surveillance. Most people in academics are now convinced where people live determines, in many ways, their health, thus the health of their communities. Life expectancies from two neighborhoods just a mile apart can differ between 15 and 20 years. The wide acceptance of this fact has taken at least 30 years in contemporary public health history, and GIS has contributed significantly to this recognition. Social physics used to rely purely on the physical distance to predict people's interaction. As modern communication technologies penetrate every corner of society, people's social networks, regardless of physical distance, can predict their interaction, health, and purchasing behaviors. Marketing professionals jumped on this wagon really quick, perhaps around 2000, whereas public health professionals are slower to act. Rather, they rarely implement network-based prevention and intervention programs. In many situations, both technology and scientific knowledge are available, and what is missing is the value of disease detection and prevention in the public health 'market'. Just like marketing, we have to value life proactively rather than post-outbreak investigations and punitive lawsuits. When the value has not been placed in the public health system, a bottom-up approach leveraging GIS for network-based surveillance is low-hanging fruit.

As a graduate student in Buffalo, I remember in a GIS advisory meeting, Jack Dangermond (ESRI president) said that GIS is infrastructure, and it needs to be built both from the ground up and the top. Then he said that ESRI had a lot of people working in the field to build GIS functionalities for industry and government sectors. When a function is common, ESRI then incorporates it into its system. I now realize that it was bottom-up GIS system building. At the time though, I had to check a dictionary to get the meaning of infrastructure. Nowadays, food labels and supply chain information in public health are for post-outbreak product tracing and recalls. This post-event paradigm could be shifted to an in-event paradigm using network distance. It is relatively easy to implement in the current GIS methodologically, but it is difficult to deploy necessary data. In addition, we need to organize our education from mostly Ford's assembly line teaching that stresses knowledge division to adaptive design learning that stresses pulling various knowledge together and adapting in time. In Case 2 above, if a point source is not directly in reference to geography, a clear knowledge division would not lead to a spatial analysis request. In Case 3, a space–time disease detection would be powerless unless we know the water supplier. However, such information is proprietary. If the public health agency could treat such data confidentially, food, and water supply companies may be willing to share. This brings to my concluding proposition in the realm of Big Data.

39.4 From Precision Medicine to Precision Neighborhood Health

In recent years, precision medicine has advanced biomedical science rapidly, where specific genes contributing to a disease can be identified and then modified through medication. Different genetic-based medications can then be administered to patients A and B as opposed to the same conventional medication to all patients. Likewise, if we want to implement precision neighborhood health, neighborhood residents, housing, health assets, transportation, and build environment would be neighborhood genes; their interaction with the outside (e.g., workplace, shopping) would be akin to gene-environment interaction. A major challenge is how to define, access, inventory, and update 'neighborhood genes'. In addition, people are mobile with wearables and other devices constantly generating data. They together with IOT sensor- and social media data become increasingly accessible forming cyber-neighborhood data.

Just like genomic sequencing of an individual, we can sequence the 'genomes' of a neighborhood. Different from human genomes, neighborhood genomes are more dynamic as residents move in and out frequently. In addition, we can treat neighborhood health assets from a disease genome perspective just like viral genome sequencing. For instance, the HIV virus had only three genes and they were fully sequenced. Their mutations and ability to resist antiviral therapy could be understood. If we identify a handful of neighborhood disease triggering 'genes' for group A patients, and another few for group B patients, we can tailor treatments to patient groups A and B more precisely. Here, we have to build from the bottom-up thousands of neighborhood solutions. GIS would serve spatial data infrastructure to inventory, update and analyze all aspects of neighborhood health. GIS would help neighborhood matching so that similar neighborhoods, regardless of geographic distance, can share effective interventions and lessons learned. GIS would provide neighborhood health images according to health providers' needs. Some of the functions might have been applied for specific diseases with specific neighborhood factors, but we are talking about enterprise and infrastructure level implementation.

Precision medicine is a buzzword, but its contribution to the health impact pyramid is relatively small (Frieden, 2010). The greatest impact is at the base—socioeconomic factors. Most of them can be sequenced and intervened at individual and neighborhood levels leading to precision neighborhood health. Human genes have evolved almost to a standstill for the last 50,000 years, while human environments are changing fast, which suppress, evoke, amplify, and mutate human genes. Both precision medicine and precision neighborhood health share computation and AI challenges in Big Data Science. GIS bridges precision medicine and precision neighborhood health so that both could be 'sequenced' together to treat disease etiology at both levels.

References

Frieden, T. R. (2010). A framework for public health action: The health impact pyramid. *American Journal of Public Health, 100*, 590–595.

Rogerson, P. (1990). Buffon's needle and the estimation of migration distances. *Mathematical Population Studies, 2*, 229–238.

Ruff, J. C., Zhang, Y., Bui, D. P., et al. (2021). Notes from the field: Acute nonviral hepatitis linked to a brand of alkaline bottled water — Clark County, Nevada and California, 2020. *MMWR Morbidity and Mortality Weekly Report, 70*, 1617–1619.

Chapter 40
Challenging Issues in Applying GIS to Environmental Geochemistry and Health Studies

Chaosheng Zhang, Xueqi Xia, Qingfeng Guan, and Yilan Liao

Abstract GIS has been widely used in geochemistry-related environmental health studies and practices to map and analyze sampling locations and spatial distribution of geochemical features and health information. In the big data era, the focus is shifting towards revealing the hidden patterns and features. This chapter explores the challenging issues of spatial analysis, machine learning, and uncertainty in such studies and practices. Spatial analysis needs to focus more on hot spot analysis and identification of spatial outliers, as well as exploration of spatially varying relationships. Machine learning can be adopted to conduct deep learning with a focus on non-linear features and their links with causal effects. The field and laboratory uncertainty of environmental geochemistry should be incorporated in GIS analysis. The analyses of the association between environment and health need to be more intelligent and accurate. GIS continues to provide useful tools to make novel findings in environmental geochemistry and health from the spatial aspect.

Keywords Environmental geochemistry · Spatial analysis · Uncertainty · Machine leaning · Environmental health

C. Zhang (✉)
International Network for Environment and Health (INEH), School of Geography, Archaeology and Irish Studies, National University of Ireland, Galway, Ireland
e-mail: Chaosheng.Zhang@nuigalway.ie

X. Xia
School of Earth Sciences and Resources, China University of Geosciences, Beijing 100083, China
e-mail: sdxqxia@163.com

Q. Guan
National Engineering Research Center of GIS, China University of Geoscience, Wuhan 430074, Hubei, China
e-mail: guanqf@cug.edu.cn

Y. Liao
The State Key Laboratory of Resources and Environmental Information System, Institute of Geographic Sciences and Natural Resources Research, Chinese Academy of Sciences, Beijing 100101, China
e-mail: liaoyl@lreis.ac.cn

B. Li et al. (eds.), *New Thinking in GIScience*,
https://doi.org/10.1007/978-981-19-3816-0_40

40.1 Introduction

Environmental geochemistry focuses on the sources and distribution of chemical elements on the Earth's surface and their relationships with plants, animals, and human health (Eby, 2016). It plays an increasingly vital role in mineral exploration, environmental management, agricultural practices, and human health. Analyses of geochemical data can help to: (1) reveal the geochemical patterns, (2) explore the relationships between these patterns and influencing factors, and (3) map environmental and health risks. With the development of geochemical mapping, massive geochemical databases have been established and environmental geochemistry has entered the era of big data. It is increasingly critical to break away from conventional methods that rely excessively on expert judgment, and instead to develop efficient methods to reveal useful information and patterns hidden in large geochemical databases and establish associations between environment and health.

40.2 Spatial Analysis of Environmental Geochemistry

The most popular applications of GIS in environmental geochemistry are mapping of sampling locations based on sample coordinates and mapping of spatial distribution of environmental parameters based on spatial interpolation. In the past, it was not easy to produce spatial distribution maps of high quality based on a limited number of samples. However, with the growing volume of data, the current focus is shifting towards revealing hidden information in the data. Meanwhile, spatial heterogeneity of environmental geochemistry makes it a challenging task to identify the hidden patterns of environmental variables and their influencing factors. Spatial statistical techniques have been applied to measure the spatial autocorrelation and spatial structure, with an aim to provide foundations for spatial interpolation. The current trend is moving towards the challenging issues of hot spot analysis and spatial outlier identification, as well as spatially varying relationships.

The analyses of hot spots and spatial outliers can provide an effective way to identify areas requiring attention, e.g., potential pollution or mineralization. The techniques for hot spot analysis include local index of spatial association (LISA) (Anselin, 1995), Getis Ord G_i^* (Getis & Ord, 1992), and others. At the regional scale, the LISA method was applied by Zhang et al (2008) to identify spatial clusters of lead in soils of Galway City and the clusters were associated with traffic. At the continental scale, Xu et al. (2019) found that the co-existence of cool spots of soil organic carbon and pH value in central eastern Europe was related to the coarse parent materials of the last glacier deposit. More hidden patterns of spatial clusters and spatial outliers in environmental geochemistry should be revealed so that their influencing factors can be better identified.

The conventional correlation analysis assumes that the relationships between variables are constant across the whole study area. This assumption has ignored the spatial heterogeneity of such relationships. The geographically weighted regression (GWR) offers a way to explore the complex spatially varying relationships between environmental variables. Yuan et al. (2020) found that anthropogenic factors weakened the relationships between Pb and Al in soils of central London, while large parks and greenspaces reserved their positive relationships. Following the discovery of both cool spots of soil organic carbon and pH in the central-eastern Europe, a further study revealed their positive relationships in this area (Xu & Zhang, 2021). Further studies should be carried out to reveal the spatially varying relationships in environmental geochemistry in order to explore more details of hidden information.

40.3 Machine Learning in Environmental Geochemistry

As a data analytical technology in the artificial intelligence area, machine learning can help intelligently discover patterns and features hidden in data and derive decisions or predictions for new data (Mahesh, 2020). In recent years, machine learning methods, such as artificial neural networks, support vector machines, logistic regression, *K*-means clustering, and autoencoder have been applied in geochemical analysis to model nonlinear systems and capture complex multi-phase geological events (Zhang et al., 2021; Zuo & Xiong, 2018). There are three main challenges in applications of machine learning and its potential future development in environmental geochemistry.

1. Most studies use shallow structured machine learning methods, which have limited ability to represent complex functions and suffer limited capability of feature extraction. Feature learning within deep learning has attracted considerable attention in the field of machine learning. Deep learning networks enhance feature learning by deepening the model structure to achieve a complex function approximation, demonstrating a powerful ability to learn the essential features of a dataset from a small sample set (LeCun et al., 2015). More deep learning methods can be adopted in environmental geochemistry to further enhance feature extraction and relationship modeling capabilities.
2. Spatial patterns have been the focus of environmental geochemical analysis. Although methods such as geographically weighted regression, Moran's *I*, and hot spot analysis have been developed and applied to quantify patterns in environmental geochemistry, their model design relies more on a priori relationships. For example, the spatial weights representing the spatial relationships between samples in these models often rely on predetermined formulas (e.g., inverse distance weighting formulas based on Tobler's First Law of Geography), resulting in these captured relationships to be relatively simplified and possibly insufficient for complex spatial patterns. Convolutional neural networks (CNN, Gu et al., 2018) and graph neural networks (GNN, Wu et al., 2020), as extended

models of deep learning with the ability to mine local spatial structure features, have been widely used in face recognition, anomaly detection, and relationship mining. The spatial structures and relationships of geochemical elements can be intelligently extracted using CNN and GNN to provide more essential information for the relationship extraction between geochemical spatial patterns and human health.

3. Many machine-learning (especially deep learning) methods are mostly "black-box" models. Although the prediction and classification results obtained by machine learning are highly accurate, they often lack interpretability of the internal principles. Moreover, most of the studies use data-driven machine learning models, and the feature relationships mined are mostly correlation rather than causation. The mining results are mostly related to the regions of the collected data, and the generalization is relatively weak (Altman & Krzywinski, 2015). Therefore, causal interpretation by experts is also required after the results are obtained, which limits the application of machine learning in environmental geochemistry. Constructing environmental geochemical knowledge graphs and using knowledge-driven models to automatically extract and reveal the existence of causations between geochemistry and the environment should be one of the key future directions (Wang et al., 2017)

40.4 Uncertainty in Environmental Geochemistry and GIS

Uncertainty from spatial prediction has been considered in GIS, and it has been well handled by the kriging theory. The kriging method of interpolation can produce the prediction map of attribute values with a map of the prediction uncertainty. But this uncertainty is only the part caused by the model, which is the so-called model uncertainty (Krivoruchko, 2011). In the traditional GIS methods, input values are assumed to be deterministic. But in environmental geochemistry, each observed value is not a true one, but a measured value with a certain error. This is not fully considered in GIS at present.

Data for the environmental variables are either measured in situ or sampled from the field and then sent to laboratory for analysis. In either case, the measured value is not the true one. The true value cannot be exactly obtained because the error of measurement always exists. The uncertainty is usually evaluated by repeated sampling and testing. But samples cannot be collected repeatedly at the exact same location, but are collected at locations near the precursor. In this case, the uncertainty, typically measured by standard deviation of the repeated measurements, is induced from two sources: laboratory measurements and microscale variation of the variable.

The uncertainty is determined by (1) the magnitude, or the scale of the measured data, and (2) the spatial characteristics of the variable. The former is determined mainly by the method or instrument of analysis. There is a range of "good" performance for most analytical methods. If the measured value is in this range, the uncertainty caused by testing is relatively small. But when the measured value is near or even lower than the detection limit, the uncertainty increases. This is also the case when the measured value is high enough relative to the upper boundary of the range. Another factor influencing the measured uncertainty is the spatial distribution of the variable and the location of the repeated sampling. In areas of high values, e.g., those related to anthropogenic pollution, the microscale variation of the element or pollutants is usually large, and large uncertainty is induced during repeated sampling. In area of low value background, the uncertainty caused by repeated sampling is small. To sum up, the uncertainty of a measured value varies according to its magnitude and its spatial location due to spatial heterogeneity.

However, the spatial and magnitude dependence of the uncertainty from measurement is neglected in traditional GIS. There are two possible solutions to include uncertainty in environmental geochemistry with GIS. (1) A model can be constructed to describe the spatial and magnitude dependence of the measurement uncertainty. In the future, the GIS system can accept repeatedly measured values at one location as the input for the model, including repeatedly tested values in the laboratory and the data of repeatedly collected samples from the field. (2) The uncertainty of input values can be incorporated into spatial modeling. In this way, the spatial predictions will take into account the uncertainty caused by both the measurement and spatial prediction. Also, the result of spatial statistics, e.g., the areal mean from block kriging, should also be accompanied by the standard error with the measurement uncertainty.

40.5 Environmental Health and GIS

Humans and animals have always been exposed to chemicals in the environment. The dramatic increases in industrialization over the past three centuries have changed both the quality and the quantity of both natural and synthetic chemicals humans expose to (Birnbaum, 2008). Human health hazards caused by chemicals include endemic diseases resulted from the deficiency or surplus of natural chemicals in local environment and health issues due to environmental pollution (Wang et al., 2002). Therefore, risk assessment of environmental chemicals aims to identify the potential hazards to human health and understand how serious such problems may be. It is composed of four components: exposure assessment, hazard identification, dose/response assessment, and risk characterization and control (Birnbaum, 2008).

The applications of GIS in studying health effects of environmental chemicals include: (1) efficient integration, manipulation, and processing of environmental health data from multi-sources; (2) flexible implementation of spatial–temporal functions for transforming, aggregating, and analyzing the data in health risk assessment; (3) effective simulation and display of effects of various prevention and control measures.

GIS provides the tools such as the Geodetector method for combining mortality and morbidity data with environmental chemical data to present a coherent picture of the role of chemicals in determining human health status (Wang et al., 2010). By explicitly linking human health outcomes to chemicals and other environmental variables, GIS allows a reorientation toward more direct, population-based, and institutional explanations for health differentials. Moreover, GIS enables a wide range of relevant non-spatial data to be integrated into a consistent framework to facilitate analysis of the geographical distributions of diseases (Tim, 1995).

However, there are some key impediments in applying GIS to studying health effects of environmental chemicals. The huge volumes of surveillance data and the rapid rise in the creation and expansion of environmental health data require GIS to analyze and display research findings more quickly and accurately. There is a challenge to couple GIS with advanced techniques such as high-performance computing and artificial intelligence algorithms. In addition, more efforts are needed to take into account the dynamic factors such as the extent of pollution and the nature of the pollutant.

References

Altman, N., & Krzywinski, M. (2015). Points of significance: Association, correlation and causation. *Nature Methods, 12*(10). https://doi.org/10.1038/nmeth.3587.

Anselin, L. (1995). Local indicators of spatial association—LISA. *Geographical Analysis, 27*, 93–115.

Birnbaum, L. S. (2008). The effect of environmental chemicals on human health—CJA. Fertility and sterility. *American Society for Reproductive Medicine, Birmingham, AL, 89*(2, Supplement 1), e31.

Eby, G.N. (2016). *Principles of environmental geochemistry*.Waveland Press.

Getis, A., & Ord, J. K. (1992). The analysis of spatial association by use of distance statistics. *Geographical Analysis, 24*(3), 189–206.

Gu, J., Wang, Z., Kuen, J., Ma, L., Shahroudy, A., Shuai, B., Liu, T., Wang, X., Wang, G., Cai, J., et al. (2018). Recent advances in convolutional neural networks. *Pattern Recognition, 77*, 354–377.

Krivoruchko, K. (2011). *Spatial statistical data analysis for GIS users* (1st ed.). Esri Press.

LeCun, Y., Bengio, Y., & Hinton, G. (2015). Deep learning. *Nature, 521*(7553), 436–444.

Mahesh, B. (2020). Machine learning algorithms—A review. *International Journal of Science and Research (IJSR), 9*, 381–386.

Tim, U. S. (1995). The application of GIS in environmental health sciences: Opportunities and limitations. *Environmental Research., 71*, 75–88.

Wang, J. F., Li, X. H., Christakos, G., Liao, Y. L., Zhang, T., Gu, X., & Zheng, X. Y. (2010). Geographical detectors-based health risk assessment and its application in the neural tube defects study of the Heshun Region, China. *International Journal of Geographical Information Science., 24*(1), 107–127.

Wang W. Y., Yang L. S., & Tan J. A. (2002). Balance and regulation of environment-health-development—Thoughts on innovation in geography (in Chinese). In: *Proceedings of the 2000–2002 annual academic conference of the Geographical Society of China.*

Wang, Q., Mao, Z., Wang, B., & Guo, L. (2017). Knowledge graph embedding: A survey of approaches and applications. *IEEE Transactions on Knowledge and Data Engineering, 29*(12), 2724–2743.

Wu, Z., Pan, S., Chen, F., Long, G., Zhang, C., & Philip, S. Y. (2020). A comprehensive survey on graph neural networks. *IEEE Transactions on Neural Networks and Learning Systems, 32*(1), 4–24.

Xu, H., Demetriades, A., Reimann, C., Jiménez., J.J., Filser, J., & Zhang, C. (2019). Identification of the co-existence of low total organic carbon contents and low pH values in agricultural soil in north-central Europe using hot spot analysis based on GEMAS project data. *Science of the Total Environmental, 678*, 94–104.

Xu, H., & Zhang, C. (2021). Investigating spatially varying relationships between total organic carbon contents and pH values in European agricultural soil using geographically weighted regression. *Science of the Total Environment, 752*, 141977.

Yuan, Y., Cave, M., Xu, H., & Zhang, C. (2020). Exploration of spatially varying relationships between Pb and Al in urban soils of London at the regional scale using geographically weighted regression (GWR). *Journal of Hazardous Materials, 393*, 122377.

Zhang, C., Luo, L., Xu, W., & Ledwith, V. (2008). Use of local Moran's I and GIS to identify pollution hotspots of Pb in urban soils of Galway, Ireland. *Science Total Environment, 398*(1–3), 212–221.

Zhang, C., Zuo, R., Xiong, Y., Shi, X., & Donnelly, C. (2021). GIS, geostatistics, and machine learning in medical geology. In: Practical Applications of Medical Geology. Springer, pp. 215–234.

Zuo, R., & Xiong, Y. (2018). Big data analytics of identifying geochemical anomalies supported by machine learning methods. *Natural Resources Research, 27*(1), 5–13.

Printed in the United States
by Baker & Taylor Publisher Services